Micro and Nano Semiconductor Devices for Digital, Analog and Sensor Design

Scrivener Publishing
100 Cummings Center, Suite 541J
Beverly, MA 01915-6106

Publishers at Scrivener
Martin Scrivener (martin@scrivenerpublishing.com)
Phillip Carmical (pcarmical@scrivenerpublishing.com)

Micro and Nano Semiconductor Devices for Digital, Analog and Sensor Design

Edited by

Suman Lata Tripathi

Vrinda Gupta

Sobhit Saxena

Shipra Upadhyay

and

Sudip Ghosh

Scrivener Publishing

WILEY

This edition first published 2026 by John Wiley & Sons, Inc., 111 River Street, Hoboken, NJ 07030, USA and Scrivener Publishing LLC, 100 Cummings Center, Suite 541J, Beverly, MA 01915, USA

For more information about Scrivener publications please visit www.scrivenerpublishing.com.

Wiley Global Headquarters
111 River Street, Hoboken, NJ 07030, USA

For details of our global editorial offices, customer services, and more information about Wiley products visit us at www.wiley.com.

The manufacturer's authorized representative according to the EU General Product Safety Regulation is Wiley-VCH GmbH, Boschstr. 12, 69469 Weinheim, Germany, e-mail: Product_Safety@wiley.com.

Library of Congress Cataloging-in-Publication Data

ISBN 9781394307319

Cover image: Generated with AI using Adobe Firefly
Cover design by Russell Richardson

Set in size of 11pt and Minion Pro by Manila Typesetting Company, Makati, Philippines

Printed in the USA

10 9 8 7 6 5 4 3 2 1

Contents

Preface

In the era of smart portable and flexible electronic devices, designers need to develop new design concepts and principles for improved performance. Advanced techniques, efficient computing algorithms, and new models help in developing efficient solutions at low cost and low power features. Most of the available analog and digital ICs are based on conventional electronic devices or architectures at different technology nodes. Although there are several new devices proposed with specific features and performances, there is a need to develop advanced applications based on these new electronic devices and architectures at the micro and nano level. The use of flexible semiconductor material and biodegradable electronic devices are the need of the hour to control electronic waste for a sustainable environment. Also, low power devices will help in reducing heating challenges in such electronic devices and ICs. Smart chips, high computational power, low voltages, and low cost are the major areas where improvement still is possible.

So, there is need of detailed discussion on design techniques, advanced semiconductor materials, and fabrication techniques intending towards the advanced application of efficient micro and nano devices. This book will fulfill these requirements through discussion on the existing and latest micro and nano devices with their applications. Design techniques with modern tools, software, and simulators in virtual environments will also be the focus of this book. In this book, design problems will be discussed with their specifications for researchers, industry professionals, and academicians working in the area of semiconductor IC design, layout and fabrication technology.

Suman Lata Tripathi
Vrinda Gupta
Sobhit Saxena
Shipra Upadhyay
Sudip Ghosh

1

Design of Advanced MOSFET Architectures

Remya Jayachandran* **and Salila Hegde**

Dept. of Electronics and Communication Engineering, The National Institute of Engineering, Mysore, Karnataka, India

Abstract

Multigate devices, widely called as FinFETs (Fin type field effect transistors), belong to a category of transistors used in modern microprocessors. They consist of a fin-shaped gate structure that surrounds the channel, due to which flow of electric current is controlled in better way. This design enables multigate devices to work at lower voltages with reduced power consumption and increased performance. The multiple gates also help to reduce leakage current, making them more reliable and efficient. Due to this, multigate devices have become a crucial component in the growth of ultraminiaturized, high-performance ICs (integrated circuits). Due to the advancement in technology especially by incorporating artificial intelligence in the gadgets, the requirement of high computation processors led to the replacement of FinFET with new device architecture beyond 3 nm technology node. The evolution of transistor from planar structure to multigate non-planar structures, tunnel FETs, reconfigurable FETs etc. are detailed in this chapter. These devices are used to build high performance and low-power circuits, such as, memory, digital and analog circuits. In digital circuits, multigate device has many advantages, which includes lower leakage current, improved sub threshold slope, and increased switching speed which are also discussed in this chapter.

Keywords: FinFET, GAAFET, TFET, RFET, analog circuits, digital IC

**Corresponding author*: remyajayachandran@nie.ac.in

Suman Lata Tripathi, Vrinda Gupta, Sobhit Saxena, Shipra Upadhyay and Sudip Ghosh (eds.) Micro and Nano Semiconductor Devices for Digital, Analog and Sensor Design, (1–30) © 2026 Scrivener Publishing LLC

1.1 Introduction

For the past five decennium, the evolution and expansion of VLSI (very-large-scale integration) circuits have been driven by the persistent scaling down of device structures in alignment with Moore's Law [1–4]. Non-planar device structures are used in recent electronic gadgets which have processors working at high frequency, embedded systems etc. instead of conventional planar transistors [5]. Since the analog circuit part is built using planar transistors, fabrication of both analog and digital circuits demand two different process technologies which makes the cost of product high. Scaling down the device dimension can increase the frequency of operation of the device with a trade-off in increased parasitic capacitances and fabrication complexity. New device architectures such as carbon nanotubes, multi-gate devices, single electron devices, tunnel FETs and reconfigurable FETs can outdo conventional planar MOSFETs in terms of lesser area, higher speed and low power consuming devices [5–10]. Among the various multi-gate transistor structures, "Gate All Around FET" (GAAFET) device has better control of gate over the channel and channel length scaled down below 5nm. One more multi-gate device in use recently is reconfigurable field-effect transistor (RFET), which can be configured either as an p-type or n-type FET by applying suitable bias. RFET device structures with single gate to multiple gates have been detailed in the published works [11–18]. Experimental demonstration of logic circuits, current mirror circuits, memory circuits, amplifiers etc. using new device architectures and its performance analysis are available in literature [19–25]. Basic digital circuits have been demonstrated using scaled-down new transistor architectures. Analog circuit and signals are more susceptible to changes in device parameters and hence makes it hard to replace conventional transistor with new device structures. In scaled-down devices, non-linear parameters become more dominant that significantly impacts analog circuit design. As device dimensions continue to scale down, the design complexity increases more significantly for analog circuits compared to digital circuits. Despite the fact that digital circuits have already made their way to the nanoscale regime, there is still research on designing analog circuits using scaled-down device architectures [17, 22]. Electrical characterization of various nanowire FET devices and the circuit simulations using these transistors are illustrated in [26–38]. The physical size of devices has been significantly reduced due to the rapid development of smart electronic devices and gadgets. As device dimensions shrink, maintaining effective gate control over the channel becomes increasingly challenging for planar transistors.

The cylindrical type GAAFET offers the lowest natural length, making it possible to scale down the device. Carrier conduction is not restricted by the Si/SiO2 interface in the GAA device, but rather by the center of the device due to its unique architecture. This alleviates interface-related issues like charge carrier mobility and interface impurities caused by the dangling bonds during development. The current technology node is 3 nm node that uses GAAFET device by name 3GAA that can replace the FinFET structure.

In this chapter, the evolution of MOSFET structures is explained in section 1.2 followed by silicon-on-insulator (SOI) MOSFET structures in section 1.3. Current multigate MOSFET structures and their challenges and future scopes are discussed in section 1.4 followed by applications of non-planar transistors in analog and digital circuits in section 1.5. The prime objective of this chapter is to introduce the new transistor architectures and challenges in existing analog and digital circuit designs using non-conventional transistors.

1.2 History of Transistors

During wartime (World wars), the need for advanced technology to create highly efficient computational devices spurred the blooming of transistors. This invention, as well as subsequent utilization of transistors in integrated circuits, marked a significant boost in the semiconductor industries. In this section, the evolution of transistors is detailed that led to the new device architectures both planar and non-planar multigate transistors.

1.2.1 Evolution of Transistors

The evolution of active electronic devices has undergone a remarkable transformation over time. The journey began with the vacuum tubes, which were superseded by the bipolar transistors (BJT) and subsequently by the advancements in the MOS (metal-oxide-semiconductor) transistor. Vacuum tubes were bulky, power-hungry, and had limited lifespan. The field of solid-state materials and devices underwent numerous explorations and breakthroughs after John Bardeen and Walter Brattain demonstrated the point-contact form of transistor in December 1947 [1]. The transistor offered a more compact, efficient, and reliable alternative to vacuum tubes, which led to the significant advancements in electronic technology. A possible alternate design was suggested by the fact that the point-contact transistor's two contacts were quite close to one another. A comparable result can be achieved with more consistent application by sandwiching an extremely thin layer of material with complimentary doping between two silicon pieces.

Building on this insight, William Shockley defined "junction" transistor. This new design, based on the p-n junction principle, offered a more robust and practical solution compared to the initial point-contact transistor. In 1956, Brattain, Bardeen and Shockley were awarded Nobel Prize for creating a device capable of generating solid-state amplification. While Shockley pioneered the bipolar transistor, Bardeen and Brattain created the point-contact transistor. William Shockley directed the efforts of Bardeen and Brattain at Bell Labs to develop a three-electrode device, which is the basis for the field-effect transistor. However, the team led by Brattain faced a significant challenge in overcoming the surface states that interfered with the gate voltage. Despite their attempts, they were unable to realize a successful insulated-gate FET. M.M.(John) Atalla and Dawon Kahng, also at Bell Labs, finally achieved the breakthrough and demonstrated the initial insulated-gate FET (IGFET) successfully in 1959. This landmark achievement paved the way for the development of the metal oxide semiconductor field effect transistor widely known as MOSFET, which has become the foundation of modern electronics and computing.

The evolution of the MOSFET has been a remarkable journey in the history of semiconductor technology. The MOSFET was first conceptualized by Julius Lilienfeld in 1925, but the first working MOSFET was demonstrated by Mohamed Atalla and Dawon Kahng at Bell Labs in 1959. Jack Kilby at Texas Instruments realized a simple (flip-flop) circuit with dual BJTs on a solo chip of germanium which led to the invention of integrated circuits in September 1958. In the 1960s, the planar MOSFET structure was developed, which allowed for the multiple transistor integration on a single chip, leading to the start of integrated circuits. Over the decades, MOSFETs have undergone continuous scaling, with the channel length being reduced from several micrometers to the nanometer scale, following Moore's law. To address the challenges of short channel effects in scaled-down planar MOSFETs, various multi-gate structures were introduced, such as the double-gate MOSFET, FinFET, and GAAFET. To address the drawbacks of conventional silicon dioxide (SiO2) gate dielectrics, metal gates and high-k dielectric materials were developed, improving the gate control and reducing leakage currents. The incorporation of strained silicon channels has been used to enhance carrier mobility and improve the performance of MOSFETs. Research has been focused on investigating non-silicon alternative channel materials such germanium, III-V compounds, and two-dimensional materials like graphene and transition metal

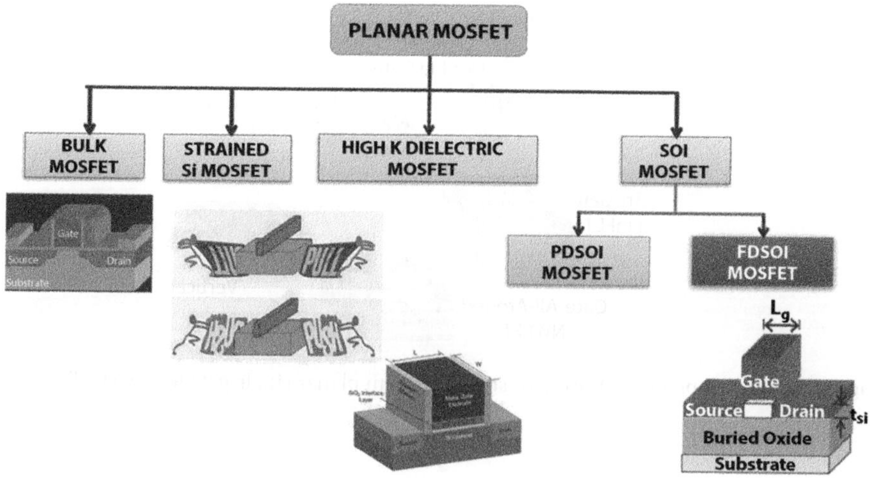

Figure 1.1 Planar MOSFET architectures.

dichalcogenides. As device dimensions continue to scale down, quantum mechanical effects become increasingly important, leading to the development of novel device structures, such as tunnel FETs, single-electron transistors, and quantum-dot-based devices. Figure 1.1 presents the different planar MOSFET devices.

1.2.2 Non-Planar Device Architectures

As the dimensions of MOSFET devices have been reduced down to smaller and smaller dimensions, a set of undesirable effects, collectively known as "short-channel effects," have become increasingly prominent. These short-channel effects arise because of the scaling of the transistor's physical dimensions, particularly the length of channel, and can degrade the device's performance and reliability. Some of the key short-channel effects in MOSFETs are roll-off of gate threshold voltage, lowering of drain induced barrier, modulation of channel length, hot carrier and punch-through effects, sub-threshold swing degradation etc.

To address these short-channel effects, various device engineering methods are developed, such as scaling of gate oxide thickness, high-k dielectric materials as metal and gate oxide, adoption of multi-gate structures (e.g. Double gate FET, FinFET, GAAFET), use of strain engineering to enhance carrier mobility, exploration of alternative channel materials (For example 2D materials, III-V compounds), etc. Continuous innovation in device design and materials is crucial in maintaining the performance and

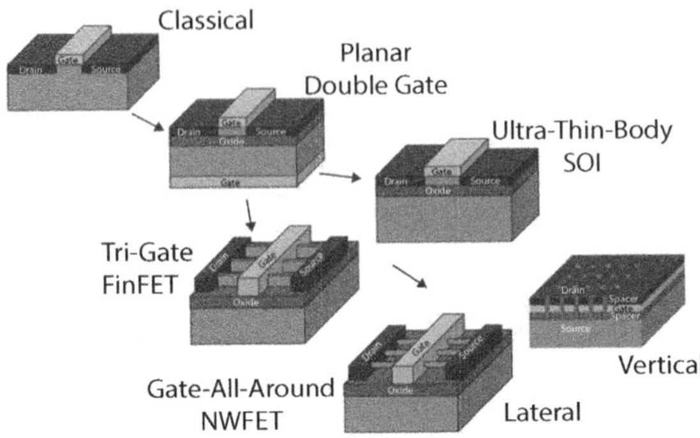

Figure 1.2 Development of transistor structure from planar single gate to GAAFET structures.

reliability of MOSFETs as they continue to scale to smaller dimensions. The transistor structure evolution from conventional MOSFET structure to non-planar structures is as in Figure 1.2. Classical structure represents the typical planar MOSFET with single gate. To obtain more gate terminal control over the channel with scaling of the device structure, SOI substrates were introduced. Due to limitations in SOI substrate with single gate in scaled down devices, planar double gate MOSFET and ultra-thin body SOI MOSFETS replaced the classical MOSFETs.

As the necessity for high speed and low power consuming processors increased, the semiconductor industry moved towards the FinFET and Trigate MOSFETs which showed remarkable performance beyond 20 nm technology node. Artificial Intelligence (AI) algorithms need processors with high capacity of computation. The demand for new transistor architectures led to the variation in FinFET structures to reduce the parasitic capacitances. Now, multiple fin FinFET structures have saturated as the performance of FinFET is not good beyond 3 nm which led to the GAAFET nanowire structures as shown in Figure 1.2. Non-planar transistor architectures have emerged as a solution to address the challenges faced by traditional planar MOSFET devices as they continue to scale to smaller dimensions. These non-planar transistor structures offer improved electrostatic control and better scaling capabilities.

1.3 SOI MOSFET: From Single Gate to Multigate

CMOS technology is the prevailing semiconductor technology and the MOS transistor is the building block of CMOS device. The dimension of the MOSFET has been reduced by half every three years in line with Moore's law. Due to the second order effects or short channel effects, the normal MOSFET architecture was replaced by new architectures for device as it was not possible to scale the dimensions of the conventional bulky MOSFET's less than 20 nanometers. Using voltage-doping transformation model (VDT), the effect of scaling is translated to electrical parameters mainly drain induced barrier lowering (DIBL) and short channel effect (SCE). The electrostatic integrity (EI) factor is the measure of how the electric field lines in the region of drain influence the region of channel, which is dependent on device geometry. From the VDT model,

$$SCE = 0.64 * (\frac{\varepsilon_{Si}}{\varepsilon_{ox}}) * EI * V_{bi} \tag{1.1}$$

and

$$DIBL = 0.80 * \left(\frac{\varepsilon_{Si}}{\varepsilon_{ox}}\right) * EI * V_{DS} \tag{1.2}$$

Where, V_{bi} is the built-in potential of drain or source, V_{DS} is the potential across source and drain.

The MOSFET threshold voltage V_{th} for a given channel length can be determined using

$$V_{TH} = V_{TH0} - SCE - DIBL \tag{1.3}$$

where V_{TH0} is the long channel MOSFET threshold voltage. Hence, as the device dimension reduces, Vth reduces due to SCE and DIBL, which is known as roll off of threshold voltage. More increase in the doping level of the drain and source area can lower down the entry of electric field lines into the channel side, limitation in the doping can be increased up to $10^{19}/cm^3$. Introducing silicon on insulator (SOI) substrate can reduce the entry of lines of electric field from drain to transistor channel area. Fully depleted and partially depleted MOSFET are two different types of SOI MOSFETs. Based on the concentration of doping, thickness of buried and gate oxides, the performance of SOI MOSFET can be better or worse than bulk MOSFET. Thin buried oxide layer beneath the channel area between the drain and source, can reduce the SCE but increases junction capacitance and body effects. To overcome these challenges, a double-gate (DG)

MOSFET structure was introduced. Both the gates are tied and drained electric field lines will end up at the back gate electrode. Hence, the SCE can be reduced in DG- MOSFET compared to the bulk and SOI MOSFETs.

As the demand increased for very compact, low-power, and high-speed electronic gadgets, the transistor structures evolved from single-gate to multiple-gate MOSFET structures which can supply high current drive, better gate voltage control over the channel, and decreased short channel effects. A single gate electrode on two opposing sides of the device is referred to as a double-gate MOSFET; a single electrode folded over three sides of the device is referred to as a triple-gate MOSFET.There are other multiple gate devices with independent gate bias, namely multiple independent gate FET (MIGFET) or four-terminal FinFET. The first double-gate SOI MOSFET to be fabricated was a fully depleted lean channel transistor (DELTA) in 1989. The structure of a FinFET is similar to that of a DELTA transistor, with the exception of a hard mask that is present on top of the silicon fin to prevent the formation of a parasitic inversion channel.Vertical channel MOSFET reduced the footprint of the multigate SOI MOSFETs. Gate all around FET (GAA-FET) is the vertical channel SOI MOSFET where the gate wraps around the channel area and the W/L (width/length of channel) ratio is near to unity. This increases the gate voltage control over the transistor channel.

EI is associated with short-channel effects, which can be enhanced in SOI devices by lowering the thickness of the silicon film or the gate oxide. The physics of the multigate FET is analyzed by using Poisson's equation. The distribution of electric potential is obtained concerning the natural length of the device structure. From the study of simulation results it is known that the effective length of gate of the MOS device should be 5 to 10 times of the natural length to avoid the SCE as we reduce the dimension.

The Natural length (λ) can be reduced by reducing t_{ox}, t_{si}, and by using high-K dielectric gate oxide along with multiple gate devices. Some of the effects that degrade the performance in multiple gate FETs are corner effects, quantum effects, volume inversion, mobility effects, threshold voltage and inter-sub band scattering effects.

Technology computer aided design (TCAD) tools facilitate detailed microlevel analysis of different FETs. These tools are instrumental in developing new device architectures and studying their characteristics. In the post-Moore's Law era, the literature proposes many worthy successors to conventional CMOS devices, including carbon nano tubes, multiple gate transistors, single electron devices, tunneling FETs and reconfigurable devices. Among these, reconfigurable devices have emerged as a particularly promising area of study. RFETs can alter functionality between p and n type transistors by altering the gate bias polarity. Tunnel FETs (TFETs) are a type of FETs that leverage quantum tunneling of charge carriers (electrons or holes) through an energy barrier to achieve switching. This mechanism is fundamentally different from the thermionic emission process used in conventional MOSFETs. Research into the implementation of analog and digital circuits using various MOSFETs like- FinFETs, GAAFETs, RFETs and TFETs can provide valuable understanding of their performance capabilities and potential circuit applications. The electrical characterization of the non-planar multigate transistors beyond 10 nm technology nodes can give awareness into the possibilities of schematic implementation employing these new devices. The next section details the new device architectures adopted by the industry and also the devices that are at the research level.

1.4 Current Non-Planar Device Architectures

1.4.1 FinFET

In 1999, Chenming Hu, a professor at Berkeley and former CTO of TSMC, and his group introduced the concept of FinFET; then in 2000, they presented UTB-SOI (FD SOI). Since a thin body is the fundamental component of both designs, the gate capacitance is closer to the channel's entire length. The body is incredibly thin less than 10 nm. Thus, a leaking channel that is distant from the gate does not exist. The leakage can be efficiently controlled by the gate. They suggested that a channel controlled by multiple channel sides would be the fundamental building block of a FinFET. The non planar or 3D transistor utilized in the construction of contemporary computers is called a FinFET or Fin Field Effect Transistor. It is constructed on SOI substrate, just like in previous planar devices. Designs of FinFETs, however, also utilize an active channel whose conductivity rises above the insulator's level, which results in a thin silicon structure known as a gate electrode that has the appearance of a fin. A single transistor can support the operation of numerous gates. Modern FinFETs are three-dimensional structures also known as tri-gate transistors which is as shown in Figure 1.3. FinFETs can be fabricated on SOI wafers or bulk silicon. This FinFET structure is made up of a substrate-mounted, vertical and thin silicon fin. Because the channel is encircled by gate, it offers superior control from three angles. Because the Si body of this device bear a resemblance to the back fin of a fish, it is known as the FinFET.

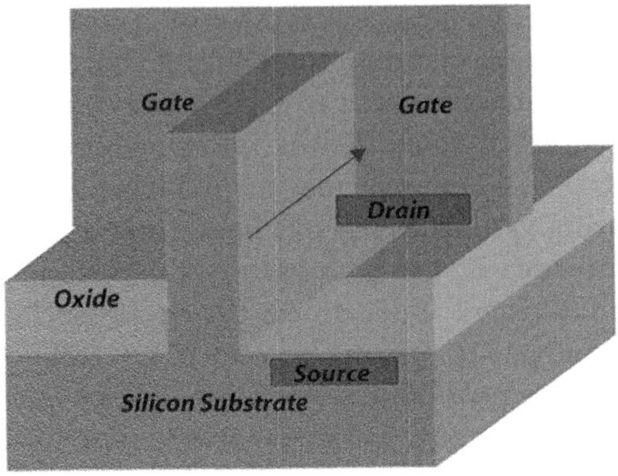

Figure 1.3 Structure of modern FinFET.

To produce FinFETs, we typically use bottom up and top-down techniques. The pros and cons of both approaches are equal. In the top-down approach, the fin must first be patterned on the active region of device by a process called E-beam lithography. Depending on the material (substrate) wet etching techniques like BOE, KOH are used, or transformer coupled plasma reactive ion etching (TCP-RIE) or ICP can also be applied to define the fin. Following the definition of the fin, the Ohmic (S/D) photo and metal deposition, quick thermal annealing, gate photo, and gate metallization are deposited. The channel in bulk-MOS (planner MOS) is horizontal, whereas it is vertical in the FinFET channel. Therefore, in a FinFET, the width of the device is determined from the channel's height (Fin). The following formula provides the channel's ideal width.

$$Fin_{Width} + Fin_{Height} = 2 \text{ x ChannelWidth} \tag{1.4}$$

One way to raise the driving current of the FinFET is to increase the Fin's height, or the width of the channel. As seen in Figure 1.4, we can also build parallel multiple fins joined together to improve the device driving current. It suggests that since the arbitrary channel width for a FinFET is always a multiple of the height (fin); it is not achievable. Consequently, the device's effective width quantizes.

1.4.1.1 FinFET Advantages

i. They need less power and outdo planar transistors in terms of performance.

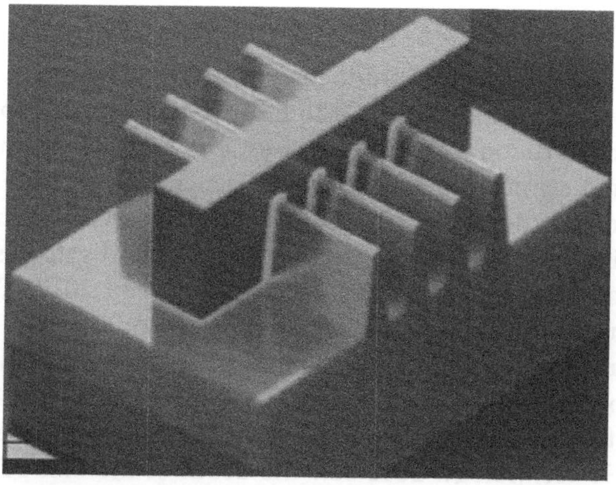

Figure 1.4 Multiple fin FINFET structure.

ii. Compared to a 28 nm technology, a 16 nm /14 nm FinFET technology may provide a 40 to 50% improvement in accomplishment or a 50% decrease in power.
iii. Can produce lower power or higher frequency values in comparison to bulk for a certain power budget.
iv. Compared to planar devices, they show higher drive current per unit area, primarily because the fin's height allows for the channel formation with a higher effective volume and still allows for the use of a wraparound gate.
v. They have a power reduction because they require fewer wide, high-drive standard cells and can function at a lower level of power supply voltage for a given leakage level.

1.4.1.2 FinFET Disadvantages

i. The primary concern is cost as the process of creating a FinFET necessitates several extra steps in the production pipeline.
ii. It is more challenging to manage fin depth on a bulk silicon process.
iii. It is difficult to control dynamic threshold voltage.
iv. More temperature sensitivity.
v. More heat dissipation.
vi. Very high capacitances.

1.4.1.3 FinFET in Semiconductor Industries

These days, FinFETs are the production platform of choice for all of the major semiconductor businesses. The FinFET platform is expected to be used in the construction of future devices. The creation of contemporary nanoelectronic semiconductor devices is dependent on FinFETs. In the early years of the decade of the 2010s, FinFET-based microchips were commercialized and went on to become the standard gate design at process nodes of 14 nm, 10 nm, and 7 nm [30, 31]. In 2012, Trigate FETs were introduced by Intel at the 22-nanometer node in the "Ivy Bridge CPU". Global Foundry, TSMC, and Samsung are among the several foundries that provide FinFET technology. TSMC declared in 2014 that it has created the first networking processor based on ARM architecture using 16nm FinFET technology. 2012 saw the debut of the first FD-SOI chips by STMicroelectronics for mobile processors at 28nm. Foundries including as Global Foundry, Samsung, and IBM are providing FD-SOI technology. Sony's PlayStation, AMD's CPU, and PowerPC microprocessors are a few items that use SOI technology. Synopsys provides the most complete and efficient FinFET solutions in the market in partnership with IDMs, foundries, and academic institutions. More than 90% of the top volume-production SoCs have been built using the Synopsys Design Family, and the majority of FinFET devices in use today are created using Synopsys' TCAD tools. Synopsys also boasts the largest library of silicon-proven IP for FinFET.

TSMC started the foundry for 5 nm FinFET (N5) in 2020 and 3nm FinFET (N3) technology in 2022 for high volume production. This is their most advanced semiconductor technology as they claim. Due to its advantages FinFETs are used in applications like smart wearables, smart gadgets, in mobile phones, laptops, in routers, in switching devices, for medical, defense and aerospace devices. Figure 1.5 depicts the comparison of SG-MOSFET and DG-MOSFET with respect to sub threshold slope (SS) and DIBL with reference to the effective channel length (Leff). As L_{eff} reduces, the DIBL and SS increases for both MOSFETs. Compared to SG-MOSFET, DG-MOSFET has better characteristics which are required for the digital circuit design.

1.4.1.4 Future of FinFET

FinFETs will eventually be more energy-efficient and sized down to 5 nm or 3 nm, but this will come at a price. The leakage problem resurfaces at lower technological nodes, such as those below 10 nm. This causes a host

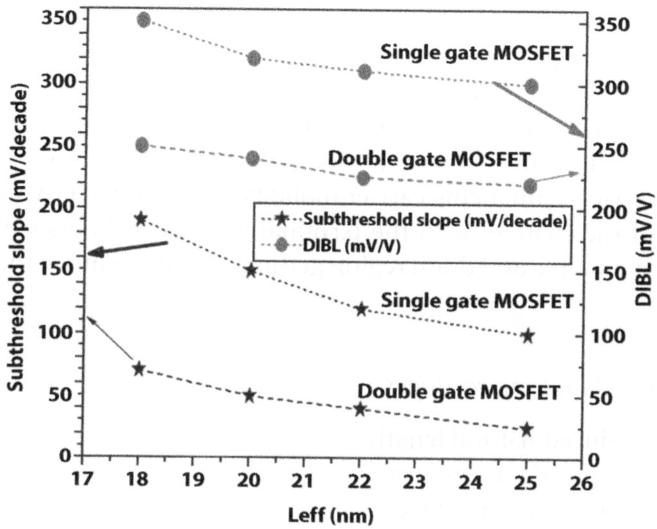

Figure 1.5 The variation of subthreshold swing and DIBL with respect to effective channel length for SG-MOSFET and DG-MOSFET.

of other problems, including heat loss, an increase in power density, and threshold flattening. Because heat can build up on the fins of a finFET construction, heat dissipation efficiency is lower. Thus, scientists are investigating novel approaches and have created the GAAFET. Its gate-like fins are many. It has a higher switching speed than a FinFET and can be shrunk down below 3 nm.

1.4.2 GAAFET

GAA FET device reported first in the late 1990s, features a unique architecture where the gate material surrounds the channel, unlike FinFETs, to achieve maximum electrostatic integrity. The cross-sectional views of square and cylindrical GAA devices demonstrate how cylindrical GAA-FETs offer the lowest natural length, facilitating further device scaling. In this configuration, carrier conduction occurs through the center of the device instead of being limited by the Si/SiO_2 interface, thereby mitigating interface related issues in device modeling. Key dimensions for device include the channel length (L_G), gate dielectric thickness (t_{ox}) and channel diameter (t_{si}). The three-dimensional view of a vertical cylindrical GAA FET device highlights its potential as the next-generation nano-structure device due to enhanced gate controllability and short channel effect (SCE) immunity. Nanowires represent the extreme case of GAA devices, and recently, cylindrical GAA FETs are created gainfully using silicon nanowire technology. This cylindrical design simplifies modeling due to the uniform field of electricity in the channel. In very thin film multiple gate SOI MOSFETs, volume inversion occurs as center of the film has high concentration of carriers, enhancing carrier mobility. The ability if GAA structure to surround the channel with the terminal gate prevents lines of electric field from the source and drain region getting into the channel, thus reducing SCEs.

1.4.2.1 *GAAFET Advantages*

i. Reduced natural length
ii. Increased current drive
iii. Reduced corner effect
iv. Reduced quantum effect

1.4.2.2 *Working of GAAFET*

In a p-MOSFET, a negative gate voltage can create an inversion channel close to the gate, forming a current path that electrically connects the p-doped regions under the drain and source contacts. Applying this hypothesis to a silicon nanowire FET, the ideal structure features an n-doped region in the gate area and p-doped regions elsewhere, as depicted in Figure 1.6.

An inversion channel can be formed along the whole length of the nanowire even if the gate-source and gate-drain distances are not too great. By applying a sufficient negative gate voltage, an inversion region is formed

Figure 1.6 3D structure of GAAFET device with circular cross-section.

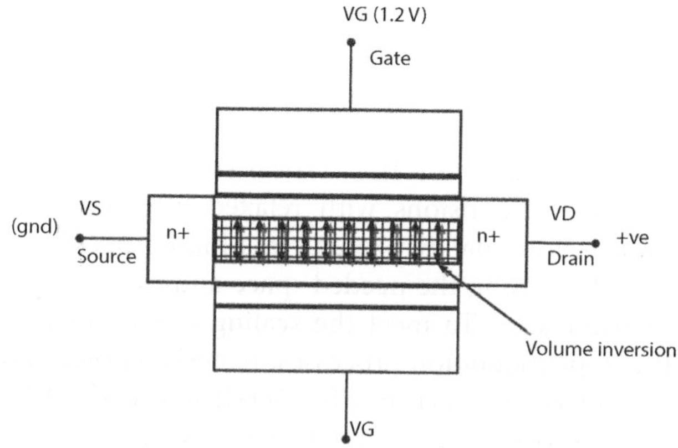

Figure 1.7 Cross-sectional view of GAAFET device.

directly under the gate, effectively inverting the nanowire. Upon applying a drain voltage, carriers can then flow through this inverted channel from the source to the drain.

Since the gate surrounds the channel in a GAAFET, the inversion region is formed at the center of the channel, forming a uniform distribution of electric field across the channel. The three-dimensional longitudinal view of an n-GAAFET in Figure 1.7 illustrates this volume inversion which occurs at the mid of channel when a gate is connected to positive voltage. Under these conditions, carrier conduction takes place through the nanowire channel from the source to the drain.

1.4.2.3 Advancement in GAA Device Design

The major challenge in GAA device design is its process complexity. However, GAA nanowire FET (GAANWFET) is considered the ultimate short-channel device for future technology. Metal gates with HKMG dielectric stacks give the required threshold voltage (V_{TH}) through tuning of work function while keeping lower gate leakage. Employing metal gates instead of poly-silicon stacks overcomes the effect of poly depletion. This benefits effective majority carrier mobility by decreasing the transverse field. In addition to that, the utilization of un-doped channels enhances low field effective carrier mobility by reducing scattering of substrate impurity. In the absence of pocket implants, mobility deterioration resulting from Coulomb scattering in short-channel devices can be further minimized.

Using III-V materials like InGaAs in place of NFET channels and greater carrier mobility semiconductors like Ge in place of PFET channels within the MOSFET channel could also potentially improve performance.

The vertical structure of GAA FETs offers advantages and challenges compared to lateral structures. The vertical structure allows for further reduction in device dimensions with relaxed lithography for channel length definition. Additionally, it results in a higher packing density and drive current by decreasing the needed space and increasing the channel width per transistor area. To meet the scaling requirements, research is ongoing on the implementation processes, material choice and ingenious unconventional device architectures. The 3D structure of GAA FET implemented in TCAD sprocess [32] is shown in Figure 1.8.

The highest level in electrostatic control is provided by the GAA FET nanowire channel body, which has the ability to push the gate length down to a few nanometers. The primary challenge is to develop large-scale manufacturing methods for these structures without introducing contamination issues. Samsung's GAA technology represents a significant breakthrough in semiconductor design, overcoming the performance limitations inherent in the traditional FinFET architecture. This innovative approach delivers tangible benefits, including improved power efficiency through reduced supply voltage levels, as well as better working by enhancing the capacity of drive current. An important turning point in the semiconductor

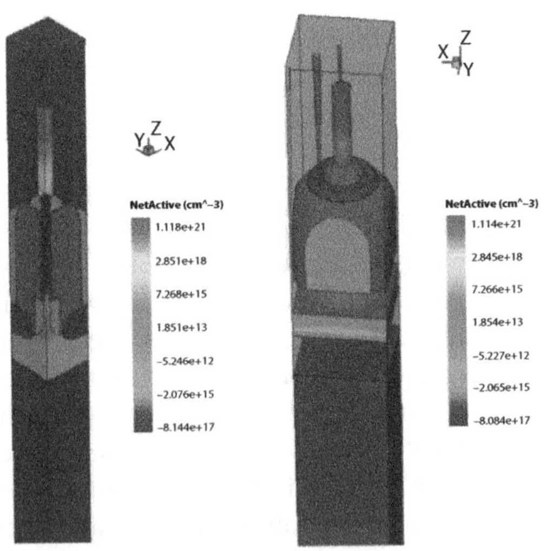

Figure 1.8 Three dimensional view of the vertical GAAFET device structure developed in Sentuarus TCAD sprocess tool.

industry has been reached with Samsung's adoption of GAA technology, which opens the door for further improvements in device performance and energy efficiency.

1.4.2.4 Future of GAAFET

Despite having many advantages over the widely used FinFET, GAAFET development has been extremely slow because of its more complicated structure, more stringent material requirements, and incompatibility of its manufacturing and design processes with existing technology platforms. Few big businesses, including Samsung and TSMC, are able to manufacture GAAFET using low-process technology. At the 2022 Technology Symposium, TSMC introduced N2, or 2nm, next-generation advanced process technology [44]. It is anticipated to go into mass production by 2025. Samsung declared the beginning of manufacturing for its 3 nm chips with the GAAFET design on the 30th of that same month [45]. There is a huge market need for low-process chips, but at the moment, only two companies worldwide are working on GAAFET research and development: Samsung and TSMC. For instance, to remain competitive in the market, tech giants like Apple and Intel, which are unable to independently develop low-process chips in the future, will require processors with more performance and smaller areas. The competition between TSMC and Samsung is still level today, but if one company can produce low-process GAAFET chips in large quantities and achieve a notable performance boost, this equilibrium will be quickly upset, and numerous tech companies will place orders.

1.4.3 Reconfigurable FET (RFET)

The rapid development of smart devices has driven a major downsizing in the actual size of electronic components. As device dimensions reduce, maintaining effective gate control on the channel in planar type transistors turns to be increasingly challenging. To address this, new device architectures have emerged that can outdo traditional planar type transistors in terms of area, power consumption/dissipation and speed. These include multiple-gate devices, tunnel FETs, carbon nanotubes, reconfigurable FETs (RFETs) and single-electron devices. Among these, the R-FET is an upcoming multigate device that can be structured as either an p or n type FET by applying a suitable voltage to the leads. RFET is an electronic device in which the mechanism of conduction is reversibly reconfigure between p-type and n-type operation modes. These devices use electrostatic doping, or the creation of mobile carriers by an external potential, rather than chemical doping brought on by impurities to enable this capability. In a nominally intrinsic semiconductor, either electrons or holes are injected, depending on the bias.

The three dimension structure of RFET is shown in Figure 1.9 which is developed in Sentuarus TCAD sprocess tool [32]. The construction of R-FET devices is less complex compared to conventional MOS transistors, as it doping is not required. Various RFET device structures have been reported, including single-gate (SG-RFET), double-gate, and triple-gate configurations. The SG-RFET, in particular, has a very simple structure that is likely to GAA FET architecture. Compared to other RFET designs, the SG-RFET offers a high-performance solution with a straightforward structure.

Digital circuits realized using RFET devices have shown good performance, comparable to other multi-gate transistors. Researchers have explored the electrostatics and performance of RFET devices realized on planar fully depleted SOI, as well as high-speed analog applications using single-gate and double-gate RFETs. Reconfigurable circuits employing polarity controlled double gate RFETs have demonstrated superior performance, including the first demonstration of a two-transistor fully functional XOR gate. Inverter circuits based on GAAFETs with elliptical cross-sections have shown improvements in delay compared to circular cross-section devices. Experimental demonstrations of current mirrors using silicon nanowire transistors, as well as analog and digital circuits implemented using tunnel FETs, have shown promising results, including low-power operation at low supply voltages. RFETs with strained silicon have been explored to improve carrier mobility.

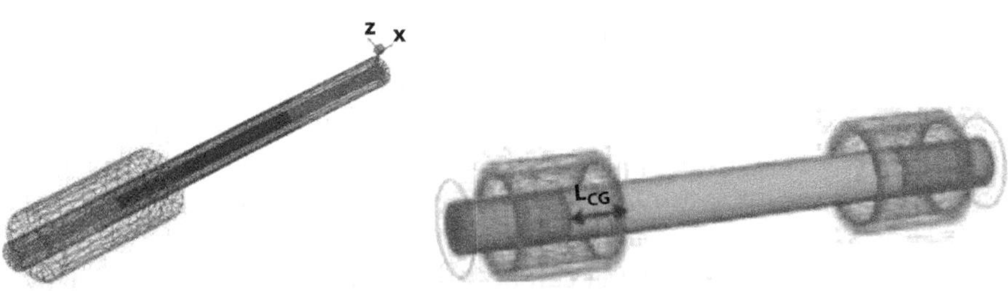

Figure 1.9 3D structure of single gate RFET and double gate RFET developed in Sentuarus TCAD device editor SDE.

1.4.3.1 Future of RFET

In addition to their advantages over CMOS counterparts, RFETs have been shown to have plenty of features at circuit level, like intrinsic XOR and wired-AND capabilities, dynamic reconfiguration, Vth control, and subduing parasitic effects of charge sharing in dynamic logic gates [42]. Emerging applications in the field of hardware security and neuromorphic computing are explained in the work of T. Mikolajick *et al.* [42].

Recent groundbreaking research has demonstrated that RFETs' increased expressiveness results in additional benefits at the circuit level as opposed to the device level. According to the simulation results in [43], Germanium nanowires can be a promising material for future reconfigurable electronic systems. Various MOSFET performance features are illustrated in [43–48].

1.4.4 Tunnel FET

TFETs known as Tunnel FETs are the recent device architecture that can potentially outperform conventional MOSFET transistors, especially in terms of power consumption and scalability [11]. The key operating principle of a T-FET relies on band to band tunneling (BTBT) rather than the traditional thermionic emission mechanism in MOSFETs. In a TFET, the channel and source regions are doped with opposite polarities, creating a p-i-n structure. When a suitable gate voltage is applied, it modulates the energy bands at the channel-source junction, enabling carriers to excavate from the source region into the channel area whose structure is depicted in Figure 1.10.

1.4.4.1 TFET Advantages

i. Steep sub threshold slope
ii. Low operating voltage

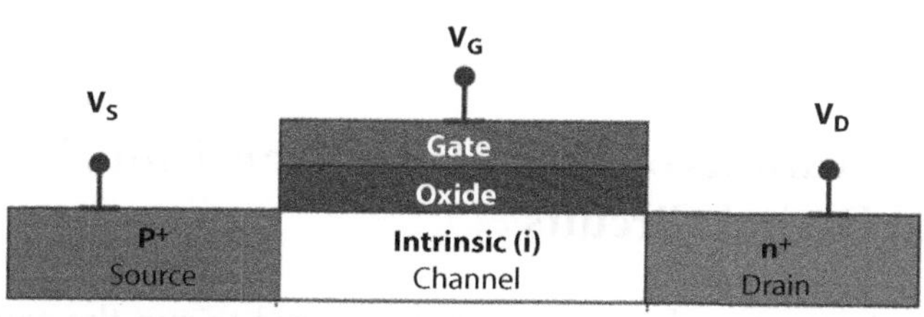

Figure 1.10 Two dimensional structure of TFET.

iii. Scalability
iv. Ambipolar behavior

However, TFETs also face challenges, such as achieving high on-current, optimizing the source-channel junction, and overcoming the inherent tradeoff between on-current and sub threshold slope. Ongoing research is focused on improving TFET performance through novel device structures, material engineering, and optimization of the tunneling mechanism. TFETs are considered a favorable candidate for future energy-efficient low-power, electronic devices and systems, particularly in applications like smart phones, IoT, and energy-restricted systems. SRAM cell using TFET for ultra-low-power IoT applications is demonstrated in literature which shows possibilities of using TFET devices in memory circuits [11].

1.4.4.2 Future of TFET

Comparing TFETs to conventional transistors, a significant reduction in power consumption is possible because of the quantum tunneling concept under which they function. Their potential for energy-efficient electronics is enhanced by their capacity to function at reduced voltages. Additionally, TFETs have better scalability and are ready to meet semiconductor manufacturing's expectations for downsizing. Researchers and major industry players are investing extensively in R&D to develop this technology, which is driving the TFET market. Potential uses include mobile devices, the Internet of Things, healthcare, and other industries. TFETs are expected to increase rapidly, which suggests that in the years to come, they may completely transform the electronics industry. Though TFETs have great potential, there are still issues to be resolved, including difficult fabrication processes and attaining high performance levels at room temperature. Nonetheless, partnerships between academic institutions and business entities, in addition to ongoing developments in nanotechnology, provide a path forward to surmount these obstacles and realize the complete potential of TFETs.

1.5 Applications of Non-Planar Transistors in Analog and Digital Circuits

High computation speed processors are required to run the complex AI/ML algorithms. The analysis of analog and digital circuits using new device architecture is inevitable. The performance analysis of the various circuits using new device architectures is reported in the literature [22–25, 32–38]. In [39], planar MOSFETs and their applications in IC design along with challenges in IC design due to the scaling down of the device dimension are explained in detail. GAA-FETs replacing FinFETs in the logic IC design, circuits designed using FinFET with multiple fins and fabrication challenges and limitations of FinFET beyond 3 nm technology in realizing digital circuits are detailed in [40].

The various applications of analog and digital circuits which require new device architectures are discussed in the coming section.

1.5.1 Applications in Digital Circuits

i. **High-Speed Microprocessors:** The improved switching speeds and reduced power consumption make non-planar transistors ideal for CPUs and GPUs.
ii. **Memory Devices:** Used in DRAM and NAND flash, these transistors help in reducing leakage and increasing storage density.
iii. **Mobile Devices:** Lower power consumption makes them suitable for battery-operated devices like smartphones and tablets.

1.5.2 Applications in Analog Circuits

i. **Analog-to-Digital Converters (ADC):** Non-planar transistors improve the performance of ADCs by offering higher linearity and lower noise.
ii. **Operational Amplifiers (Op-Amps):** Benefits include lower offset voltages and reduced thermal noise, enhancing the precision of analog circuits.
iii. **RF Circuits:** Improved high-frequency performance makes them suitable for radio frequency applications, improving the efficiency and effectiveness of RF amplifiers and mixers.

1.5.3 Design Considerations

i. **Layout and Fabrication:** Non-planar transistors require advanced fabrication techniques and careful layout to ensure performance and yield.

ii. **Thermal Management:** 3D structures can lead to hotspots, necessitating efficient thermal management solutions.
iii. **Simulation and Modeling:** Increased complexity of non-planar devices requires advanced simulation tools to accurately predict their behavior in circuits.

1.5.4 Challenges and Future Trends

i. **Complex Fabrication Processes:** The transition from planar to non-planar technology involves significant changes in fabrication processes, which can be complex and costly.
ii. **Integration with Existing Technologies:** Non-planar transistors need to be integrated with existing planar technologies, requiring compatibility at various levels of the design and fabrication process.
iii. **Continued Scaling:** While non-planar transistors provide a path for continued scaling, future innovations will need to address limitations as we approach atomic scales.
iv. **Hybrid Approaches:** Combining non-planar transistors with emerging technologies like 2D materials and quantum dots could unlock new potentials in both analog and digital circuits. In [39], planar MOSFETs and their applications in IC design is discussed in detail. The challenges in IC design due to the scaling down of the device dimension are explained in detail. In [40, 41], the emerging device GAAFET that can replace FinFET in logic IC design is explained in detail with the challenges in fabrication processes. NWFET features a nanoscale wire as its channel, allowing for increased mobility and scalability.

GAA-FET, on the other hand, has a gate electrode surrounding the channel, providing better control over the flow of current. Both types of FETs offer lesser power consumption, increased integration density and improved performance, making them suitable for applications in emerging technologies such as IoT devices, 5G networks, and next-generation computing. Table 1.1 depicts the comparison of FinFET SRAM circuit and nanowire FET SRAM circuit [41]. N type Nanowire FET SRAM has better performance compared to FinFET in terms of saturation current (I_{Dsat}), leakage current (I_{off}), threshold voltage (V_t), DIBL and SS. Along with SRAM, logic gates, ring oscillators and comparison of various circuit parameters are also detailed in [41].

Table 1.1 Comparison of SRAM FinFET and Nanowire FET for L_G = 16 nm, Leff = 14 nm, Wspacer = 5nm.

	FinFET		**Nanowire FET**	
	nMOS	**pMOS**	**nMOS**	**pMOS**
I_{Dsat} (μA/fin)	24.2	18.9	33.8	25.01
I_{off} (μA/fin)	4.6	3.93	2.72	1.78
V_t (V)	0.301	-0.357	0.275	-0.317
DIBL (mV/V)	69.24	69.24	21.55	24.63
SS (mV/Decade)	75.30	75.70	64.212	65.019

GAAFETs are suitable for high power and high frequency applications, like power management, analog-to-digital converters, and radio-frequency transmitters. In IC design, GAAFETs offer a reliable and efficient means of handling high-power signals while minimizing power consumption which makes it suitable to replace the current FinFET. Multigate devices are particularly useful in applications requiring low power consumption, high density and high speed, justifying them a popular choice for modern integrated circuit design.

The performance comparison of various MOSFET devices is summarized in Table 1.2. and Table 1.3 based on application and structure respectively. RFET, TFET are emerging structures which can replace the existing devices in future. Table 1.2 presents the vertical structures which are used in high power and communication applications.

1.6 Conclusion

Non-planar transistors represent a significant advancement in semiconductor technology, with widespread implications for both analog and digital circuits. They offer advantages in terms of performance, power efficiency, and scalability, making them a crucial component in the continued development of modern electronic devices. As fabrication techniques evolve and new materials are explored, the role of non-planar transistors is expected to expand, driving further innovations in the semiconductor industry. An overview of the evolution of transistors, various device architectures and the challenges involved in the implementation are discussed

Table 1.2 Performance comparison of various MOSFET architectures based on application.

MOSFET architecture	Planar MOSFET	Trench MOSFET	Superjunction MOSFET	FinFET	GaN MOSFET	SiC MOSFET
Key Features	Standard 2D structure	3D structure with vertical channels	Multiple layers for improved performance	3D Fin structure	Gallium nitride material	Silicon carbide material
Advantages	Simple fabrication, mature technology	Reduced on-resistance, better performance at high frequencies	Higher breakdown voltage, lower RDS(on)	Reduced leakage, improved electrostatics	High efficiency, fast switching	High thermal conductivity, high voltage operation
Applications	General switching applications	Power supplies, motor drives	High-voltage applications, automotive	Advanced ICs, high-performance applications	RF amplifiers, power conversion	Electric vehicles, industrial power systems
On-Resistance (RDS(on))	$< 100\ \Omega$	Low ($\leq 10\ m\Omega$)	Very low ($\leq 5\ m\Omega$)	Very low (dependent on fin height)	Very low ($\leq 3\ m\Omega$)	Low ($\leq 10\ m\Omega$)

(Continued)

Table 1.2 Performance comparison of various MOSFET architectures based on application. (*Continued*)

MOSFET architecture	Planar MOSFET	Trench MOSFET	Superjunction MOSFET	FinFET	GaN MOSFET	SiC MOSFET
Breakdown Voltage (V(BR))	> 50 V	Moderate (≤ 600V)	High (up to 1200V)	Moderate to high (up to 1kV)	High (up to 1kV or more)	Very high (up to 3kV)
Switching Speed	Fast (up to 100 kHz)	Fast (up to 100 kHz)	Fast (up to 100 kHz)	Very fast (GHz range)	Extremely fast (up to MHz)	Moderate to fast (up to 100 kHz)
Thermal Performance	average	Good due to vertical structure	Excellent, can handle high power	Excellent, better thermal scaling	Excellent, high thermal conductivity	Excellent, good thermal handling
Key Advantages	Compact, low power	High efficiency, compact design	Very low conduction losses	Reduced leakage, improved drive current	High efficiency, high frequency operation	High temperature operation, ruggedness
Typical Applications	Digital designs, analog designs, low power circuits	Power supplies, motor drives	High-voltage power conversion, automotive	High-performance ICs, CPUs, GPUs	RF applications, power converters	Electric vehicles, industrial applications

Table 1.3 Performance Comparison of various MOSFET architectures based on structure.

Feature/technology	Structure	Scalability	Short channel effects	Drive current	Subthreshold swing	Power consumption	Device variability	Fabrication complexity	Applications	Market maturity
Planar MOSFET	2D planar	Limited beyond 5nm	Significant	Moderate	60 mV/decade	High at small nodes	High	Moderate	Digital circuits	Mature
FinFET	3D fin structure	Better scalability	Reduced	High	60 mV/decade	Lower than MOSFET	Moderate	Higher	High-performance CPUs	Emerging
GAAFET	3D gate-all-around	Excellent scalability	Further reduced	Higher than FinFET	< 60 mV/decade	Very low	Low	High	Emerging applications	Experimental
Carbon Nanotubes	1D nanostructures	High potential	Minimal effects	Extremely high	< 60 mV/decade	Very low	Very low	Very high	Nanoscale devices	Research stage
Reconfigurable FET	Flexible architectures	Highly scalable	Variable	Moderate to high	Variable	Low	Low	Moderate to high	Adaptive circuits	Emerging
Tunnel FET	3D layered	Limited by tunneling	Low due to tunneling	Moderate	< 60 mV/decade	Very low	Moderate	Moderate	Low-power devices	Emerging

in this chapter. Circuits using multigate devices have gained attention in the field of electronics due to their potential for improved performance and power efficiency. These devices feature multiple gate electrodes, allowing them to control the flow of current with greater precision and flexibility. This added control enables the creation of complex digital circuits with reduced power consumption, making them suitable for applications namely mobile devices, IoT devices, and AI systems. Analog as well as digital circuits using multigate transistors and future of non-planar transistors are overviewed in this chapter.

References

1. Iwai, H. and Misra, D., The transistor was invented 75 years ago: A big milestone in human history. *Electrochem. Soc. Interface*, 31, 4, 65, 2022.
2. Maloberti, F. and Davies, A.C. (Eds.), *A short history of circuits and systems*, CRC Press, 2022.
3. Cao, W., Bu, H., Vinet, M., Cao, M., Takagi, S., Hwang, S., Ghani, T., Banerjee, K., The future transistors. *Nature*, 620, 7974, 501–515, 2023.
4. Bhol, K., Jena, B., Nanda, U., Journey of Mosfet from planar to gate all around: A review. *Recent Pat. Nanotechnol.*, 16, 4, 326–332, 2022.
5. Kajal, and Sharma, V.K., FinFET: A Beginning of Non-Planar Transistor Era, in: *Nanoscale VLSI: Devices, Circuits and Applications*, pp. 139–159, 2020.
6. Sun, Y., Yu, X., Zhang, R., Chen, B., Cheng, R., The past and future of multi-gate field-effect transistors: Process challenges and reliability issues. *J. Semicond.*, 42, 2, 023102, 2021.
7. Samal, A., Tripathi, S.L., Mohapatra, S.K., A journey from bulk MOSFET to 3 nm and beyond. *Trans. Electr. Electron. Mater.*, 21, 5, 443–455, 2020.
8. KhademHosseini, V., Dideban, D., Ahmadi, M.T., Ismail, R., An analytical approach to model capacitance and resistance of capped carbon nanotube single electron transistor. *AEU Int. J. Electron. Commun.*, 90, 97–102, 2018.
9. Lin, Y.-M., Appenzeller, J., Knoch, J., Avouris, P., High-performance carbon nanotube field-effect transistor with tunable polarities. *IEEE Trans. Nanotechnol.*, 4, 5, 481–489, 2005.
10. Heinzig, A., Slesazeck, S., Kreupl, F., Mikolajick, T., Weber, W.M., Reconfigurable Silicon nanowire transistors. *Nano Lett.*, 12, 1, 119–124, 2011.
11. Ahmad, S., Alam, N., Hasan, M., Robust TFET SRAM cell for ultralow power IoT applications. *AEU Int. J. Electron. Commun.*, 89, 70–76, 2018.
12. Weber, W.M., Geelhaar, L., Lamagna, L., Fanciulli, M., Kreupl, F., Unger, E., Riechert, H., Scarpa, G., Lugli, P., Tuning the polarity of Si-nanowire transistors without the use of doping, in: *8th IEEE Conference on Nanotechnology, NANO'08*, pp. 580–581, 2008.

13. Wessely, F., Krauss, T., Schwalke, U., CMOS without doping: Multi-gate silicon-nanowire field-effect-transistors. *IEEE J. Solid-State Circuits*, 70, 33–38, 2012.
14. Sacchetto, D., Leblebici, Y., De Micheli, G., Ambipolar gate-controllable SiNW FETs for configurable logic circuits with improved expressive capability. *IEEE Electron Device Lett.*, 33, 2, 143–145, 2012.
15. Darbandy, G., Claus, M., Schröter, M., High-performance reconfigurable Si nanowire field-effect transistor based on simplified device design. *IEEE Trans. Nanotechnol.*, 15, 2, 289–294, 2016.
16. Navarro, C., Barraud, S., Martinie, S., Lacord, J., Jaud, M.-A., Vinet, M., Reconfigurable field effect transistor for advanced CMOS: Advantages and limitations. *IEEE J. Solid-State Circuits*, 128, 155–162, 2017.
17. Darbandy, G., Schröter, M., Claus, M., Optimization of reconfigurable SiNWFETs for analog high-frequency applications, in: *International Conference on Simulation of Semiconductor Processes and Devices (SISPAD)*, pp. 245–248, 2016.
18. De Marchi, M., Sacchetto, D., Frache, S., Zhang, J., Gaillardon, P.-E., Leblebici, Y., De Micheli, G., Polarity control in double-gate, gate-all-around vertically stacked silicon nanowire FETs, in: *IEEE International Electron Devices Meeting (IEDM)*, IEEE, pp. 8–4, 2012.
19. Kumar, S. and Jha, S., Impact of elliptical cross-section on the propagation delay of multi-channel gate-all-around MOSFET based inverters. *Microelectron. J.*, 44, 9, 844–851, 2013.
20. De Marchi, M., Zhang, J., Frache, S., Sacchetto, D., Gaillardon, P.-E., Leblebici, Y., De Micheli, G., Configurable logic gates using polarity-controlled silicon nanowire gate-all-around FETs. *IEEE Electron Device Lett.*, 35, 8, 880–882, 2014.
21. Gaillardon, P.-E., Amaru, L., Zhang, J., De Micheli, G., Advanced system on a chip design based on controllable-polarity FETs, in: *Proceedings of the Conference on Design, Automation & Test in Europe*, European Design and Automation Association, p. 235, 2014.
22. Huang, R., Zou, J., Wang, R., Fan, C., Ai, Y., Zhuge, J., Wang, Y., Experimental demonstration of current mirrors based on silicon nanowire transistors for inversion and subthreshold operations. *IEEE Trans. Electron Devices*, 58, 10, 3639–3642, 2011.
23. Wang, D., Sheriff, B.A., Heath, J.R., Complementary symmetry silicon nanowire logic: Power-efficient inverters with gain. *Small*, 2, 10, 1153–1158, 2006.
24. Qiu, X., Chen, D., Wang, Z., Response of ring oscillator to periodic interference on the power supply. *AEU Int. J. Electron. Commun.*, 82, 383–390, 2017.
25. Strangio, S., Settino, F., Palestri, P., Lanuzza, M., Crupi, F., Esseni, D., Selmi, L., Digital and analog TFET circuits: Design and benchmark. *Solid State Electron.*, 146, 50–65, 2018.

26. Heinzig, A., Mikolajick, T., Trommer, J., Grimm, D., Weber, W.M., Dually active silicon nanowire transistors and circuits with equal electron and hole transport. *Nano Lett.*, 13, 9, 4176–4181, 2013.
27. Moura, R., Tiencken, N., Mothes, S., Claus, M., Blawid, S., Reconfigurable nanofets: Performance projections for multiple-top-gate architectures. *IEEE Trans. Nanotechnol.*, 17, 3, 467–474, 2018.
28. Simon, M., Liang, B., Fischer, D., Knaut, M., Tahn, A., Mikolajick, T., Weber, W.M., Top-down fabricated reconfigurable fet with two symmetric and high-current on-states. *IEEE Electron Device Lett.*, 41, 7, 1110–1113, 2020.
29. Fuchs, F., Gemming, S., Schuster, J., Electron transport through nisi2–si contacts and their role in reconfigurable field-effect transistors. *J. Phys.: Condens. Matter*, 31, 35, 355002, 2019.
30. Rikabi, A.L., Salim, H.TH., Ali, G.M., A survey on the latest FET technology for samsung industry. *AIP Conf. Proc.*, 2977, 1, AIP Publishing, 2023. https://doi.org/10.1063/5.0181953
31. Samsung Begins Chip Production Using 3nm Process Technology with GAA Architecture (Press release). Samsung. Archived from the original on 30 June 2022.
32. Jayachandran, R., Komaragiri, R.S., Dhanaraj, K.J., Simulation of Reconfigurable FET Circuits Using Sentaurus TCAD Tool, in: *Sub-Micron Semiconductor Devices*, pp. 165–180, CRC Press, 2022.
33. Jayachandran, R., K J, D., Subramaniam, P.C., Planar CMOS and multigate transistors based wide-band OTA buffer amplifiers for heavy resistance load. *Facta Univ. Ser. Electron. Energ.*, 35, 1, March 2022.
34. Jayachandran, R., Komaragiri, R.S., Subramaniam, P.C., Ranjith, R., High Performance Reconfigurable FET for a Simple Variable Gain Buffer Amplifier Design. *Int. J. Electron.*, 109, 1, 100–118, Jan. 2022.
35. Ranjith, R., Jayachandran, R., Suja, K.J., Komaragiri, R.S., Two dimensional analytical model for a reconfigurable field effect transistor. *Superlattices Microstruct.*, 114, 62–74, February 2018.
36. Jayachandran, R., Komaragiri, R.S., Subramaniam, P.C., Reconfigurable circuits based on single gate reconfigurable field effect transistors, in: *6th IEEE International Conference on Electronics, Computing and Communication Technologies (CONECCT)*, GITAM University, Bangalore, July 2020.
37. Jayachandran, R., Komaragiri, R.S., Subramaniam, P.C., Study of Circuits Based on SOI-Vertical Gate-All-Around FET, in: *15th IEEE India Council International Conference, INDICON*, Amrita Vishwa Vidyapeetham Coimbatore, December 2018.
38. Jayachandran, R., High performance OTA buffer amplifiers for driving resistive loads, http://hdl.handle.net/10603/449954.
39. El Ghouli, S. and Rideau, D., Planar MOSFETs and Their Application to IC Design, in: *Springer Handbook of Semiconductor Devices*, pp. 391–421, Springer International Publishing, Cham, 2022.

40. Kalinin, S.V., Ostertak, D.I., Poteryaev, D.A., Kuznetsov, M.A., A New Revolution in Logic Silicon IC Technology: GAA FETs are Replacing FinFETs, in: *2023 IEEE XVI International Scientific and Technical Conference Actual Problems of Electronic Instrument Engineering (APEIE)*, IEEE, pp. 160–165, 2023.
41. Vashishtha, V. and Clark, L.T., Comparing bulk-Si FinFET and gate-all-around FETs for the 5 nm technology node. *Microelectron. J.*, 107, 104942, 2021.
42. Mikolajick, T., Galderisi, G., Rai, S., Simon, M., Böckle, R., Sistani, M., Cakirlar, C., Bhattacharjee, N., Mauersberger, T., Heinzig, A., Kumar, A., Weber, W.M., Trommer, J., Reconfigurable field effect transistors: A technology enablers perspective. *Solid State Electron.*, 194, 2022, ISSN 0038-1101, https://doi.org/10.1016/j.sse.2022.108381.
43. Trommer, J., Heinzig, A., Heinrich, A., Jordan, P., Grube, M., Slesazeck, S., Mikolajick, T., Weber, W., Material Prospects of Reconfigurable Transistor (RFETs) – From Silicon to Germanium Nanowires. *MRS Proc.*, 1659, 225–230, 2014, https://doi.org/10.1557/opl.2014.110.
44. Wannan, End FinFET transistors! TSMC officially announced a 2nm power consumption reduction of 30%, https://news.mydrivers.com/1/839/839766.htm, 2022.
45. Samsung, Samsung Begins Chip Production Using 3nm Process Technology With GAA Architecture, https://semiconductor.samsung.com/news-events/news/samsung-begins-chip-production-using-3nm-process-technology-with-gaa-architecture/?wcmmode=disabled, 2022.
46. Tripathi, S.L., Pathak, P., Kumar, A., *et al.*, Improved Drain Current with Suppressed Short Channel Effect of p+Pocket Double-Gate MOSFET in Sub-14 nm Technology Node. *Silicon*, 14, 10881–10891, 2022, https://doi.org/10.1007/s12633-022-01816-2.
47. Tripathi, S.L., Mishra, R., Narendra, V., Mishra, R.A., High performance Bulk FinFET with bottom spacer. *2013 IEEE International Conference on Electronics, Computing and Communication Technologies*, Bangalore, India, pp. 1–5, 2013, doi: 10.1109/CONECCT.2013.6469282.
48. Bhattacharya, S., Tripathi, S.L., Kamboj, V.K., Design of tunnel FET architectures for low power application using improved Chimp optimizer algorithm. *Eng. Comput.*, 39, 1415–1458, 2023, https://doi.org/10.1007/s00366-021-01530-4.

2

Multi Gate MOSFET Architectures

Vrinda Gupta[1], L.G. Naveen Kumar[2*] and K.N.S. Santosh[2]

[1]*Dept. of Electronics & Communication Engineering, NIT Kurukshetra, Kurukshetra, India*
[2]*School of VLSI Design and Embedded Systems, NIT Kurukshetra, Kurukshetra, India*

Abstract
With conventional planar MOSFETs, the ongoing transistor downsizing in integrated circuits (ICs) has reached a physical limit. Device performance deteriorates as a result of the increased short-channel effects (SCEs) caused by scaling. Comparing multi-gate MOSFETs to traditional transistors is an interesting idea, as it greatly enhances electrostatic control and lowers SCEs. Fin-FETs and trigate devices, which have gates on three sides of the channel, are examples. This chapter investigates the concept of multi-gate MOSFET architecture.

With better gate control and channel electrostatics, the multi-gate MOSFETs particularly the FinFETs and nanowire design has become a key development in semiconductor technology. This architecture uses several gates surrounding the channel to achieve improved scalability and performance, propelling advancements in computing power and energy efficiency. The underlying ideas, Classifications of multi-gate MOSFETs, MOSFET architectures, manufacturing processes, Applications and future work on multi-gate MOSFETs are examined.

***Keywords*:** FinFET, trigate, multi-gate MOSFET, scaling, integrated circuits, short-channel effects

**Corresponding author*: naveenkumarlg1998@gmail.com

Suman Lata Tripathi, Vrinda Gupta, Sobhit Saxena, Shipra Upadhyay and Sudip Ghosh (eds.) Micro and Nano Semiconductor Devices for Digital, Analog and Sensor Design, (31–50) © 2026 Scrivener Publishing LLC

2.1 Introduction

In multi-gate control, short-channel effects are still present in advanced MOSFET structures, despite the fact that silicon thickness scaling is essential for mitigating their impacts [1]. Here, the extremely thin SOI layer and BOX thickness are studied to gain better understanding of the source of the short-channel effects for independently controlled multi-gate MOSFET. Multi-gate MOSFETs, which have many advantages over conventional planar MOSFETs, have completely changed the semiconductor industry. Many gates encircle the semiconductor channels of these transistors, significantly changing the transistor's electrical properties and performance metrics. By enhancing electrostatic control over the channel and reducing short-channel effects, multi-gate MOSFETs answer important problems about transistor size reduction without sacrificing efficiency or performance.

Since Sekigawa and Hayashi's 1984 [2] proposal, the development of a self-aligned double gate MOSFET has been the ultimate objective for scientists and engineers. The first modern self-aligned vertical multi-gate MOSFET was presented in 1989 by D. Hisamoto *et al.* [3] and was known as a DELTA (totally depleted channel resistor). Multi-Gate transistors, also referred to as multiple gate field effect transistors, are MOSFETs that include numerous gates into a single device. Transistors, the microscopic engines that drive our gadgets, have always shrunk in size, which has allowed technology to improve.

Figure 2.1 illustrates how transistors evolved starting in the early 1990s in accordance with Moore's law. However, as transistors get smaller and smaller, they start to exhibit short-channel effects (SCEs), which are

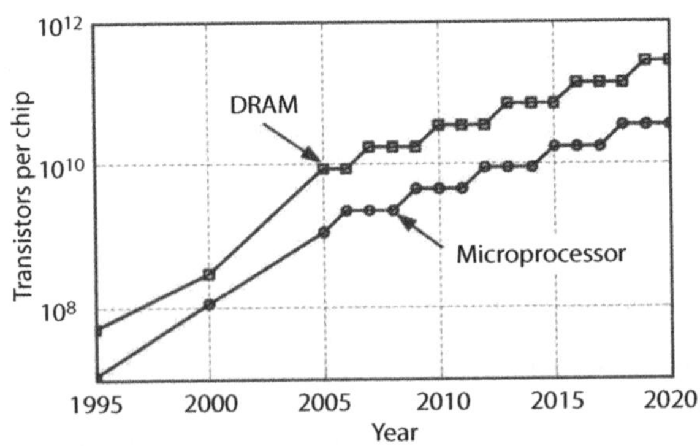

Figure 2.1 Evolution of the number of transistors per chip (Moore's law).

undesired electrical side effects. By interfering with transistors' regular operations, these SCEs result in leakage currents and reduce transistor performance [5].

Take a normal transistor and imagine it as a switch where one gate controls the electrical current. This idea takes on a completely new meaning with MuG-FET. Because MuG-FETs have many gates encircling the channel instead of just one, they function as a team to precisely control the current flow. Moore's law asserts that non-ideal effects occur when MOSFET dimensions are continuously dropped below the 10 nm node [4]. However, by the late 1990s, it is clear that a novel type of substrate called SOI (Silicon-On-Insulator) may greatly improve performance. This deviates from the ideal values of the device's characteristics and publishes a report detailing the various technologies, design tools, and other products that need to be created to keep up with the exponential growth of semiconductor devices. The FinFET, having a semiconductor channel with gates wrapped around in a form of fin shape, and nanowire MOSFETs, which use vertical nanowires as channels, are two well-known uses of multi-gate MOSFET technology. These designs can greatly lower leakage currents and improve transistor energy efficiency and switching speed by expanding the surface area accessible for gate control. The typical planar MOSFET, which has supported semiconductor devices for decades, is beginning to demonstrate its limitations as transistor widths approach nanoscale levels.

Transistor performance is compromised, and further downsizing is limited by short-channel phenomena including leakage currents and decreased gate control [1]. In response, these difficulties have led to the development of multi-gate MOSFET topologies like Fin-FETs and nanowire MOSFETs. Multi-gate MOSFET fabrication calls for sophisticated methods of semiconductor production, including epitaxial growth, exact etching procedures, and creative material integration. These procedures make it possible to precisely create complex three-dimensional transistor layouts at the nanoscale, guaranteeing dependable performance and scalability for the next semiconductor devices. The heightened short-channel effects (SCEs) brought on by this scaling impair device performance. It is interesting to consider a comparison between multi-gate MOSFETs (MuG-FETs) and conventional transistors.

A multi-gate-managed channel in these devices improves electrostatic control and reduces SCEs significantly [6]. Fin-FETs and Trigate devices which feature gates on three sides of the channel are common examples of semiconductors with fin-shaped channels. This research delves into the concept of multi-gate MOSFET architecture. The limitations of planar MOSFETs with respect to scalability are examined, along with how MuG-FETs get around them. But the Multi-Gate MOSFET (MuG-FET), a ground-breaking alternative, is just around the corner. MuG-FETs have an innovative architecture that addresses the constraints of miniaturization, in contrast to their traditional single-gate equivalent. Let's take a closer look at MuG-FETs and see how they are affecting the advancement of transistor technology. The total current flowing along all the interfaces covered by a multi-gate FET's gate electrode is nearly identical to the current drive. Since planarization lessens the workload for gate etch and photolithography especially if the gate stack is over the fin topography it is the recommended technique for flattening the gate surface [5]. To clear the underside of the fins, a significant over etching of the gate material is required. Therefore, the gate etch needs to have a high selectivity to the gated dielectric on top of the fin if one wants to shield it during gate etching. We then create the sidewalls of the gate and fin using S/D offset spacers. After the sidewall spacers are removed from the fins, selective epitaxy is used to expose the fins to grow, raised source, and drain. The parasitic resistance linked to thin fins can be decreased by strengthening the source and drain structure. Consequently, if carriers have the same mobility at each interface, it is equivalent to the current in a single-gate device multiplied by the corresponding number of gates. Each fin in a triple- or vertical-gate structure is the same width and thickness. To optimize power and delay, thin-film multi-gate MOSFETs are adaptively threshold-voltage regulated *via* distinct back-gate biassing. For a given gate length, the current drive can be fixed to a single discrete value. The current drive of a multi-fin MOSFET can be calculated by multiplying the value of the current flowing through a single fin with the total number of fins. Multi-Gate Field Effect Transistors (Mu-GFET) as depicted in Figure 2.2.

Realizing analog building blocks is essential for next-generation SoC solutions in these cutting-edge technologies. Therefore, using standard

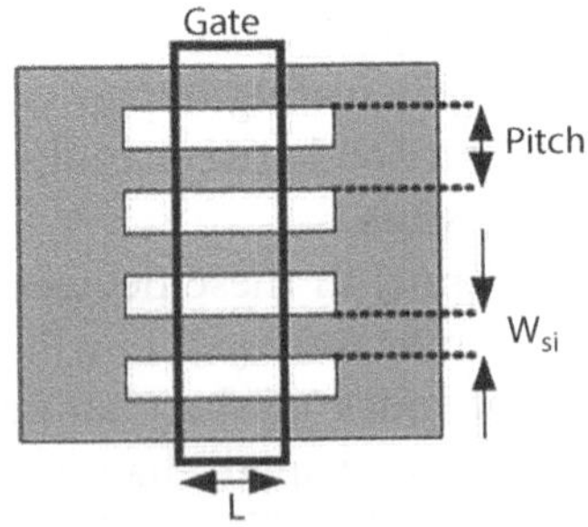

Figure 2.2 Multi-fin multi-gate FET layout.

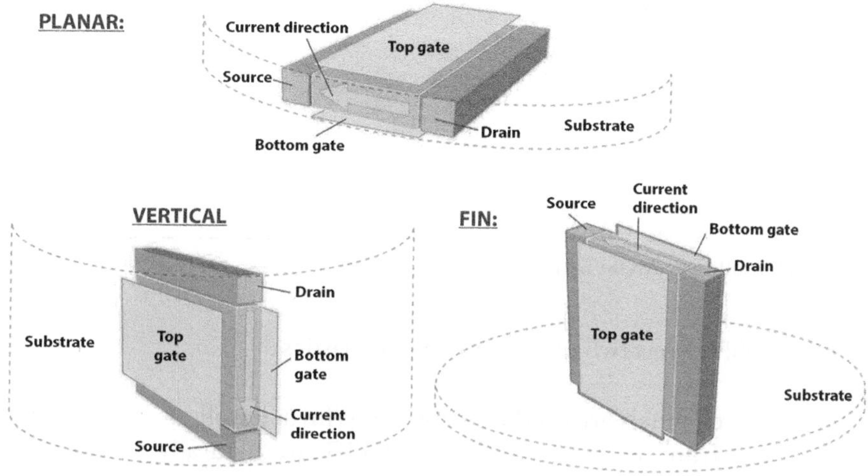

Figure 2.3 Structures of double gate MOSFET.

unbind SOI wafers, basic analog and digital circuits were built in an experimental Mu-GFET CMOS technology. This work investigates the link between digital and analog circuit design and technology, whereas the bulk of published Mu-GFET results to date have focused on the devices' performance and fabrication.

Figure 2.3 shows the Double Gate MOSFET's PLANAR, VERICAL, and FIN structures. A chapter summary is provided at the end of the chapter, which follows extensive talks on the benefits of double-gate MOSFETs, their applications, and their types.

2.2 About Multi-Gate MOSFETs

A multi-gate MOSFET is a device that has a unique structure. In a normal MOSFET, the current flowing through a channel is controlled by a single gate electrode, but in a multi-gate MOSFET it allows for more control over the current because it has numerous gates surrounding the channel. A single gate electrode can control several gates simultaneously, in which case the many gate surfaces function electrically as a single gate. Alternatively, separate gate electrodes can control multiple gates simultaneously. Different gate electrodes are part of the multi-gate layout of devices referred to as multiple-independent-gate field-effect transistors (MIGFETs). Multi-gate devices that are not planar, or three-dimensional, are the most commonly utilized.

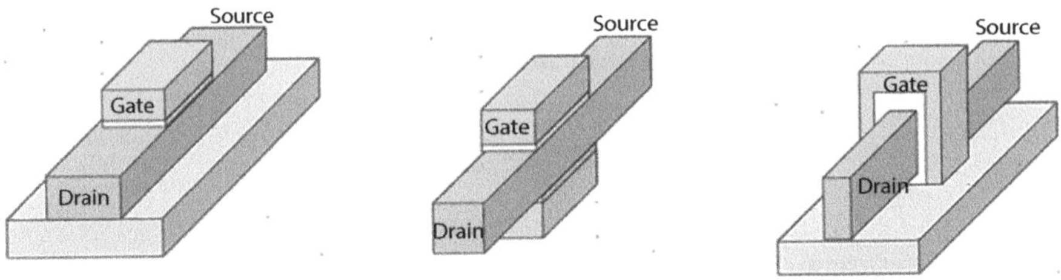

Figure 2.4 Multi-gate MOSFET and its modes.

The MOSFET has two important characteristics:

1) The channels are on the silicon body/fin sidewall
2) The front and rear gates of MOSFET are naturally in line with one other.

Figure 2.4 shows how Fin-FETs, Trigate, and Π/gate devices have all advanced from this initial idea.

The current in a single-gate device multiplied by the same number of gates is what happens if carriers have the same mobility at each interface. In a vertical double-gate or triple-gate configuration, each fin has the same width and thickness. This indicates that the driving current is fixed to a single discrete quantity for a given gate length. Larger currents are driven by the use of multi-fin device.

A comparison of single- and double-gate MOSFETs may be seen in Table 2.1 below.

2.3 Types of Multi-Gate MOSFETs

2.3.1 Double-Gate MOSFETs

The limitations of conventional single-gate MOSFETs were addressed by the development of double-gate MOSFETs (DG-MOSFETs), an advanced kind of field-effect transistor that becomes more and more relevant as device dimensions approach the nanoscale. This section provides a thorough analysis of several aspects of double-gate MOSFETs, such as their construction, benefits, manufacturing process, difficulties, and uses. Mobility degradation and short channel effects (SCEs) have a significant influence on CMOS technology below 100 nm [8].

Table 2.1 Comparison related to single gate and multi gate MOSFET.

Feature	Single-gate MOSFET	Multi-gate MOSFET
Structure	Single gate electrode controls a channel	Multiple gates surround the channel
Electrostatic Control	Limited control, especially as devices shrink	Superior control due to multiple gates
Short-Channel Effects	More susceptible to leakage currents as devices miniaturize	Reduced leakage due to better channel control
Current Drive	Lower current flow	Higher current flow
Subthreshold Swing	Less sharp transition between "on" and "off" states	Steeper subthreshold slope for sharper switching
Leakage Current	Higher leakage current in "off" state	Lower leakage current, leading to lower power consumption
Scalability	Scaling limitations as devices shrink due to short-channel effects	Improved scalability for continued miniaturization
Manufacturing Complexity	Simpler fabrication process	More complex and expensive to manufacture due to intricate structures
Applications	General-purpose electronics, mature technology	Nanoelectronics, low-power electronics, high-performance electronics (emerging)

2.3.1.1 *Structure of DG MOSFET*

In a dual-gate MOSFET, movement of carriers in the channel is regulated by two gates which is shown in Figure 2.5. Compared to traditional single-gate MOSFET structure, DG MOSFET structure has a number of benefits, such as better channel control, a decrease in short-channel effects, and better scalability. Poisson's three-dimensional (3D) charge density equation has been solved with the proper boundary conditions, yielding the surface potential for the whole channel [10].

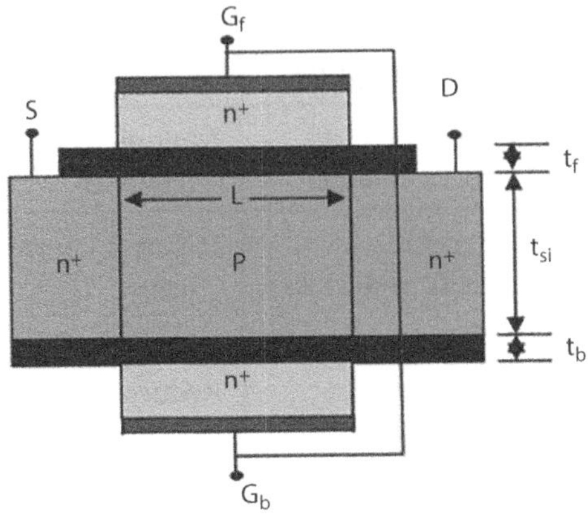

Figure 2.5 Cross sectional view of double gate (DG) MOSFET [5].

For n-channel MOSFETs, the substrate is often n-type silicon and for p-channel MOSFETs, it is typically p-type silicon. Carriers pass through severely doped zones that form on either side of the channel (n+ for n-channel, p+ for p-channel) to enter and exit the channel. The space between the source and drain is where current flows. A key component in determining the MOSFET's characteristics is the channel's width and length. An insulating layer, usually made of silicon dioxide (SiO2), separates the channel from the gate electrodes. The control of the carrier flow in the channel is contingent upon this layer. These gate electrodes can be independently operated, making them more efficient than a single gate.

2.3.1.2 Applications of DG MOSFETs

- **Microprocessors and CPUs:** DG MOSFETs may deliver greater driving currents and faster switching speeds, which improve overall performance and efficiency; they are employed in the production of microprocessors.
- **Graphics Processing Units (GPUs):** GPUs benefit from the enhanced control and decreased leakage currents of DG MOSFET because high-speed and high-efficiency transistors are needed for graphic rendering.
- **Mobile Devices:** DG MOSFETs are perfect for mobile devices like smartphones and tablets because of their lower leakage currents and lower power consumption, which increase battery life and improve energy efficiency.

- **Wearable Technology:** In wearable devices, where DG MOSFETs are used, low power consumption and high performance are necessary for operation and user experience.
- **Analog-to-Digital Converters (ADCs):** The performance of ADCs, which are essential for translating analog signals to digital form in variety of applications, is improved by the accuracy and lower noise of DG MOSFETs.

2.3.1.3 Design Challenges of DG MOSFETs

The intricate structure of Double Gate MOSFETs and the strict specifications for good performance and dependability make designing DG MOSFETs difficult [11–16]. The following are a few of the main design obstacles:

- **Lithography:** The top and bottom gates must be precisely aligned, which calls for sophisticated lithography methods. The performance of the device might be severely impacted by any misalignment.
- **Etching and Deposition:** High precision and technical difficulty are required when etching narrow fins or producing thin silicon layers, as well as when depositing homogeneous gate oxides and metal gates.
- **Channel Length Scaling:** Short-channel effects (SCE) are harder to regulate as the channel length is lowered. SCE can reduce performance; hence, DG MOSFETs must be constructed with it as little as possible.
- **Doping Profile:** To control the threshold voltage and lower SCE, precise control over the doping profile and concentration in the source/drain and channel sections is needed.

2.3.2 Tri-Gate MOSFETs

Tri-Gate MOSFETs also known as Fin-FETs, are a technological advance in transistor design that overcomes the drawbacks of conventional planar MOSFETs. They provide exceptional control over the channel because of their distinctive three-dimensional structure, which greatly enhances performance and lowers power consumption [7].

2.3.2.1 *Structure of Tri-Gate MOSFET*

The construction of a Tri-Gate MOSFET, commonly known as a Fin-FET, is depicted below in Figure 2.6. Fin-FETs, because of their distinctive three-dimensional structure, provide better channel control, which enhances performance and reduces power consumption. A Tri-Gate MOSFET's fin, gate, source and drain regions, and oxide layer are its main structural components.

Fins are tiny, vertical silicon protrusions that rise above the substrate. The path that current travels to go from the source to the drain is made by it. The fin's three-dimensional shape gives the gate greater surface area to interact with, enhancing the channel's electrostatic control. The three sides that the gate encircles are the top and the two vertical sidewalls of the fin. This architecture gives rise to the moniker Tri-Gate MOSFET. For the gate insulator, high-k dielectric materials are frequently utilized to increase gate control and decrease leakage. Metal gates are typically used to increase performance even further. Through the creation of an electric field that alters the fin's conductivity, the gate controls the amount of current that flows through the channel. At each end of the fin are the severely doped portions called Source and Drain. As a result, current enters the source and exits the drain. The drain and source provide low-resistance pathways that allow current to flow into and out of the channel. MOSFETs are typically doped with materials to produce n-type or p-type regions (NMOS or PMOS), depending on the type. A thin insulating layer, usually made

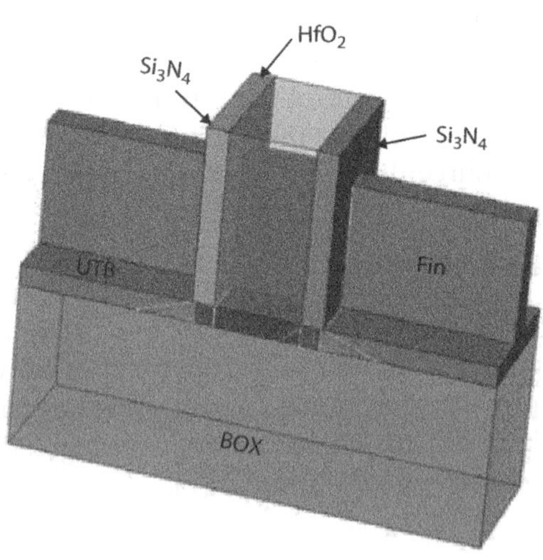

Figure 2.6 Tri-gate MOSFET [8].

of silicon dioxide (SiO2) or a high-k dielectric material, separates the gate and fin. The oxide layer prevents direct current flow, allowing the gate to electrostatically control the channel.

2.3.2.2 *Applications of Tri-Gate MOSFETs*

- **CPUs and GPUs:** Tri-gate MOSFETs find widespread use in CPUs and GPUs. Their increased switching speed and high drive current enable faster and more effective data processing, which is needed for high-performance computing tasks including artificial intelligence, scientific simulations, and gaming.
- **Smartphones and Tablets:** In smartphones and tablets, Tri-Gate MOSFETs are integrated with other integrated circuits (ICs) and processors. They have minimal power consumption, which prolongs battery life, yet they have the performance needed for fast data processing and multitasking.
- **Laptops and Ultrabook's:** They are basically used in the CPUs and various other components of laptops and ultra books. Better thermal management and longer battery life are the results of improved performance and lower power consumption, which are essential for portable devices.
- **Analog Signal Processing:** Tri-Gate MOSFETs are used for processing and signal amplification in analog circuits. Their improved electrostatic control and less leakage result in higher performance in applications involving analog signal processing.
- **RF Communication Circuits:** They are used in RF circuits for wireless communication devices. Enhanced channel control and decreased noise levels enhance the dependability and effectiveness of radio frequency communication systems.

2.3.2.3 *Design Challenges of Tri-Gate MOSFETs*

Tri-gate MOSFETs, also called Fin-FETs, provide a number of special design issues because of their three-dimensional structure and the requirement to balance a number of performance criteria. Below are some of the design challenges faced:

- **Fin Formation and Uniformity:** For the gadget to work consistently, fins must be created with consistent height, width, and spacing. Differences in driving current and threshold voltage may result from variations.
- **Gate Wrap-Around:** It's difficult to make sure the gate encircles the fin evenly. Any anomalies may have an impact on the channel's gate control, which could result in inconsistent performance.
- **Lithography and Etching:** Sophisticated lithography methods are needed to precisely characterize the microscopic details of tri-gate systems. This raises the fabrication process's complexity and expenses.
- **Threshold Voltage Control:** It is difficult to maintain a constant threshold voltage throughout all fin dimensions, particularly as the fins get smaller. Deviations in the threshold voltage can cause devices to work less well than they should.
- **Short Channel Effects:** Although tri-gate MOSFETs are developed to reduce short channel effects, it can be difficult to keep this control as device dimensions get smaller. Precise engineering of the fin size and doping profiles is necessary for this.
- **Self-Heating Effects:** Tri-gate MOSFET performance can be deteriorated by self-heating, especially at high current densities. This calls for meticulous thermal planning and maybe the use of cutting- edge cooling methods.

2.3.3 Gate-All-Around (GAA) MOSFETs

Newest development in transistor technology is the Gate-All-Around (GAA) MOSFET, which offers even more channel control than Fin-FETs, perhaps extending Moore's Law further. GAA MOSFETs' distinct structure provides notable gains in scalability, power efficiency, and performance. This paper examines the design, functionality, benefits, manufacturing process, difficulties, and uses of GAA MOSFETs.

2.3.3.1 Structure of Gate-All-Around (GAA) MOSFET

The channel illustrated in Figure 2.7 is fully enclosed by a state-of-the-art transistor architecture called Gate-All-Around (GAA) MOSFETs, which comprise a Source, Drain, and Gate nanowire stack of 5x, providing the

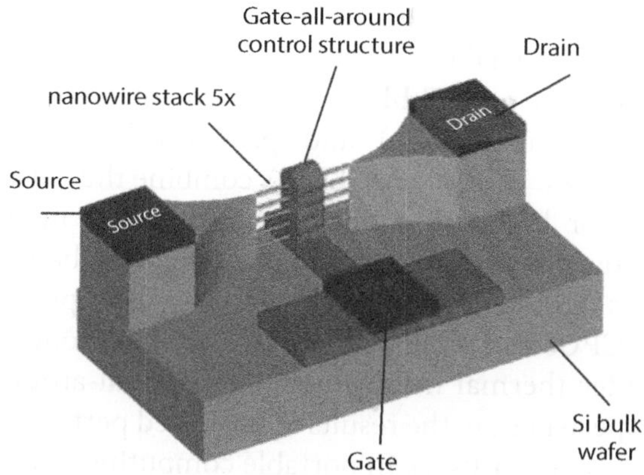

Figure 2.7 Gate-all-around (GAA) MOSFET.

user with excellent control over it. This unique arrangement improves the device's overall performance [9], and minimizes short-channel effects.

In lower technology nodes, this work uses a quantum transport model to illustrate the performance of a cylindrical gate all around (GAA) MOSFET based on silicon nanowire (Si NW) A round or rectangular nanowire made of silicon or another semiconductor material makes up the channel. The typical range for the width (for rectangular nanowires) or diameter (for cylindrical nanowires) is a few nanometers. The nanowire has a flatter, wider structure and a higher current carrying capacity. Because of its greater width and thinner thickness, a nano sheet can manage current more skilfully than a nanowire. Complete control over the channel in all directions is provided by the gate material, which completely encloses the nanowire or nanosheet. often built with high-k dielectric materials and a metal gate to reduce leakage and increase efficiency. The gate length is the distance measured along the channel by the gate electrode. The source and drain are heavily doped regions located at either end of a nanowire or nanosheet. Both the source and the drain serve as the locations of entry and exit for the current. N-type or p-type areas are produced by doping these regions.

2.3.3.2 *Applications of Gate-All-Around (GAA) MOSFET*

- **CPUs and GPUs:** Tri-gate MOSFETs are frequently used in CPUs and GPUs. Tri-Gate MOSFETs' high drive current and enhanced switching performance enable faster data processing and efficient parallel processing for high-performance

computer workloads including gaming, scientific simulations, and artificial intelligence (AI).
- **Smartphones and Tablets:** They are integrated into the system-on-chips (SoCs) and processors of tablets and smartphones. Tri-Gate MOSFETs combine the performance needed for high-speed data processing and multitasking with a low power consumption that lengthens battery life.
- **Laptops and Ultrabook's:** Utilized in laptops and ultra books' CPUs and other vital parts, They have longer battery life, better thermal management, and an all-around better user experience are the results of increased performance and decreased power usage in portable computing devices.
- **Power Management ICs:** For effective power distribution and management in electronic devices, they are incorporated into power management integrated circuits (PMICs). Power management in consumer electronics and industrial applications requires more efficiency and lower power loss.
- **Voltage Regulators:** They are utilized in voltage regulators to keep electronic components powered on a steady basis. The overall power efficiency of electronic systems is improved *via* improved voltage control performance and efficiency.
- **Advanced Driver-Assistance Systems (ADAS):** Utilized in ADAS for data management and real-time processing, the superior performance and dependability of Tri-Gate MOSFETs are important for safety-critical automotive applications like as collision detection systems, adaptive cruise control, and lane-keeping assistance.

2.3.3.3 Design Challenges of Gate-All-Around (GAA) MOSFET

There are significant challenges in the design. These challenges are mostly brought on by the complex three-dimensional structure and the advanced production methods:

- **Gate Stack Formation:** It is difficult to evenly deposit the gate material around the whole channel (nanowire or nanosheet). For the best possible device performance, conformal coverage of metal gate materials and high-k dielectrics must be achieved.
- **Etching Precision:** To produce the nanowires/nanosheets and remove the sacrificial layers employed during manufacture

without causing damage to the device structure, sophisticated etching procedures are needed.

- **Subthreshold Swing:** For low power operation, a low subthreshold swing is essential, and achieving these calls for careful design and material selection to improve gate control.
- **Leakage Currents:** A careful balance in device design and material selection is required to maintain strong on-state driving currents while managing off-state leakage currents.
- **Self-Heating Effects:** The performance of GAA MOSFETs can be greatly impacted by self-heating, especially at high current densities. It is imperative to design with efficient thermal management in mind.

2.3.4 Omega-Gate MOSFET

The goal of the Omega-Gate MOSFET (Ω-Gate MOSFET) is to offer improved electrostatic control over the channel in comparison to traditional planar MOSFETs. This structure has several advantages that make it suitable for high-end semiconductor applications. It gets its name from its distinctive gate form, which resembles the Greek letter Ω.

2.3.4.1 Structure of Omega-Gate MOSFET

When examined in cross-section, the distinctive gate structure of the Omega-Gate MOSFET (Ω-Gate) omega (Ω) shape around three sides of the channel. By strengthening electrostatic control over the channel, this design increases scalability and performance.

The below Figure 2.8 shows the Omega Gate MOSFET with PolySi Electrode. The base layer of the MOSFET is typically made of silicon.

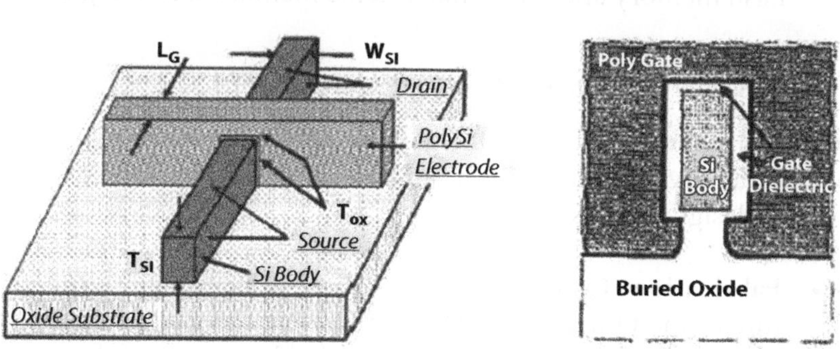

Figure 2.8 Omega Gate MOSFET.

Over three sides, the gate forms an omega (Ω) shape enclosing the channel. The channel's charge carrier flow is regulated by the control terminal. More control is given to the channel by the Omega-Gate MOSFET's gate, which surrounds it from three angles. The objective is to increase capacitance and reduce leakage by placing an insulating layer between the gate and the channel, usually made of silicon dioxide (SiO_2) or high-k dielectric materials.

2.3.4.2 *Applications of Omega-Gate MOSFET*

- **High-Performance Microprocessors:** Omega-Gate MOSFETs are perfect for high-performance microprocessors found in servers and computers because they can function at greater speeds due to their higher driving currents and decreased short-channel effects. Moore's Law can be further supported by further scaling down device sizes made possible by the improved channel control.
- **Portable Devices:** Because of their lower leakage currents, Omega-Gate MOSFETs are a good fit for battery-operated products like wearables, tablets, and smartphones where power efficiency is essential.
- **Internet of Things (IoT):** Low-power operation is essential for IoT devices that need to have long battery lives and consume little electricity.
- **DRAM and SRAM:** Improved subthreshold slope and reduced leakage currents increase the energy efficiency and performance of memory applications like SRAM [17], and DRAM.
- **Non-Volatile Memory:** Omega-Gate MOSFETs offer faster access times and improved data retention when utilized in flash memory and other non-volatile memory technologies.

2.3.4.3 *Design Challenges of Omega-Gate MOSFET*

To fully utilize the potential of Omega-Gate MOSFETs while maintaining manufacturability and reliability, a number of hurdles must be overcome during the design process.

- **Fabrication Complexity:** The three-dimensional structure of the Omega-Gate MOSFET, where the gate encircles the channel, calls for sophisticated fabrication methods.

When comparing this to conventional planar MOSFETs, the complexity increases.

- **Precision Lithography:** It is difficult to achieve the lithography precision needed to construct the complex gate architectures; this requires the use of cutting-edge tools and procedures.
- **Material Selection and Interface Quality:** Enhancing gate control with high-k dielectric materials may bring up problems with dielectric reliability, interface quality, and material compatibility. To keep the carrier mobility high and minimize scattering, the channel and gate dielectric must have a low density of interface states.
- **Parasitic Capacitances and Resistances:** The device's switching speed and power consumption may be impacted by additional parasitic capacitances introduced by the three-dimensional gate structure.

2.4 Advantages of Multi-Gate MOSFETs

Multi-Gate MOSFETs (Mu-GFETs) have a unique structure that encircles the channel with several gates, providing better control than traditional single-gate MOSFETs (Sg-MOSFETs). Below is a summary of the primary benefit.

Enhanced Electrical Control:

1. **Sharper Switching:** Because of their improved gate control over the channel, MuG-FETs show a steeper transition between the "on" and "off" states. This results in enhanced signal integrity and quicker switching times in digital circuits. One of the main benefits of multi-gate MOSFETs (Mu-GFETs) over conventional single-gate MOSFETs (Sg-MOSFETs) is sharper switching.
2. **Lower Leakage Current:** In the "off" state, more precise channel management reduces unwanted current flow. Leakage current greatly increases total power consumption, hence this is important for low-power electronics.

Diminished Effects of Short Channels:

a. **A Higher Degree of Scalability:** Electronic devices can continue to be miniaturized because Mu-GFETs, as opposed to Sg-MOSFETs, maintain good performance even at reduced dimensions.
b. **Diminished Leakage Subthreshold:** The amount of leakage current that happens in MuG-FETs when the gate voltage is not high enough to completely turn the transistor "on" is greatly decreased.

Enhanced Output:

1. **Increased Current Drive:** Electronic devices can operate more quickly, thanks to Mu-GFETs' ability to tolerate higher currents, thanks to more efficient gate control.
2. **Improved Analog Features:** MuG-FETs perform better in analog applications like amplifiers and sensors because they have lower channel length modulation and higher inherent gain (amplification).

2.5 Conclusion

In conclusion, because of short-channel effects, the trend of IC downsizing has reached a critical point with planar MOSFETs. Due to their enhanced electrostatics and superior gate control, multi-gate MOSFETs especially Fin-FETs and nanowire designs offer an appealing alternative. Better overall performance, decreased leakage currents, and increased scalability are obtained from multi-gate MOSFETs. MuG-FETs are a breakthrough in semiconductor technology that will drive improvements in computing power and energy efficiency as well as future IC design innovation. The future of electronics holds great promise for further investigation of their underlying theories, classifications, production methods, uses, and future research avenues. Here are some important fields for further research:

- Sustaining the scaling, the trend of electronics shrinking will persist, driving down the size of Mu-GFETs, investigating new materials for the channel and gate in an effort to boost efficiency and get around the drawbacks of silicon and other existing materials and creating novel techniques for the accurate creation of these complex structures at the nanoscale.

- In order to achieve enhanced functionality, efforts are being made to integrate MuG-FETs with other device technologies, arranging MuG-FETs in many vertical layers to build circuits that are even denser and more potent and investigating the use of MuG-FETs in non-CMOS logic applications, like sensors and memory devices.
- Novel devices have a gate that encircles the channel entirely, providing the highest level of control but posing serious manufacturing difficulties, Using incredibly thin nanowires as the conduit for superior gate control and flexible electronics possibilities.

All things considered, multi-gate MOSFETs have a bright future ahead of them. MuG-FETs have the potential to completely transform the electronics industry by tackling present issues and creating novel opportunities. This will make it possible to create next-generation devices with previously unimaginable power.

References

1. Ghosh, S., Miura-Mattausch, M., Iizuka, T., Rahaman, H., Mattausch, H.J., Operating-condition optimization of MG-MOSFETs for low-voltage application. *Proc. Elect. Dev. Technol. Manuf. Conf. (EDTM)*, pp. 97–99, Mar. 2022.
2. Sekigawa, T. and Hayashi, Y., Calculated threshold-voltage characteristics of an XMOS transistor having an additional bottom gate. *Solid-State Electron.*, 27, 8–9, 827–828, Aug.–Sept. 1984. doi: 10.1016/0038-1101(84)90005-0.
3. Hisamoto, D., Kaga, T., Kawamoto, Y., Takeda, E., A fully depleted lean-channel transistor (DELTA) – a novel vertical ultra thin SOI MOSFET. *International Technical Digest on Electron Devices Meeting (IEDM)*, 833–836, Dec. 1989. doi: 10.1109/IEDM.1989.74182.
4. Herrera, F.A., Miura-Mattausch, M., Iizuka, T., Kikuchihara, H., Hirano, Y., Mattausch, H.J., Modeling of short-channel effect on multi-gate MOSFETs for circuit simulation. *Proc. Int. Symp. Dev. Circuits Syst. (ISDCS)*, pp. 1–4, 2020.
5. Pandey, S.P. and Kushwah, R.S., Analytical Modeling and Comparison of Triple Gate MOSFET with Double Gate MOSFET. *2013 International Conference on Control, Computing, Communication and Materials (ICCCCM)*, 10.1109/ICCCCM.2013.6648901.
6. Herrera, F.A., *et al.*, Leading-edge thin-layer MOSFET potential modeling toward short-channel effect suppression and device optimization. *IEEE J. Elect. Dev. Soc.*, 7, 1293–1301, 2019.

7. Arun Samuel, T.S., Venkatesh, Karthigai Pandian, M., Vimala, P., Investigation of ON Current and Subthreshold Swing of InSb/Si Heterojunction Stacked Oxide Double Gate TFET with Graphene Nanoribbon. *J. Electron. Mater.*, 50, 7037–7043, 2021.
8. Pradhan, K.P. and Sahu, P.K., Investigation on Asymmetric Dual-k Spacer (ADS) Trigate Wavy FinFET: A Novel Device. *2016 3rd International Conference on Devices, Circuits and Systems (ICDCS)*, 10.1109/ICDCSyst.2016.7570643.
9. Matioli, E. and Nela, L., High-performance normally-off tri-gate GaN power MOSFETs. *2019 31st International Symposium on Power Semiconductor Devices and ICs (ISPSD)*, 10.1109/ISPSD.2019.8757690.
10. Das, S.K., Nanda, U., Biswal, S.M., *et al.*, Performance Analysis of Gate-Stack Dual-Material DG MOSFET Using Work-Function Modulation Technique for Lower Technology Nodes. *Silicon*, 14, 2965–2973, 2022.
11. Bhol, K. and Nanda, U., Nanowire Array-Based MOSFET for Future CMOS Technology to Attain the Ultimate Scaling Limit. *Silicon*, 14, 1169–1177, 2022.
12. Ganapati, R., Samoju, V.R., Jammu, B.R., Analytical Modeling of Threshold Voltage for Dual-Metal Double-Gate Gate-All-Around (DM-D G-GAA) MOSFET. *Silicon*, 13, 2869–2880, 2021.
13. Mendiratta, N. and Tripathi, S.L., A Review on Performance Comparison of Advanced MOSFET Structures Below 45nm Technology Node. *J. Semicond.*, IOP science, 41, 6, 1–10, 061401, 2020, DOI:10.1088/1674-4926/41/6/061401.
14. Samal, A., Tripathi, S.L., Mohapatra, S.K., A Journey from Bulk MOSFET to 3 nm and Beyond. *Trans. Electr. Electron. Mater.*, 21, 443–455, 2020, https://doi.org/10.1007/s42341-020-00222-y.
15. Singh Patel, G., Lata Tripathi, S., Awasthi, S., Performance Enhanced Unsymmetrical FinFET and its Applications. *2018 IEEE Electron Devices Kolkata Conference (EDKCON)*, Kolkata, India, pp. 222–227, 2018, doi: 10.1109/EDKCON.2018.8770411.
16. Tripathi, S.L., Mishra, R., Narendra, V., Mishra, R.A., High performance Bulk FinFET with bottom spacer. *2013 IEEE International Conference on Electronics, Computing and Communication Technologies*, Bangalore, India, pp. 1–5, 2013, doi: 10.1109/CONECCT.2013.6469282.
17. Kumar, T.S. and Tripathi, S.L., Process evaluation in FinFET based 7T SRAM cell. *Analog Integr. Circuits Signal Process.*, 109, 545–551, 2021, https://doi.org/10.1007/s10470-021-01938-4.

3

Design and Comparative Analysis of Hybrid DG-MOSFET with Conventional CMOS Using Visual TCAD

Kushagra[1], Suman Lata Tripathi[2]* and Balwinder Raj[3]

[1]VLSI Design LAB, SEEE, Lovely Professional University, Phagwara, India
[2]Department of E & TC, Symbiosis Institute of Technology (SIT), Symbiosis International (Deemed University), Pune, India
[3]Department of ECE, NIT, Jalandhar, India

Abstract

Steep subthreshold performance and low leakage current are major objectives of MOSFET design. The conventional MOSFET performance degrades with the shrinking dimensions. Multi gate MOSFET architectures shows improved subthreshold performance with increased gate control. 15nm n-channel and p-channel DG-MOSFET are designed and their performance are compared with similar dimension. This work presents the design and analysis of a hybrid design DG-MOSFET compared to conventional DG-MOSFETs, utilizing Visual TCAD software. Focused on the 15nm technology node, the study aims to evaluate the performance superiority of the hybrid design over the conventional counterparts. The analysis involves a comprehensive examination of both n and p-channel DG-MOSFETs, investigating key parameters such as leakage current, ON/OFF current ratio and compatibility with CMOS logic circuit. Electric field and channel potential is also plotted at different bias conditions for the evaluation of electrostatic behavior. The proposed hybrid DG-MOSFET shows I_{on}/I_{off} current ration up to 10^9 with leakage current in the range of 10^{-14} A.

Keywords: Hybrid DG-MOSFETs, visual TCAD, 15nm technology, leakage current, CMOS

**Corresponding author*: tri.suman78@gmail.com

Suman Lata Tripathi, Vrinda Gupta, Sobhit Saxena, Shipra Upadhyay and Sudip Ghosh (eds.) Micro and Nano Semiconductor Devices for Digital, Analog and Sensor Design, (51–72) © 2026 Scrivener Publishing LLC

3.1 Introduction

Designers are continuously working on exploration of new transistor with different material level, doping and architecture level analysis to meet the desired performance even after scaling down the dimensions of transistors [1, 2]. Muti-gate transistors were popular option at smaller dimension supporting to scaling down to pack more transistor per unit IC area. Double-gate (DG) architectures have more control over the channel but it is more complex to fabricate back gate. Also, there is a challenge of symmetry of top and bottom gate [3]. Asymmetric junctionless DGMOSFET have demonstrated better switching performance with high value of transistor ON-to-OFF current ratio along with lesser effect on transistor performance with fabrication related variations or defects [4–6]. To avoid the challenges of design planner DGMOSFET, vertical channel FinFETs are used with self-aligned feature [7, 8]. FinFET provide 3-fold drain current compared to the conventional planner single-gate MOSFET of similar dimensions. DGMOSFETs and FinFET were explored in designing of logic circuit, SRAM/DRAM cell and analog circuit applications [9–13]. FinFET are available in heterojunction with different semiconductor material and multi fin architectures with superior performance than regular Si based FinFETs [14, 15]. TFET architecture are also going popular option with steep subthreshold curve and high transconductance [16–20]. The major transistor performance analysis includes the parameters such as subthreshold parameters, switching, delay, power and energy. The device power is calculated including both dynamic $(fC_L V_{DD}^2)$ and static power consumptions $(V_{dd} \cdot I_{static})$. Low voltage, reduced parasitic capacitance and reduces frequencies are important constraints to overcome the issues related to low dimension transistors. The subthreshold parameters like leakage current, Vth, SS (subthreshold slope) and DIBL (drain induced barrier lowering) are also major focus of designers to control static power consumptions. Performance optimization is based on circuit level or type of application such digital or analog circuit.

Technology-Driven-Computer-added (TCAD) design is a suitable mean to design transistors at different feature size and use the same for logic, memory and analog circuit to verify their performance at circuit level. In EDA tool, circuit are implemented through available transistor model at fixed technology and transistor architecture. Cadence or HSpice tools are popular in such back VLSI circuit design at different technology model files.

This chapter demonstrated a design of hybrid DGMOSFET including both n-channel and p-channel together, and its performance is evaluated for the CMOS based logic circuit implementations. In this work transistor output and transfer characteristics are plotted to evaluate the switching performance. Electric and surface potential are also plotted to evaluate the device electrostatic behavior.

3.2 Design Methodology

Technology based transistor design follows the steps as:

(i) Gate Engineering
In this, different gate architectures are designed from single-gate to multiple-gate, rectangular to circular gate with different gate-material from polysilicon to high work-function metals. Multi-gate architectures are used to improve the gate control over the channel regions for e.g., DGMOSFET and FinFET. FinFETs are available in both double gate and triple gate format with variation in top gate oxide thickness compared to side gates.

(ii) Channel Engineering
Channel shape is one of the important design parameters. Conventional MOSFET has planner channel region while FinFETs have vertical channels with increased conducting channel width. Channel is made of semiconductor material with doping. Si as channel material is used at different doping level from lightly to moderately doped. High doping improves the transistor on current but leads to increase in more subthreshold current due to the increased scattering of carriers in presence of dopant atoms. Multi-channel devices are also going popular.

(iii) Heterojunctions
Different semiconductor materials are used to make heterojunctions such as $Si\text{-}Si_xGe_{1-x}$ Si-Ge or $Ge\text{-}Si_xGe_{1-x}$. Heterojunctions are used to improve subthreshold performance by reducing leakage current. Heterojunctions are more popular in optically sensitive devices. Heterojunctions were shown in literature for TFET, DGMOFETs and FinFETs etc.

3.3 Device Architecture and Materials Description

The proposed n-channel 15nm DG MOSFET and p-channel 15nm DG MOSFET are designed and extended to 15nm DGMOSFET based CMOS architecture. Platinum metal and HfO2 oxide are used gate contact and oxide material for both n- and p-DGMOSFET. The architecture design is same for both n- and p-DGMOSFET with small difference in Y-characteristics for similar current voltage plot. Figure 3.1 shows the architecture of DG-MOSFET. Table 3.1 shows the properties of n-channel and p-channel DG-MOSFET.

3.4 Results and Discussion

This section includes the designing of n-channel and p-channel MOSFET designing and their performance comparison using TCAD tool.

3.4.1 N-Channel DG-MOSFET

A 2D n-channel DG-MOSFET simulation set up is shown in Figure 3.2 designed on Visual TCAD. The simulation is carried out for different bias condition of gate and drain with respect to source. Both the gate of DG-MOSFETs is internally shortened and applied to similar gate voltage.

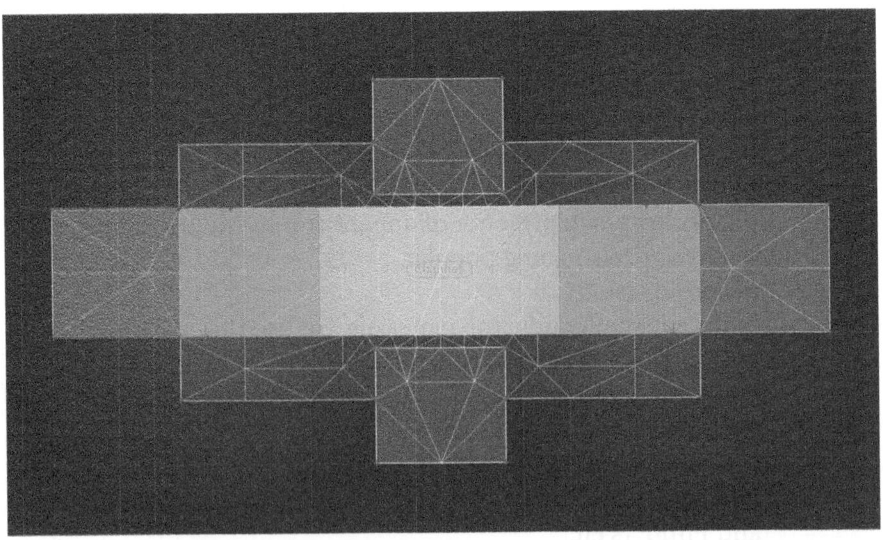

Figure 3.1 DG-MOSFET MOSFET architecture.

Table 3.1 Device properties.

Parameter	N-channel DG-MOSFET	P-channel DG-MOSFET
Source/Drain Doping type	Gaussian doping	Gaussian doping
Source/Drain Doping species	Donar Concentration 1e+18	Acceptor Concentration 1e+18
Y characteristics	0.133	0.11
Channel Region doping	Uniform	Uniform
Channel doping	A Acceptor Concentration 8e+17	Donar Concentration 5e+18

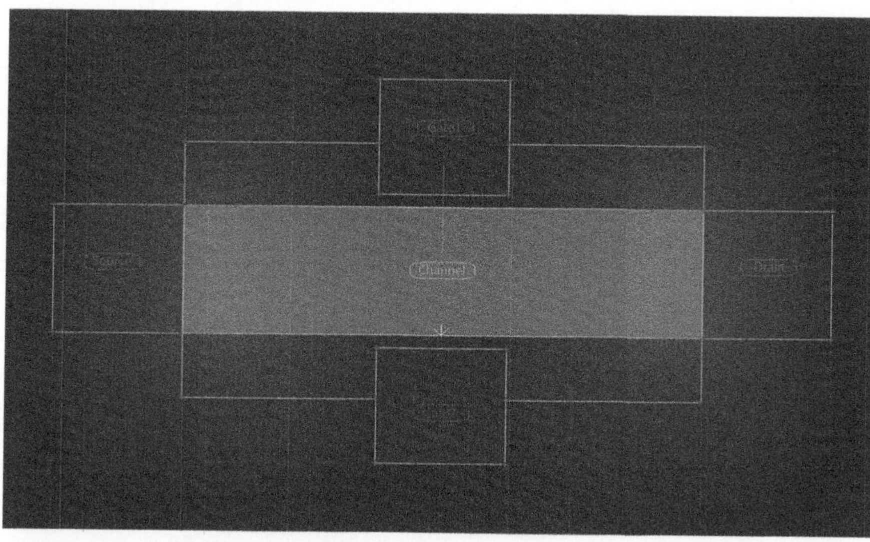

Figure 3.2 N-channel DG-MOSFET simulation on TCAD.

For gate voltage higher than threshold voltage channel inversion layer created both top and bottom of channel under the gates. Table 3.1 shows the properties of DGMOSFET.

i) Drain current versus Drain voltage plot

Figures 3.3 and 3.4 present drain current changes with varying drain voltage at linear and log scale.

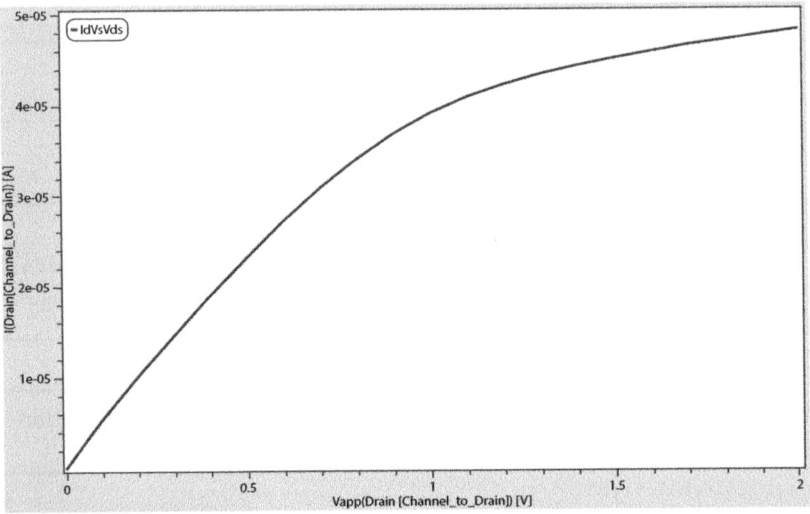

Figure 3.3 Id vs Vds plot (linear scale).

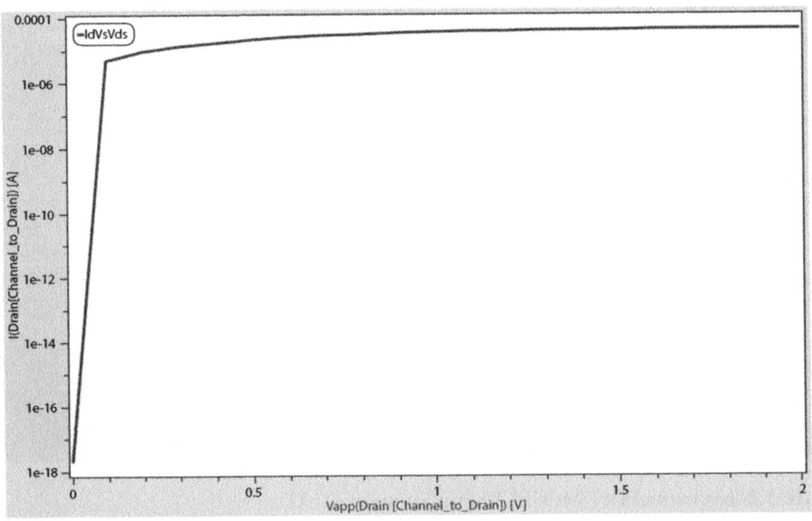

Figure 3.4 Id vs Vds plot (log scale).

The curve shows a sharp change in ON and OFF-state current which is suitable for logic circuit applications.

ii) Drain current versus Gate voltage plot

Figures 3.5 and 3.6 show a sharp change in drain current with varying gate voltage at linear and log scale. Table 3.2 shows the I_{ON} and I_{OFF} values with suitable drain and gate bias conditions.

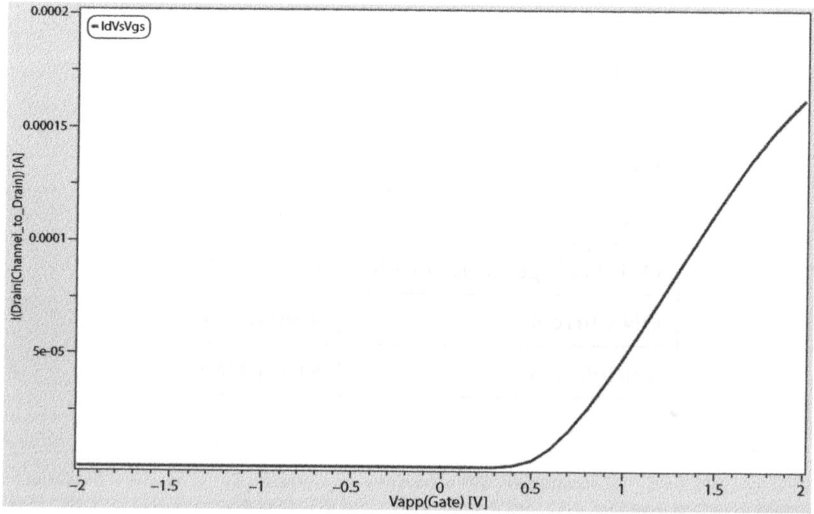

Figure 3.5 Id vs Vgs plot (linear scale).

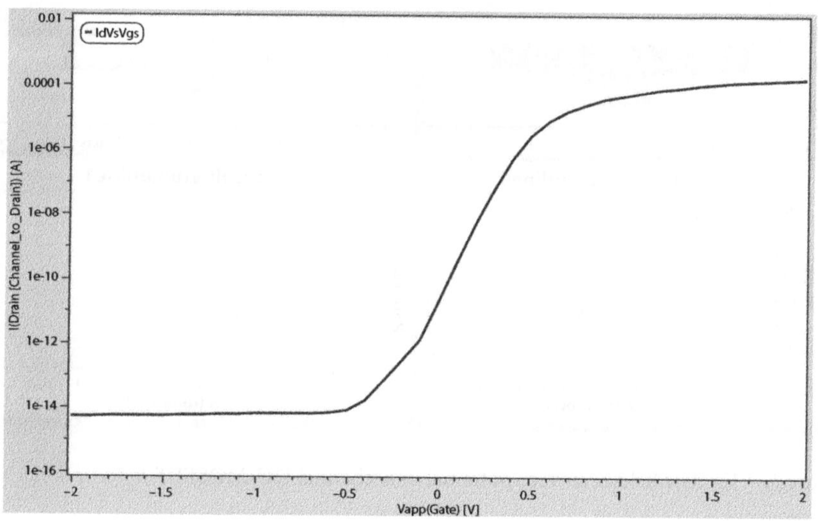

Figure 3.6 Id vs Vgs plot (log scale).

iii) Electric Field and Potential

Figure 3.7 shows electric field and surface potential plots p-channel devices obtained in 2D visuals of TCAD. The curve shows a sharp change in electric filed and surface potential at different bias conditions.

Table 3.2 N-channel DG MOSFET analysis.

Parameter	Values
Vth	0.5 V
Vds(lin)	0.2 V
Vds (Sat)	1.2 V
OFF/Leakage Current (Ioff)	1.23*10^-11 v
ON Current	9.86*10^-5 v
Ion/Ioff ratio	8.01*10^6 v

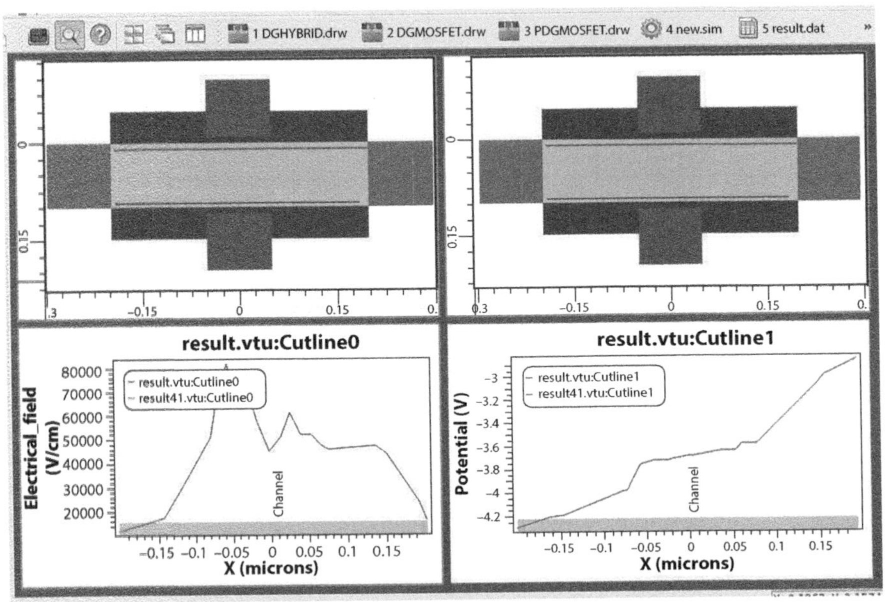

Figure 3.7 Electric field & surface potential of n-channel DG-MOSFET.

3.4.2 P-Channel DG-MOSFET

A 2D p-channel DG-MOSFET simulation set up is shown in Figure 3.8 designed on Visual TCAD.

i) Drain current versus Drain voltage plot

Figures 3.9 and 3.10 present drain current changes with varying drain voltage at linear and log scale.

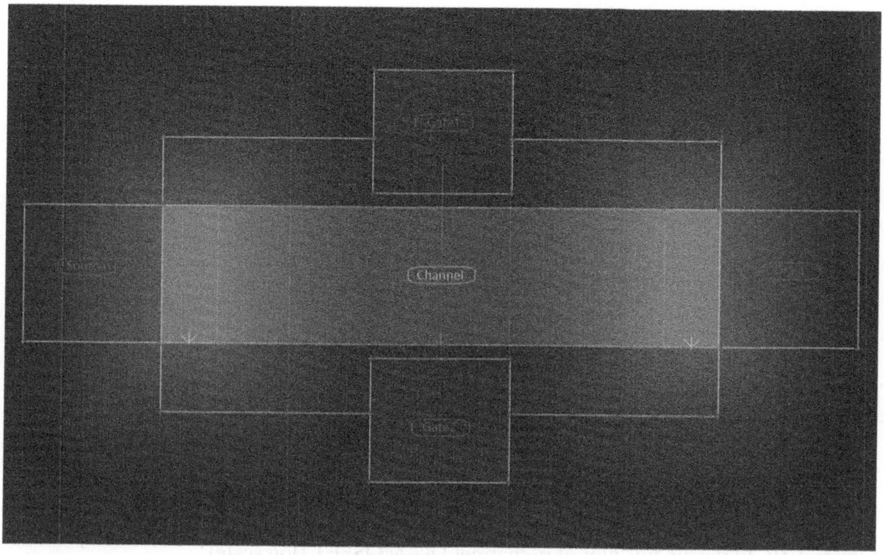

Figure 3.8 P-channel DG-MOSFET simulation on TCAD.

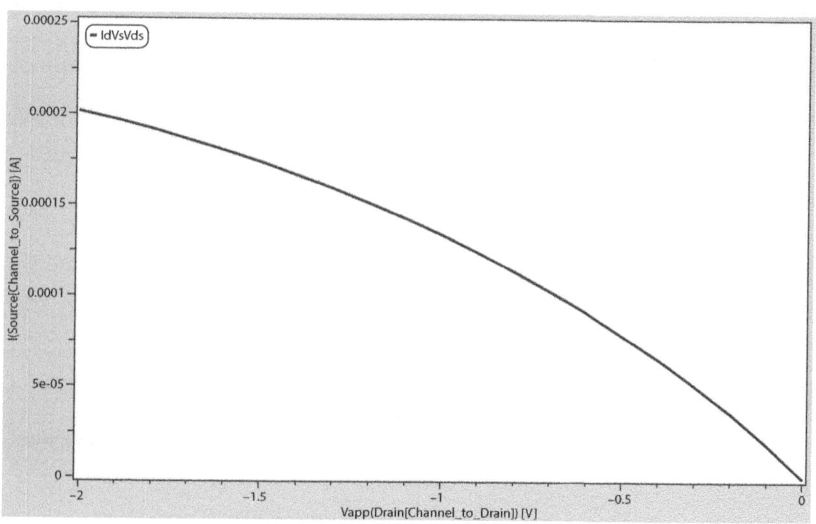

Figure 3.9 Id versus Vds plot of p-channel DG-MOSFET (linear scale).

ii) Drain current versus Gate voltage plot

Figures 3.11 and 3.12 present drain current changes with varying gate voltage at linear and log scale. Table 3.3 shows the I_{ON} and I_{OFF} values with suitable drain and gate bias conditions.

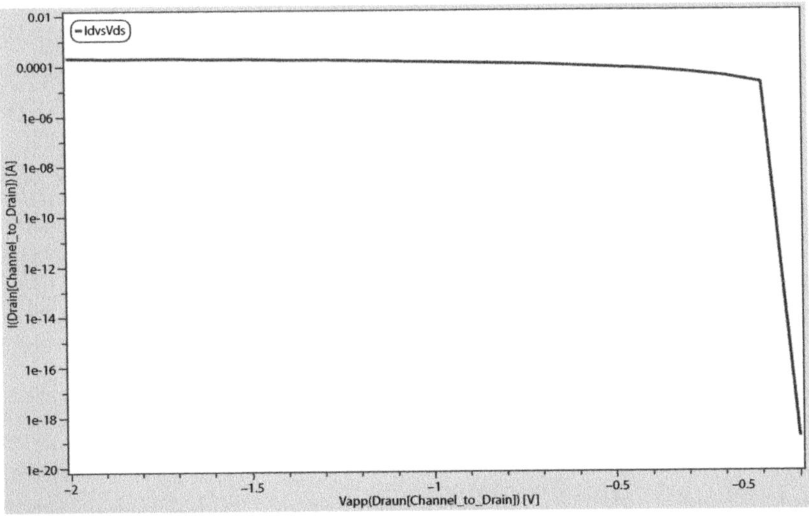

Figure 3.10 Id versus Vds plot of p-channel DG-MOSFET (log scale).

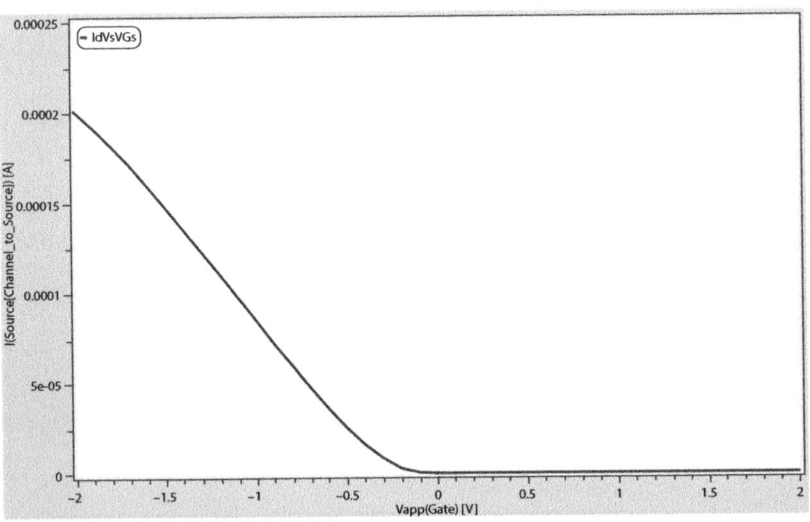

Figure 3.11 Id versus Vgs plot of p-channel DG-MOSFET (linear scale).

iii) Electric Field and Potential plot

Figure 3.13 shows electric field and surface potential plots p-channel devices obtained in 2D visuals of TCAD. Also, the drain current curve shows a sharp change in electric filed and surface potential at different bias conditions.

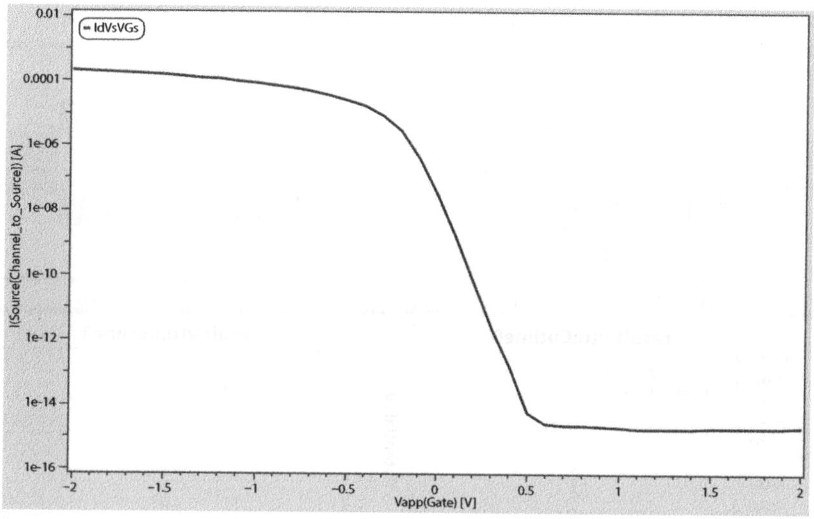

Figure 3.12 Id versus Vgs plot of p-channel DG-MOSFET (log scale).

Table 3.3 P-channel DG MOSFET analysis.

Parameter	**Values**
Vth	0.5 V
Vds(lin)	-0.2 V
Vds (Sat)	-1.2 V
OFF/Leakage Current (Ioff)	3.57*10^-15 v
ON Current	9.78*10^-5v
Ion/Ioff ratio	2.73*10^10

3.5 CMOS Compatibility of Proposed n- & p- Channel DG-MOSFET

Figure 3.14 shows a comparative drain current performance with varying gate voltage in range of -2 to +2 V of p-channel and n-channel DGMOSFET that are designed as separately using TCAD.

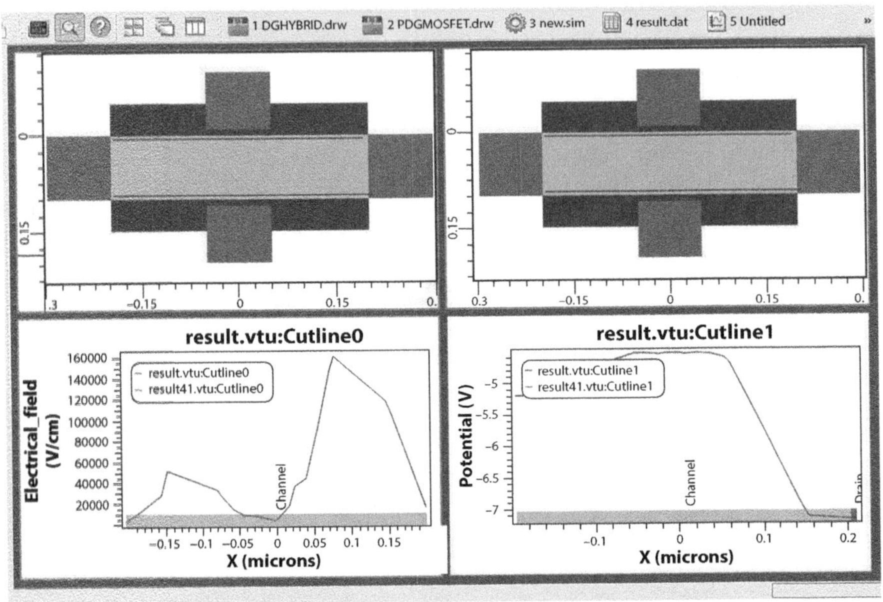

Figure 3.13 Electric field & surface potential of p-channel DG-MOSFET.

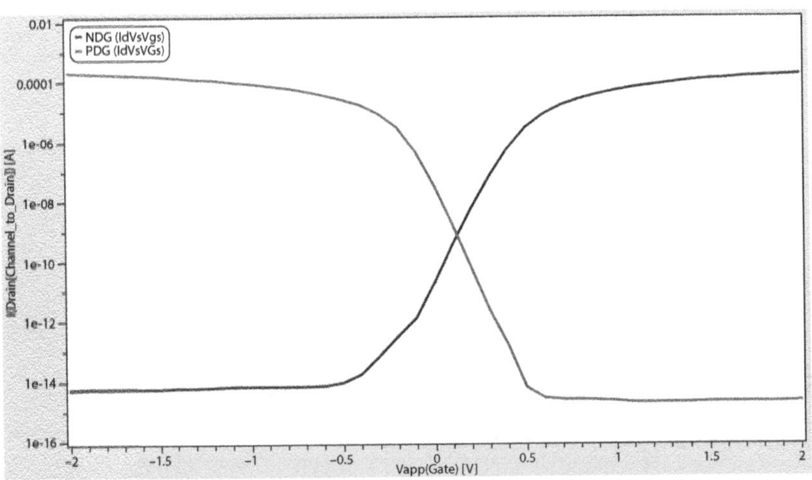

Figure 3.14 Graph analysis of PDG MOSFET and NDG MOSFET together.

3.6 Hybrid DG-MOSFET

A hybrid DG-MOSFET is a configuration where both n-channel and p-channel are designed vertically one after the other to check the suitability of it with CMOS based circuit design. In hybrid architecture the bottom gate of n- and p-channel DGMOSFET is common and dielectric layers of both devices are separated by spacer (SI3N4) material.

3.6.1 Device Simulation on TCAD

Figure 3.15 describes the hybrid DG-MOSFET architecture designed using TCAD 2D graphical interface. Tabel 3.4 describes the device properties with dimension, region material and doping.

3.6.2 Hybrid DG-MOSFET Response

The hybrid DG-MOSFET is simulated and the results are plotted between input and output parameters.

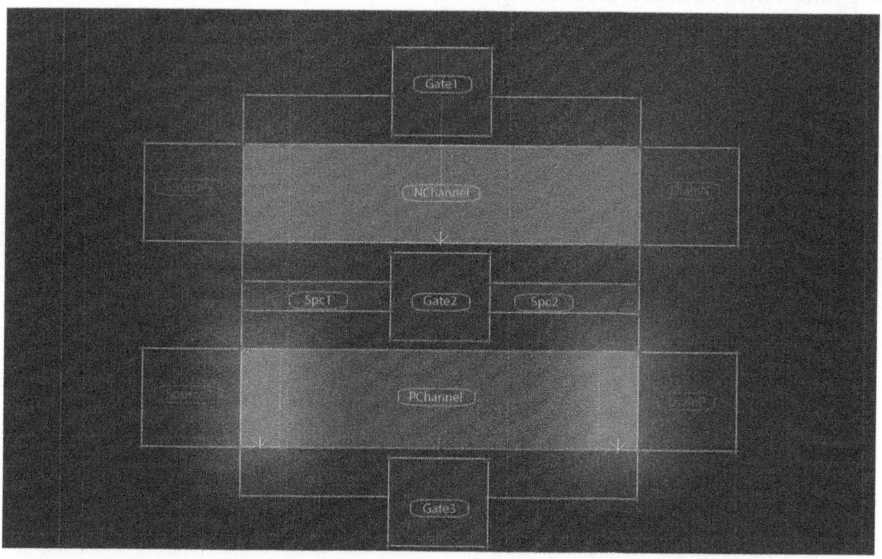

Figure 3.15 Hybrid CMOS DG-MOSFET including both n-and p-DG-MOSFET on TCAD.

Table 3.4 Device properties.

Parameter	N-channel DG-MOSFET	P-channel DG-MOSFET
Source/Drain Doping type	Gaussian doping	Gaussian doping
Source/Drain Doping species	Donar Concentration 1e+18	Acceptor Concentration 2e+18
Y characteristics	0.133	0.07
Channel Region doping	Uniform	Uniform
Channel doping	A Acceptor Concentration 5e+17	Donar Concentration 1e+15

Table 3.4 shows the comparative device properties of n-channel and p-channel DG-MOSFETs.

i) Id vs Vds plot

Figures 3.15 and 3.16 present drain current changes with varying drain voltage at linear and log scale for hybrid DG-MOSFET. For matching the drain characteristics, the optimization in bias conditions and doping levels are taken into considerations. Figure 3.17 shows Id versus Vds variation for the hybrid DG-MOSFET.

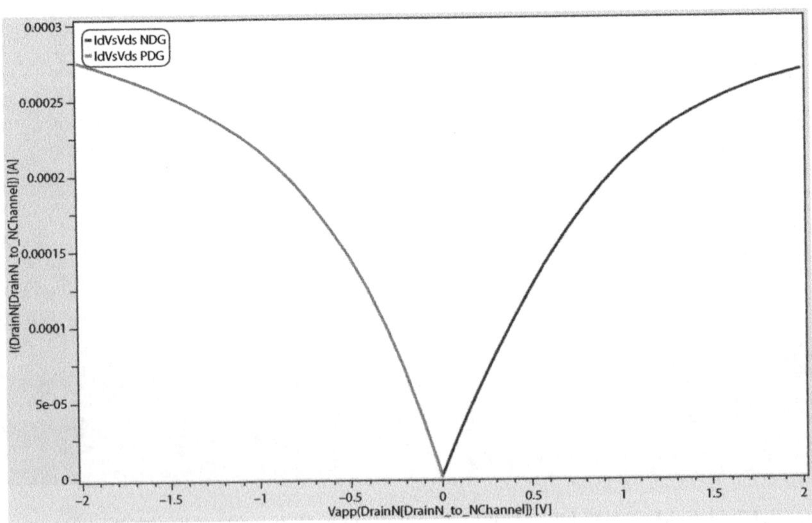

Figure 3.16 Id versus Vds of Hybrid DG-MOSFET (linear scale).

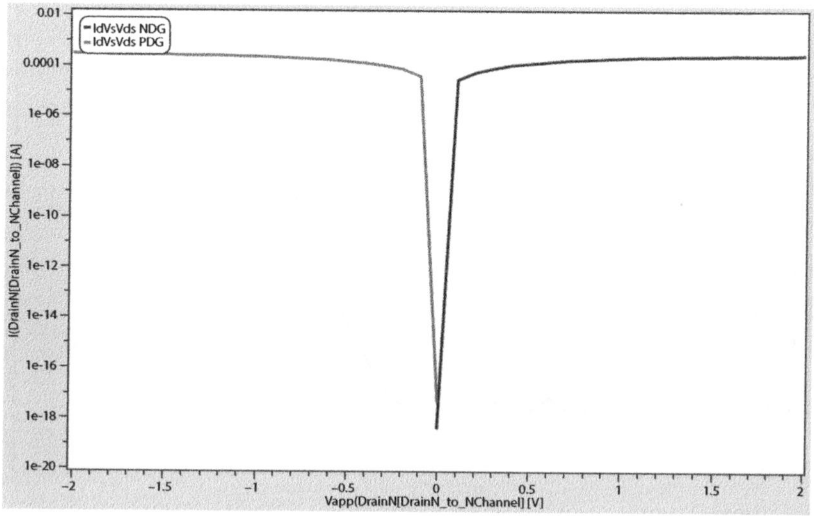

Figure 3.17 Id versus Vds of Hybrid DG-MOSFET (log scale).

ii) Id vs Vgs plot

Figures 3.18 and 3.19 present drain current changes with varying gate voltage at linear and log scale for hybrid DG-MOSFET. Table 3.5 depicts the performance of hybrid DG-MOSFET at suitable bias conditions. The drain

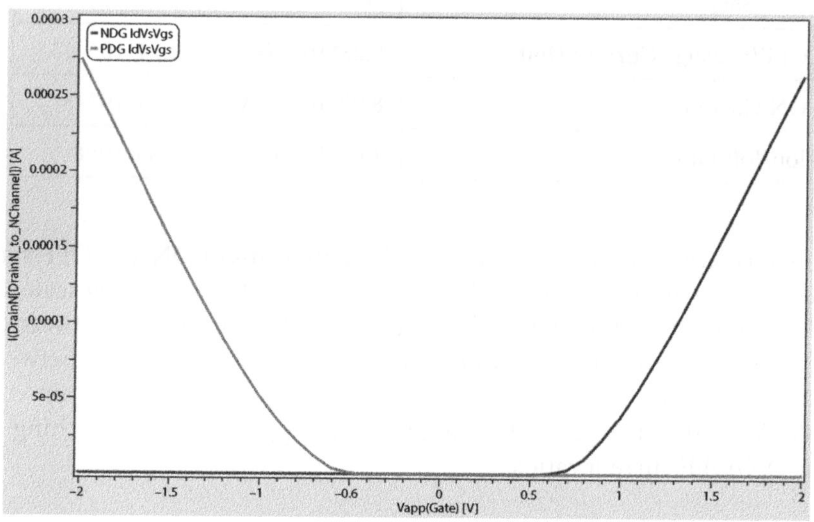

Figure 3.18 Id versus Vgs of hybrid DG-MOSFET (linear scale).

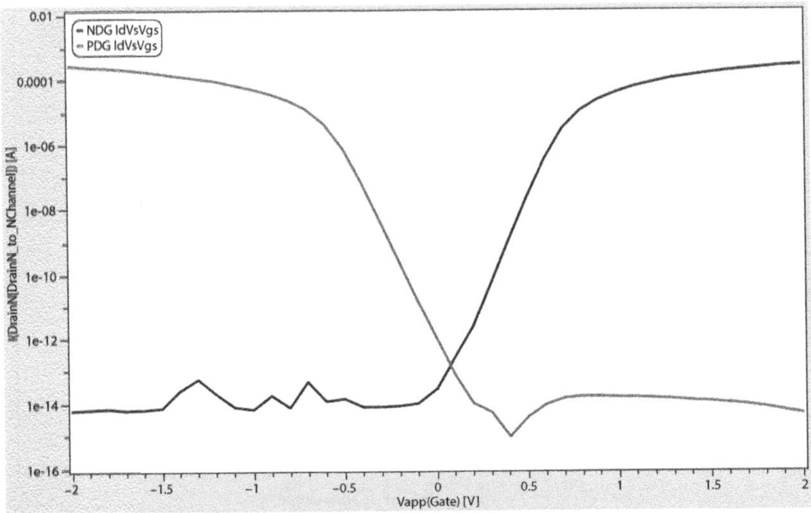

Figure 3.19 Id versus Vgs of hybrid DG-MOSFET (log scale).

Table 3.5 Hybrid DG-MOSFET response.

Parameter (Hybrid DG-MOSFET)	P-side	N-side
Vth	-0.5 V	0.7 V
Vds(lin)	-0.2 V	0.5 V
Vds (Sat)	-1.2 V	1.5 V
OFF/Leakage Current (Ioff)	8.28*10^-14 A	2.74*10^-13
ON Current	8.93*10^-5 A	7.84*10^-5
Ion/Ioff ratio	1.07*10^9	2.86*10^8

characteristics of both devices are matching in terms of ON and OFF state values along with their subthreshold slope on both linear and log scale.

P-channel DG-MOSFET is showing a suitable range of V_{gs} between -2 to 0 V while n-channel DG-MOSFET is showing suitable range between 0-2V with approximately similar values of ON and OFF state drain currents. Also, the n- and p-DG-MOSFET are approximately matching in their ON/OFF current ratios.

3.6.3 Electric Field & Potential Plot

Figure 3.20 shows electric field plot of n-channel and p-channel devices obtained in 2D visuals of TCAD. Figure 3.21 shows surface potential of n-channel and p-channel devices obtained in 2D visuals of TCAD. The curve shows a sharp change in electric filed and surface potential at different bias conditions for n- and p-channel hybrid DG-MOSFETs.

3.7 Applications & Future Scope

DG-MOSFETs are used for various application specifically in combinational logic circuits such adder/subtractor, multiplexer/decoder etc. [21, 22]. SRAM and DRAM cells are design with DG-MOSFETs and FinFETs showing stable read and write operations [23–26]. Additionally, DG-MOSFETs depicted a very good sensitivity for biological species as biosensor by adding nanogap cavity under both the gates [27, 28]. Such muti-gate MOSFET architectures can be explored for data converters like ADC, DAC's, analog comparators and amplifiers [29–31]. Researches are still going on for exploration of DG-MOSFET, FinFETs and TFETs for analog circuit design.

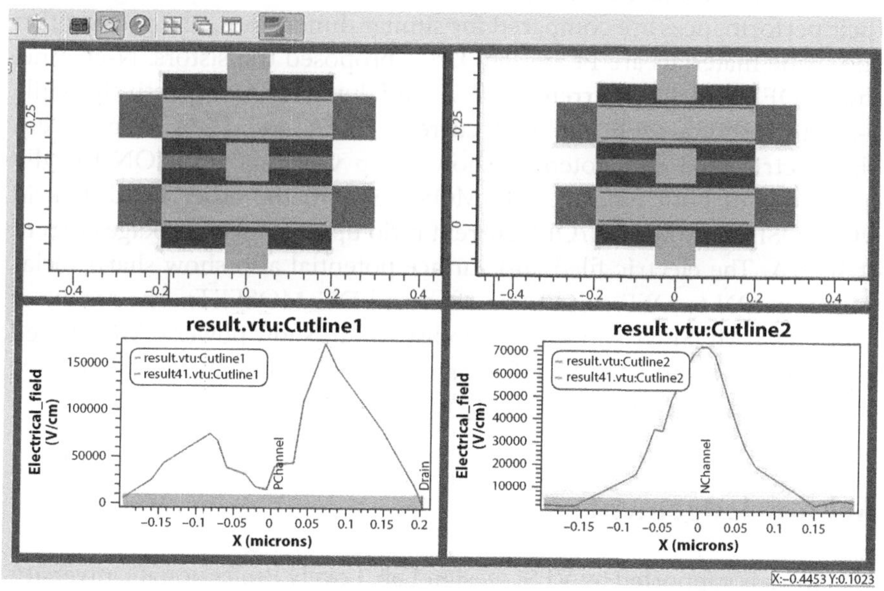

Figure 3.20 Electric field plot on TCAD.

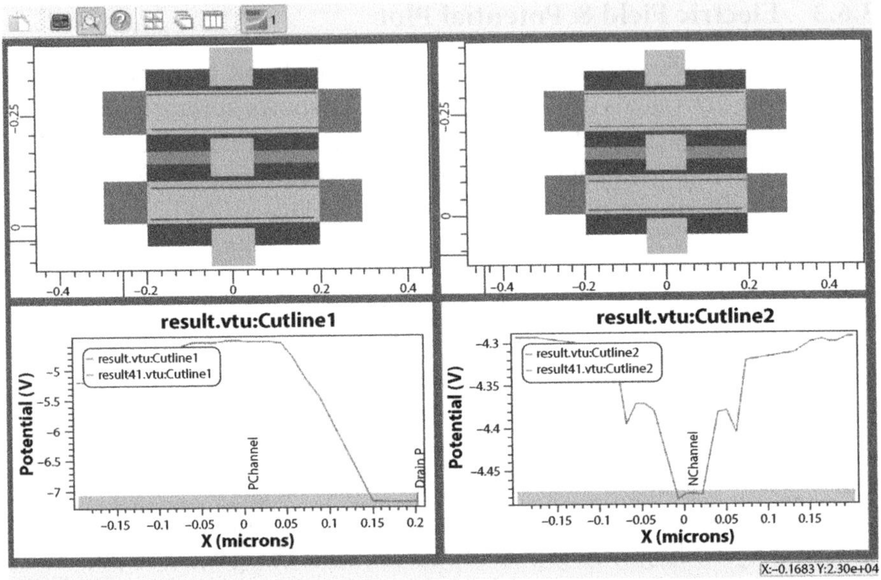

Figure 3.21 Surface potential plot on TCAD.

3.8 Conclusion

In this work, n-channel and p-channel DG-MOSFETs are designed and their performances are compared for similar dimensions. The gate contact and oxide materials are Pt and HfO2 for proposed transistors. N-channel shows OFF and ON current 10^{-11} A and Ion 10^{-5} A respectively while p-channel shows OFF and ON current 10^{-15}A and 10^{-5}A respectively. The electric filed and potential show sharp variation from ON to OFF conditions of both types of DG-MOSFETs. On the other hand, hybrid DG-MOSFET shows ON/OFF current ratio up to 10^9 with leakage current of 10^{-14} A. The electric filed and surface potential also show sharp variation from ON to OFF conditions of hybrid DG-MOSFETs. The results of hybrid DG-MOSFET show its suitability towards low power CMOS based logic circuit. Hybrid DGMOSFET can be used to design logic, memory, and analog circuit due its suitability towards CMOS based designs.

Acknowledgement

This work is supported by VLSI Design Lab, Lovely Professional University, NIT Jalandhar and DST SERB TAR/000325.

References

1. Mendiratta, N. and Tripathi, S.L., A Review on Performance Comparison of Advanced MOSFET Structures below 45nm Technology node. *J. Semicond.*, IOP science, 41, 6, 1–10, 061401, 2020, DOI:10.1088/1674-4926/41/6/061401.
2. Samal, A., Tripathi, S.L., Mohapatra, S.K., A Journey from Bulk MOSFET to 3 nm and Beyond. *Trans. Electr. Electron. Mater.*, 21, 443–455, 2020, https://doi.org/10.1007/s42341-020-00222-y.
3. Tripathi, S.L., Pathak, P., Kumar, A., *et al.*, Improved Drain Current with Suppressed Short Channel Effect of p+Pocket Double-Gate MOSFET in Sub-14 nm Technology Node. *Silicon*, 14, 10881–10891, 2022, https://doi.org/10.1007/s12633-022-01816-2.
4. Mendiratta, N. and Tripathi, S.L., 18nm n-channel and p-channel Dopingless Asymmetrical Junctionless DG-MOSFET: Low Power CMOS Based Digital and Memory Applications. *Silicon*, 14, 6435–6446, 2022, https://doi.org/10.1007/s12633-021-01417-5.
5. Mendiratta, N., Tripathi, S.L., Adhikari, M.S., Design and Process Variation Analysis of High-performance n and p-channel Insulated-gate Asymmetric-DG MOSFET. *Silicon*, 15, 8019–8029, 2023, https://doi.org/10.1007/s12633-023-02642-w.
6. Mendiratta, N., Tripathi, S.L., Chander, S., Analytical Model of Dopingless Asymmetrical Junctionless Double Gate MOSFET. *Silicon*, 14, 10765–10774, 2022, https://doi.org/10.1007/s12633-022-01819-z.
7. Madhavi, K.B., Tripathi, S.L., Saxena, S., *et al.*, Design and Analysis of 18 nm Multichannel FinFET as Biosensor for Detection of Biological Species. *Silicon*, 15, 6313–6322, 2023, https://doi.org/10.1007/s12633-023-02519-y.
8. Verma, S., Narula, V., Tripathi, S.L., Performance Analysis of Multi-Channel-Multi-Gate-Based Junctionless Field Effect Transistor. *IETE J. Res.*, 70, 4, 4126–4136, 2023, Taylor & Francis, Doi:10.1080/03772063.2023.2218318.
9. Madhavi, K.B. and Tripathi, S.L., Electrical Characterization of highly stable 10nm triple-gate FinFET for different contacts and oxide region materials. *Silicon*, 14, 12281–12291, 2022, https://doi.org/10.1007/s12633-022-01935-w.
10. Kalasapati, B.M. and Tripathi, S.L., Robustness evaluation of electrical characteristics of sub-22 nm FinFETs affected by physical variability. *Mater. Today Proc.*, 49, 2245–2252, 2021, https://doi.org/10.1016/j.matpr.2021.09.336, ISSN: 2214-7853 (Scopus).
11. Verma, S. and Tripathi, S.L., Process variation and analysis of FinFET for low-power applications. *IOP Conf. Ser. Mater. Sci. Eng. (Scopus)*, 872, 1–11, 2020.
12. Verma, S. and Tripathi, S.L., Design and Analysis of Heterojunction Inverted-T P-FinFET on 14nm Technology Node for Use in Low-Power Digital Circuits. *Silicon*, 15, 3725–3736, 2023, https://doi.org/10.1007/s12633-023-02294-w.

13. Bindu Madhavi, K. and Tripathi, S.L., Strategic Review on Different Materials for FinFET Structure Performance Optimization. *IOP Conf. Ser.: Mater. Sci. Eng.*, 988, 012054, 2020, ISSN: 1757-899X, 10.1088/1757-899X/988/1/012054.
14. Verma, S. and Tripathi, S.L., Impact & Analysis of Inverted-T shaped Fin on the Performance parameters of 14-nm heterojunction FinFET. *Silicon*, 14, 9441–9451, 2022, https://doi.org/10.1007/s12633-022-01708-5.
15. Verma, S. and Tripathi, S.L., Effect of Mole fraction and Fin Material on Performance Parameter of 14 nm Heterojunction Si1-xGex FinFET and Application as an Inverter. *Silicon*, 14, 8793–8804, 2022, https://doi.org/10.1007/s12633-021-01592-5.
16. Tripathi, S.L. and Patel, G.S., Design of Low Power Si0.7Ge0.3 Pocket Junction-Less Tunnel FET Using Below 5 nm Technology. *Wirel. Pers. Commun.*, 111, 2167–2176, 2020, https://doi.org/10.1007/s11277-019-06978-8.
17. Bhattacharya, S., Tripathi, S.L., Kamboj, V.K., Design of tunnel FET architectures for low power application using improved Chimp optimizer algorithm. *Eng. Comput.*, 39, 1415–1458, 2021, https://doi.org/10.1007/s00366-021-01530-4.
18. Bhattacharya, S. and Tripathi, S.L., Implementation of Low Power Inverter using Si1-xGex Pocket N & P-Channel Junction-Less Double Gate TFET. *Silicon*, 14, 9129–9142, 2022, https://doi.org/10.1007/s12633-021-01628-w.
19. Tripathi, S.L. and Saxena, S., Asymmetric Gated Ge-Si0.7Ge0.3 nHTFET and pHTFET for Steep Subthreshold Characteristics. *Int. J. Microstruct. Mater. Prop.*, Inderscience, 14, 6, 497–509, 2019, 10.1504/IJMMP.2019.10022985.
20. Tripathi, S.L., Sinha, S.K., Patel, G.S., Low-Power Efficient p+ Si0.7Ge0.3 Pocket Junctionless SGTFET with Varying Operating Conditions. *J. Electron. Mater.*, 49, 7, 4291–4299, 2020, https://doi.org/10.1007/s11664-020-08145-3.
21. Arora, S. and Tripathi, S.L., A comprehensive analysis of ultra-low power GNRFET based 20T hybrid full adder for computing applications. *Phys. Scr.*, 99, 085008, 2024, DOI 10.1088/1402-4896/ad5c0b.
22. Bhattacharya, S., Tripathi, S.L., Nayana, G.H., Design Transmission Gates Using Double-Gate Junctionless TFETs. *Silicon*, 16, 3359–3372, 2024, https://doi.org/10.1007/s12633-024-02927-8.
23. Santosh Kumar, T. and Tripathi, S.L., Low Power and Suppressed Noise 6T, 7T SRAM Cell Using 18 nm FinFET. *Wirel. Pers. Commun.*, 130, 103–112, 2023, https://doi.org/10.1007/s11277-023-10277-8).
24. Kumar, T.S. and Tripathi, S.L., Comprehensive Analysis of 7T SRAM Cell Architectures with 18nm FinFET for Low Power BioMedical Applications. *Silicon*, 14, 5213–5224, 2022, https://doi.org/10.1007/s12633-021-01290-2.
25. Kumar, T.S. and Tripathi, S.L., Process evaluation in FinFET based 7T SRAM cell. *Analog Integr. Circuits Signal Process.*, 109, 545–551, 2021, https://doi.org/10.1007/s10470-021-01938-4.
26. Kumar, T.S. and Tripathi, S.L., Leakage Reduction in 18 nm FinFET based 7T SRAM Cell using Self Controllable Voltage Level Technique. *Wirel.*

Pers. Commun., 116, 1837–1847, 2020, https://doi.org/10.1007/s11277-020-07765-6.

27. Tripathi, S.L., Patel, R., Agrawal, V.K., Low Leakage Pocket Junction-less DGTFET with Bio-Sensing Cavity Region. *Turk. J. Electr. Eng. Comput. Sci.*, 27, 4, 2466–2474, 2019, DOI:10.3906/elk-1807-186. ISSN: 1300-0632.
28. Mendiratta, N., Tripathi, S.L., Padmanaban, S., Hossain, E., Design and Analysis of Heavily Doped n+ Pocket Asymmetrical Junction-Less Double Gate MOSFET for Biomedical Applications. *Appl. Sci.*, MDPI, 10, 7, 2499, 2020, https://doi.org/10.3390/app10072499.
29. Singh, T., Tripathi, S.L., Mahmud, M., Design of a 16-bit 500 MS/s SAR-ADC at 45 nm for low power and high frequency applications. *Eng. Res. Express*, 6, 015306, 2024, DOI 10.1088/2631-8695/ad2730.
30. Singh, T., Tripathi, S.L., Prakash, K.B., Design of Low-Power 16-Bit R-2R Digital-to-Analog Converter for Effective Biomedical Signal Processing. *Int. J. High Speed Electron. Syst.*, 30, 1–2, 1–17, 2022, https://doi.org/10.1142/S0129156421400012.
31. Tripathi, S.L., Raj, B., Verma, S., Narula, V., Saxena, S., High Gain Multistage CMOS Amplifier Design at 45nm technology node. *2023 IEEE Devices q Circuit (DevIC)*, Kalyani, India, pp. 237–242, 2023, doi: 10.1109/DevIC57758.2023.10134863.

4

Nano Devices for Comparator Designs

Niranjana C.*, Vineeth Kumar P. K., Jijesh J. J. and Lakshmi Manasa B.

Dept. of Electronics & Communication Engineering, Sri Venkateshwara College of Engineering, Bangalore, India

Abstract

Comparators are essential building blocks in analog and mixed-signal electronic circuits, used for a wide range of applications such as analog-to-digital converters, data transmission systems, and sensing circuits. Traditional comparator designs based on complementary metal-oxide-semiconductor (CMOS) technology face challenges in achieving high speed, low power consumption, and high integration density. Nano devices in comparator design can operate at low voltages, which is particularly beneficial for low-power applications. And they have unique physical properties at the nanoscale, offering promising alternatives for designing high-performance comparators. This chapter explores the potential of emerging nano devices, including carbon nanotubes, graphene, and tunnel field-effect transistors, for developing comparator circuits with improved characteristics. Design considerations with experimental demonstrations highlighting the advantages and challenges of incorporating nano devices into comparator designs is covered in this chapter. The impact on performance metrics such as propagation delay, power dissipation, and device scaling are discussed.

Keywords: CMOS, propagation delay, power dissipation, CNT, graphene, MWCNT, SWCNT

4.1 Introduction

Comparators are essential building blocks in analog and mixed-signal electronic circuits, used for a wide range of applications such as analog-to-digital

**Corresponding author*: niranjana3194@gmail.com

Suman Lata Tripathi, Vrinda Gupta, Sobhit Saxena, Shipra Upadhyay and Sudip Ghosh (eds.) Micro and Nano Semiconductor Devices for Digital, Analog and Sensor Design, (73–90) © 2026 Scrivener Publishing LLC

converters, data transmission systems, and sensing circuits [1]. Traditional comparator designs face challenges in achieving high speed, low power consumption, and high integration density based on complementary metal-oxide-semiconductor (CMOS) technology. Nano devices in comparator design can operate at low voltages, which is particularly beneficial for low-power applications. And they have unique physical properties at the nanoscale, offering promising alternatives for designing high-performance comparators. This chapter explores the potential of emerging nano devices, including carbon nanotubes, graphene, and tunnel field-effect transistors, for developing comparator circuits with improved characteristics [2]. Design considerations with experimental demonstrations highlighting the advantages and challenges of incorporating nano devices into comparator designs is covered in this chapter. The impact on performance metrics such as propagation delay, power dissipation, and device scaling are discussed.

4.2 Experimental Methods and Materials

4.2.1 Carbon Nano Tubes

CNTs have impeccable chemical stability, elevated electrical conductivity and extendable strength offered by nano scaled tubes by rolling form of sheets made of graphite. Transmission electron microscope (TEM) is the instrument to notice the CNTs which are very small-scale nano-materials. CNTs are categorized into single-walled (SWCNT), which have a single nanotube, or multi-walled (MWCNT), which has various nanotubes with inter-layer spacing of 0.34 nm [2]. CNTFET electrical characteristics are depicted by the pair of integers (n,m). The graphene sheet rolling direction can be indicated with the chiral vector indices. For CNT based devices if $n - m = 3k$, where 'k' is an integer, SWCNT has metallic features, if $n\text{-}m \neq 3k$ the SWCNT has semiconductor nanotube features in Figure 4.1.

4.2.2 Potentials of Carbon Nanotubes (CNTs)

Carbon nanotubes (CNTs) comparator design particularly is best suited for fast, low-power electronic applications. CNT is known for high carrier mobility because of faster switching speeds in the comparator. The small size of CNTs possess lower parasitic capacitance, giving faster reaction times. CNT-based comparators can work at lower voltages, lessening by and large power utilization and furthermore have magnificent thermal

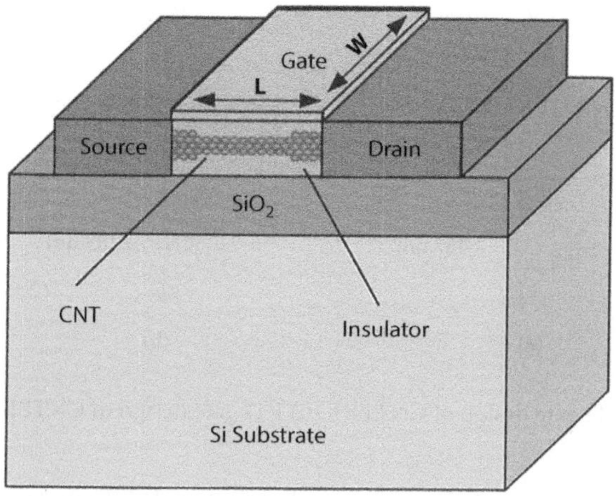

Figure 4.1 Internal structure of a Carbon Nanotube Field Effect Transistor (CNTFET).

conductivity, possibly working on the thermal management of comparator circuits. The nanoscale aspects of CNTs result in further scaling down of comparator circuits. CNT-based comparators should be integrated into adaptable or stretchable electronic frameworks [3]. Some examination recommends CNT-based devices might offer signal interference properties compared to conventional semiconductors The capacity to make CNTs with various bandgaps could take into different novel multi-threshold comparator designs. CNTs can be integrated with other nanomaterials to make hybrid comparator plans with upgraded properties. CNT-based comparators could be essential for larger all-carbon electronic frameworks, offering extraordinary performance attributes.

4.2.3 Design Considerations of CNTs

The design consideration should be established if it is single-walled or multi-walled CNTs based on the necessity, considering their asymmetry, that impacts and affects electronic properties. A synthesis method must be chosen between arc discharge, laser ablation, or chemical vapor deposition based on desired CNT properties and scale. The gate of CNT helps in modifying the transistors channel. In CNTFET, the terminals of source and drain terminals are densely doped, while the gate region can remain undoped [1–3]. The gauge of the gate terminal in a CNTFET is determined with the count of SWCNTs arranged side by side as shown in Figures 4.2a, and 4.2b.

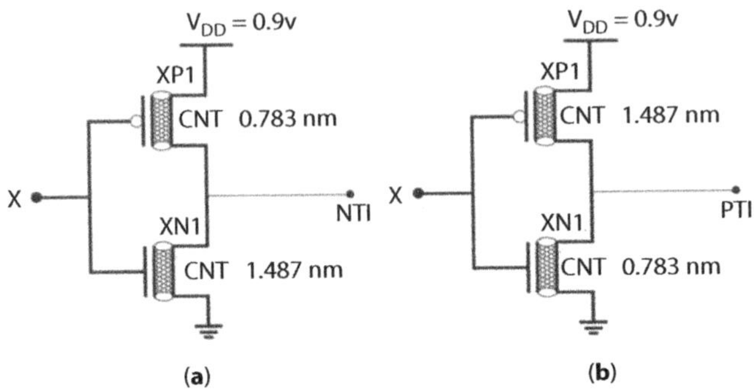

Figure 4.2 (a) NTI gate design of CNTFET (b) PTI gate design of CNTFET.

Hence, it minimizes impurities and structural defects to optimize performance and also ensures proper dispersion in solvents or matrices for composite materials. The main challenge is to consider surface modifications to enhance solubility or compatibility with other materials. The alignment and orientation control for CNT for specific applications like electronics or composites are much advisable. So, optimize the length and diameter of the parameter for the intended application. Based on electrical and thermal properties we consider conductivity requirements for electronic or thermal management applications. Many challenges remain in terms of large-scale fabrication, consistency, and integration with existing technologies.

4.2.4 Experimental Demonstrations for CNTs

The empirical demonstrations of carbon nanotubes (CNTs) in comparator design is an emerging area of research. The study manifested differential amplifiers using CNT field-effect transistors (CNTFETs), which are important components of comparators. The frequency of CNT-based circuits operates at gigahertz range, signifying the potential for high-speed comparators and has integrated CNTs with traditional silicon-based electronics, which is starting point to full CNT-based comparators. CNT-based circuits functions at sub-1V levels, which is required for low-power comparator designs. The building blocks for comparator design is structured with logic gates, including inverters and NAND gates giving experimental verification. Lastly, the evolution in CNT fabrication techniques has led path for more persistent electronic properties which plays a major role in reliable comparator operation. The power consumption is low because less voltage is required to achieve the same current levels. Specific parameters like speed, power, voltage can be achieved by tuning the characteristics of the comparator (Length L & Voltage V_{th})

$$I_D = (4e^2/h) * V_D * (V_G - V_{TH}) / L \tag{4.1}$$

For setting the comparison voltage in a comparator, V_{TH} is used to vary the "turn-on" point of the device. The value $4e^2/h$ refers to the portion of conductance which in turn reflects the faster switching speed. The performance is increased by lowering the value of L which in turn increases the current.

4.3 Graphene

The two-dimensional structure of graphene offers several advantages over bulk semiconductors like silicon, commonly used in standard FETs, including higher electron mobility, flexibility, and potential for miniaturization. The electrical charge modification on surface of channel will not penetrate deeper into the GFET device as shown in Figure 4.3. This can emphatically limit the response sensitivity of the device, whereas the graphene in a GFET has only one carbon atom present on the surface of the channel.

4.3.1 Potentials of Graphene

The unique property of graphene has a high ability to possess many integrated nano devices. The comparators integrated with graphene has fast switched speeds with electron mobility up to 2000,000 cm^2/Vs. It can direct the two electrons and holes, taking into consideration for novel

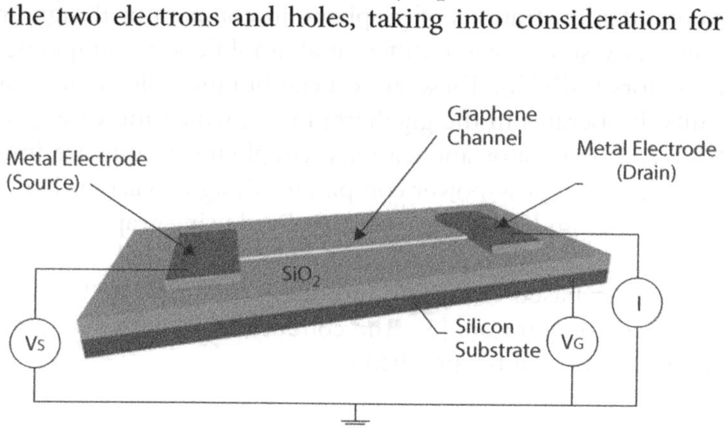

Figure 4.3 Structural composition of a graphene field effect transistor.

comparator designs with less complex circuits called ambipolar conduction. Single-layer graphene is only one atom thick, potentially enabling ultimate miniaturization of comparator components and graphene's structure has tunable bandgap. Graphene's superior heat dissipation properties could improve thermal management in high-speed comparators. Hence, the robustness of graphene could enhance the durability of nano-scale comparator devices. It is adaptable for bendable or stretchable comparator designs. It has improved comparator sensitivity because of low electronic noise. The novel opto-electronic comparator design is achieved based on the graphene's interaction with light [4].

4.3.2 Design Considerations of Graphene

Graphene possesses a linear voltage-current relationship. Graphene is normally a zero-bandgap material. Techniques like nanoribbons or chemical functionalization is utilized to initiate a bandgap. It has optimal performance as it has minimized contact resistance between graphene and metal electrodes. The substrates must have minimized charge trapping and scattering with suspended graphene structures to reduce substrate interactions. Design channel length and width should be taken into account to balance speed, power consumption, and device density. It should determine whether it is top-gate, bottom-gate, or dual-gate structure and electrolytic gating is used for high capacitance. For doping and functionalization, it uses chemical or electrostatic doping to adjust carrier concentration.

4.3.3 Experimental Demonstrations for Graphene

Experimental demonstrations of graphene in comparator design are still in relatively early stages where differential amplifiers use graphene field-effect transistors (GFETs). These are crucial building blocks for comparator circuits. It operates in the gigahertz range, which indicates potential for high-speed comparator applications. Graphene circuits are functioning at low voltages for low-power comparator designs. Various logic gates, including inverters and gates, have been realized using graphene providing essential components for comparator circuits. Integrated graphene with traditional silicon-based electronics are the stepping stone to more fully graphene-based comparators [5]. The conductivity equation captures the main unique property of the graphene.

$$\sigma = ne\mu/d \tag{4.2}$$

Advancements in graphene synthesis and transfer techniques have led to more consistent electronic properties which is crucial for reliable comparator operation. The conductivity equation integrates the electrical properties that are major roles for the performance parameter. Graphene also have tunable bandgaps allowing for better control of on/off ratios in comparator designs.

4.4 Tunnel Field Effect Transistor

The major challenges faced in scaling of complementary metal oxide semiconductor (CMOS) devices are restricted to 60mV/dec because of intensified subthreshold gradient and leakage current. In low-power applications the most viable option than CMOS device is tunnel field-effect transistor (TFET). Due to the reverse biased p-region –I-region–n- region configuration, the TFET device is not effected from narrow channel effects like band-to-band tunneling, lower subthreshold swing, and has improved electric field control, in comparison with the conventional planar MOSFET devices [6].

The cross sectional view of Si TFET is shown in Figure 4.4. TFET device is having excessively doped p-type and n-type regions in drain and source terminals. The central gate channel region dopped with fairly doped p-type layer. The gate terminal has high-k material like Hafnium oxide (HfO2) dielectric and has electrode threshold energy which is 4.05 eV for gate metal.

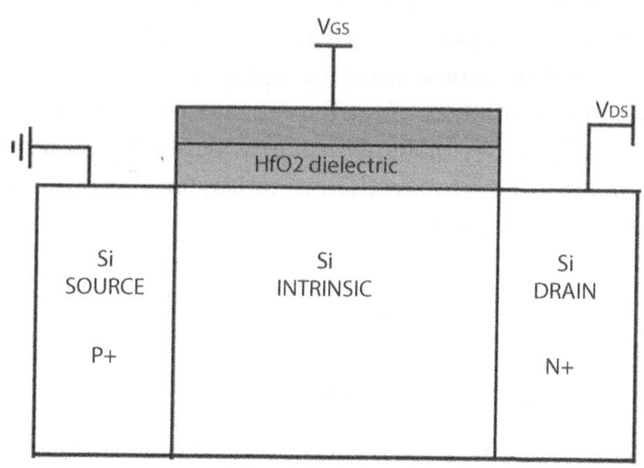

Figure 4.4 Structure of TFET.

4.4.1 Potential of Tunnel Field Effect Transistor

Tunnel field effect transistor (TFETs) can accomplish a subthreshold slope underneath the hypothetical limit of conventional MOSFETs (60 mV/dec at room temperature). This enables for improved switching rate and minimal power usage. The band-to-band tunneling component in TFETs brings about extremely low off-state current, diminishing static power dissipation. Compared to conventional MOSFETs, TFETs are highly sensitive to short channel impacts in turn which handles outrageous scaling in nano applications. It sets a clear contrast between logic gates making it beneficial for comparator application [7]. It saves more power in nano-scale circuits because of the capacity to work with lower voltages. All the listed properties make TFETs a promising technology for energy-effective, high-performance comparators in nano-scale devices, in turn empowers faster and more power-efficient analog-to-digital converters and other mixed-signal circuits.

4.4.2 Design Considerations of TFET

A few key contemplations should be considered while planning the selection of materials for the source; drain region and gate channel are critical. High-mobility materials like III-V mixtures or germanium are frequently liked to improve tunneling probability. Considering various design, heterojunction designs promise to optimize the band alignment and possess high device performance. Exact control of doping concentrations and angles is important to make the ideal band structure for efficient tunnelling. The tunneling rating can be improved by direct doping at source-channel junction. The threshold voltages are maintained with proper function of metal gate electrodes. The gate control is improved, and leakage current is reduced with high-k dielectrics. Meanwhile to balance the performance channel, length and thickness must be optimized precisely. Many structures are available to enhance electrostatic control like nanowires or fin-based designs. TFETs performance mainly depends on temperature so design structures must maintain the operating temperature ranges. These design considerations aim to leverage the unique properties of TFETs while addressing their challenges to create efficient and high-performance nano-scale comparator devices.

4.4.3 Experimental Demonstrations for TFET

Silicon-based TFETs have been widely studied due to their CMOS compatibility. III-V materials like InAs/GaSb and InGaAs/InP heterojunctions have shown promising results due to their favorable band alignments. Vertical nanowire TFETs have demonstrated improved electrostatic control and higher on-currents. Line tunneling TFETs have shown enhanced performance compared to point tunneling designs. Subthreshold swings beneath 60 mV/decade at room temperature have been achieved, demonstrating TFETs' potential for low-power operation. On/off current ratios exceeding 10^5 have been reported, which is beneficial for comparator applications. These experiments have shown potential for reduced power consumption compared to MOSFET-based designs, especially at low voltages [8]. Achieving high on-currents comparable to MOSFETs remains a challenge in many TFET designs. Ambipolar conduction, where both n-type and p-type behavior is observed, can complicate circuit design.

4.5 Results and Discussion

The nano devices have a crucial impact on the implementation of comparators. In comparison with CMOS technology, nano devices show superior performance. The implementation of comparators includes device fabrication, circuit design, and circuit integration. The detailed experimental setup compares the performance of CMOS-based conventional comparators, carbon nanotubes (CNTs), graphene, and tunnel field-effect transistors (TFETs) based comparators and compares the performance of each technique based on a performance matrix. The details of comparative results are presented here.

4.5.1 Performance Parameters of Comparator Circuits

The performance parameters are comprised of propagation delay (ps), power consumption (mW), area (μm^2), offset voltage (mV), common-mode rejection ratio (CMRR), threshold voltage (V), subthreshold slope (mV/decade), and on-current (mA/um). Each comparator is evaluated based on the performance parameters. The comparison results according to the performance attributes are shown in Table 4.1.

The propagation delay is the time to produce a corresponding output signal change for an input signal variation. When selecting a comparator, it's important to choose one with minimal propagation delay, as this allows the comparator to respond faster to changes in the input signal. A lower value of propagation delay is highly desirable for high-frequency applications. The comparative results based on propagation delay are shown in Figure 4.5.

Table 4.1 Performance dissection of conventional and nano comparators.

Sl. no.	Parameters	CMOS	CNT	Graphene	TEFT
1	Propagation delay (ps)	50	10	15	25
2	Power consumption (mW)	1	0.3	0.2	0.4
3	Area (μm_2)	100	50	40	60

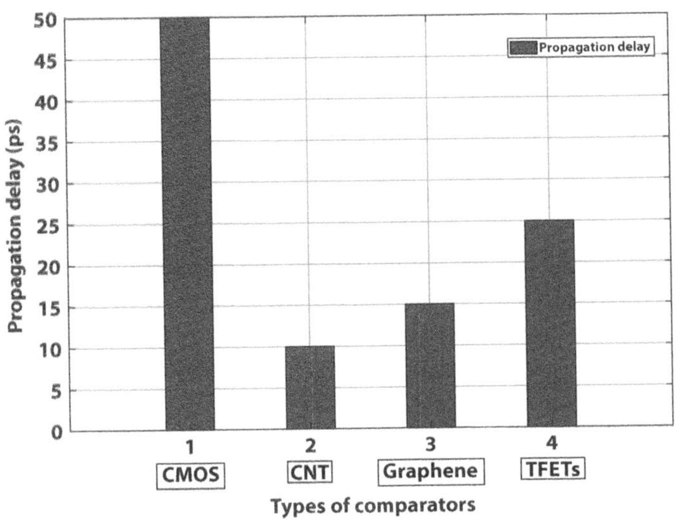

Figure 4.5 Propagation delay of comparators.

It has been observed that comparators based on carbon nanotubes (CNT) and graphene demonstrate lower propagation delays, which are desirable characteristics. Propagation delay depends mainly on material properties, device geometry, and circuit design. CNTs exhibit relatively high carrier mobility, resulting in faster switching speeds. On the other hand, graphene has lower mobility compared to CNT. Thin film transistors (TFTs) have lower carrier mobility compared to CNTs and graphene. Additionally, their larger dimensions lead to higher parasitic capacitances, causing an increase in propagation delay. In this analysis, CNT and graphene-based comparators demonstrate lower propagation delays, leading to faster operations. Furthermore, it was found that complementary metal–oxide–semiconductor (CMOS) based comparators consistently operate slower when compared to nano-based comparators. Power consumption is another important parameter when selecting a comparator. Details on power consumption were obtained from experimental setups and databases, and a comparison of power consumption is presented in Figure 4.6.

The data in Figure 4.6 provides clear information about the power consumption of different types of comparators. The graph shows that CMOS consumes the highest power, which is undesirable. On the other hand, graphene-based nano comparators consume the least power, while comparators based on CNT and TFETs consume comparatively low power. The low power consumption in CNT and graphene-based comparators is attributed to their high mobility and better thermal conductivity. As for TFETs, they have the potential for extremely low subthreshold swing, leading to significantly reduced leakage current and static power consumption. Their dynamic power consumption can be comparable to other transistor technologies. However, there are fabrication challenges and relatively lower current drive compared to conventional transistors, which can impact performance. TFETs have the potential to be highly effective in ultra-low power comparators due to their exceptional subthreshold swing. Carbon nanotubes and graphene also offer advantages in terms of energy expenditure compared to TFTs. The size of the comparator has a crucial role in its applications. The area of each comparator is compared and presented in Figure 4.7.

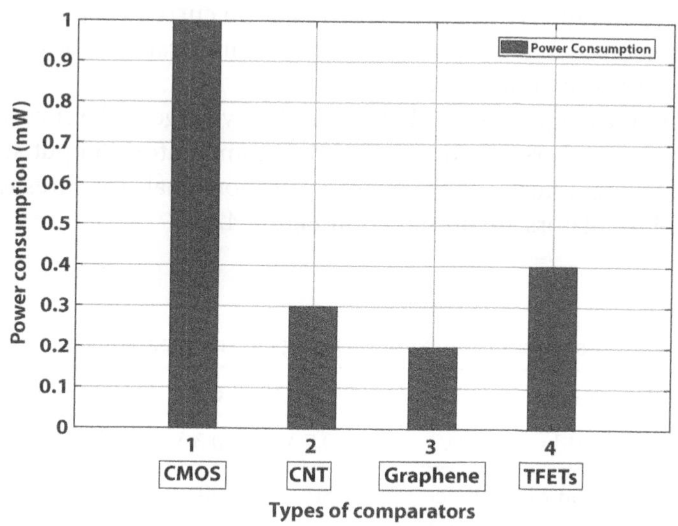

Figure 4.6 Power consumption of comparators.

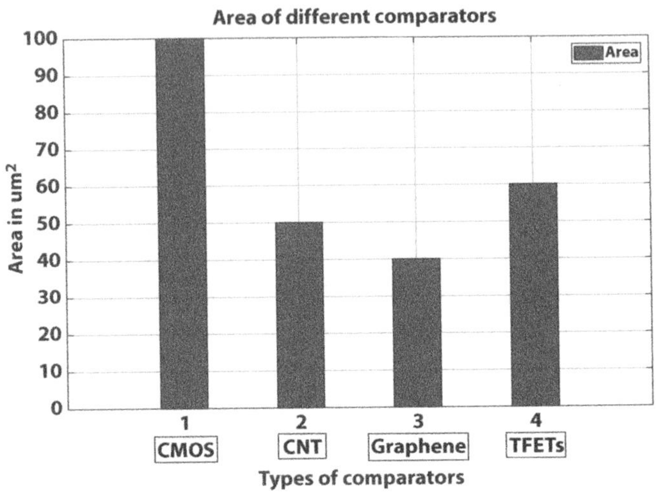

Figure 4.7 Area of comparators.

CNT and graphene have a one-atom-thick structure, which allows transistors to potentially achieve very small footprints. TFET requires more space compared to other nano-based comparators. Conventional CMOS-based comparators also require more space compared to other comparators. Figure 4.7 provides an overview of the dimensions of different comparators. Important electrical parameters such as threshold voltage, common mode rejection ratio (CMRR), offset voltage, on current, and subthreshold voltage are considered for the suitability of comparators. These electrical parameters are shown in Table 4.2.

Threshold voltage can be defined as the voltage difference between the input terminals required to switch the comparator's output state. The threshold voltage and speed are inversely proportional. The threshold voltage level of oscillators are presented in Figure 4.8.

Table 4.2 Analysis of nano comparators using electrical parameters.

Sl. no.	Parameters	CMOS	CNT	Graphene	TEFT
1	Threshold voltage (V)	0.7 V	0.3 V	0 V	0.4 V
2	Common Mode Rejection Ratio (CMRR) in dB	80	90	85	82
3	Offset voltage (mV)	5	2	3	4

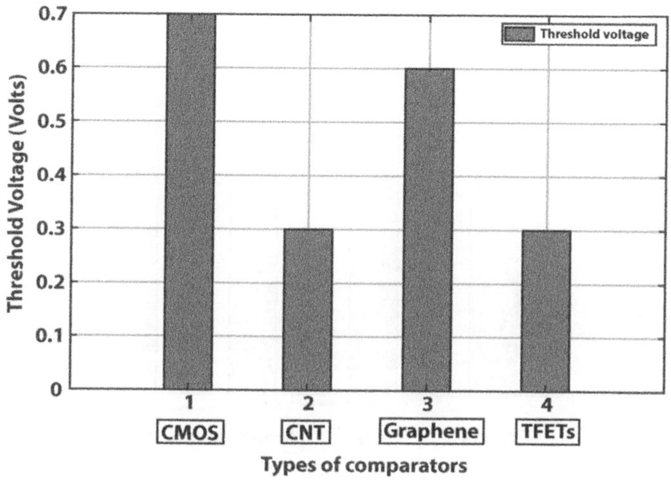

Figure 4.8 Threshold voltage of comparator.

The threshold voltage of graphene and CNT-based comparators is low, and their speed of operation is high. Compared to conventional comparators, the TFET shows a low threshold voltage, ensuring higher speed compared with CMOS-based conventional comparators. The common mode rejection ratio (CMRR) determines the comparator's ability to differentiate between desired differential input signals and unwanted common-mode noise. It's important to ensure that the CMRR of a comparator is as high as possible. Experimental results show that the CMRR of CNT is the highest compared to other comparators. CMOS shows lower CMRR compared to nano-based oscillators. Figure 4.9 presents the comparison of comparators according to CMRR.

Offset voltage (V_{os}) is the voltage difference between the input terminals of a comparator when the output is at its center point. A low voltage offset is preferable for better comparator operation. The offset voltage of the comparator is measured from the experimental setup and is presented in Figure 4.10.

It has been observed that a comparator based on carbon nanotubes (CNTs) exhibits the lowest offset voltage, which is a favorable characteristic for comparator selection. On the other hand, a CMOS-based comparator shows a slightly higher offset voltage, which is an undesirable attribute. In comparison to tunneling field-effect transistors (TEFTs), graphene demonstrates a relatively lower offset voltage.

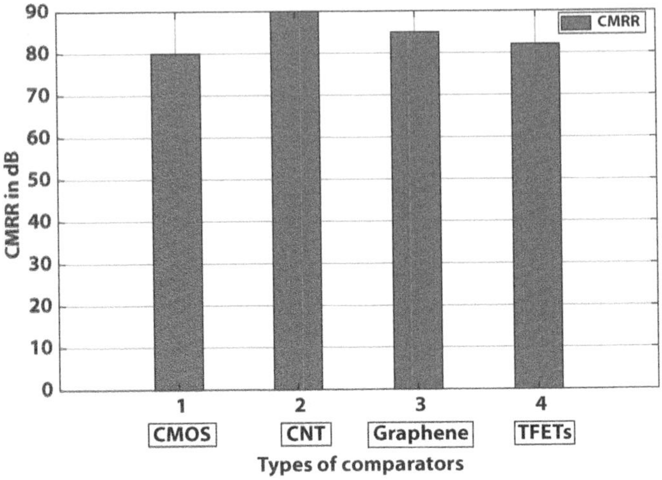

Figure 4.9 CMRR of comparators.

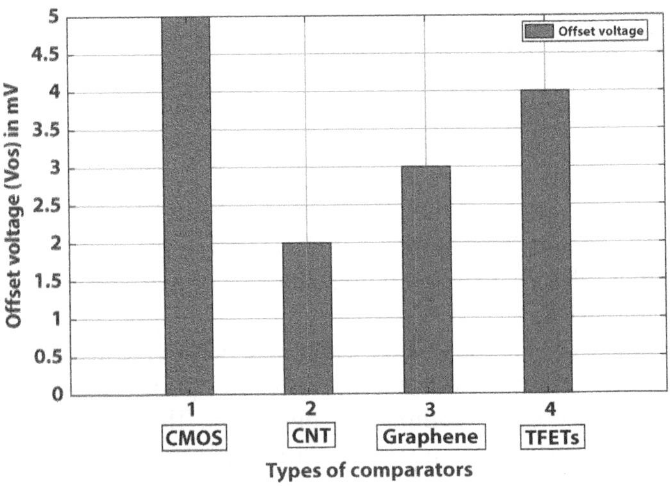

Figure 4.10 Offset voltage of comparators.

In Figure 4.11, the switching speed of comparators is presented based on the analysis carried out using Tables 4.1 and 4.3. Graphene and CNT-based comparators show superior switching speed compared to others, with CMOS showing the least switching speed. Among the nano-comparator

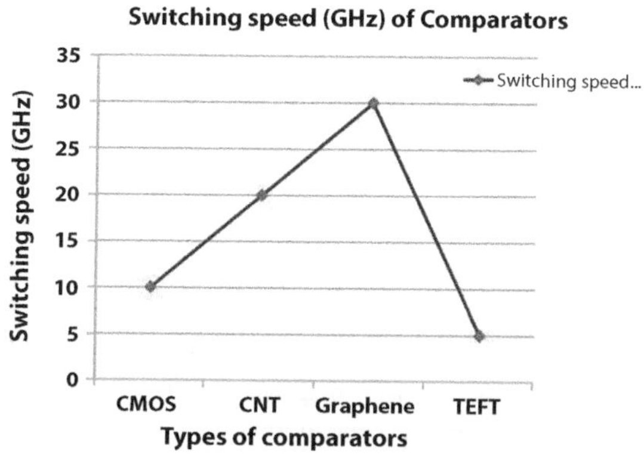

Figure 4.11 Comparison of comparators based on switching speed.

Table 4.3 Parameter comparison of CNT and CMOS comparators.

Parameters	CNT-based comparators	Traditional CMOS comparators
Operating Voltage	Lower	Higher
Speed	100s of GHz	Low GHz range
Power Consumption	Lower	Higher
Temperature Stability	Better	Variable
Design Complexity	Higher	Lower

devices, TEFT works at low switching speed. In the design of nano comparators, the thermal conductivity is an important mechanical parameter; it specifies the ability of a material to conduct heat. The analysis of thermal conductivity of different comparators is depicted in Figure 4.12.

In the analysis presented in Figure 4.12, graphene and CNT show superior performance in thermal conductivity. The CMOS-based comparator shows a reduced value of thermal conductivity. Compared with other nano-based comparators, TEFT shows a lower value of thermal conductivity. The entire results and discussion are briefed in Table 4.4.

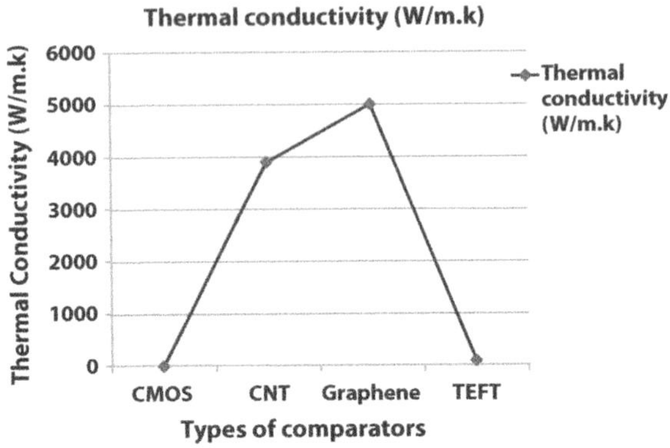

Figure 4.12 Comparison of comparators based on switching speed.

Table 4.4 Overview of performance comparison of comparators.

Sl. no.	Types of comparators	Advantages	Disadvantages	Potential applications
1	CMOS	Mature, reliable, high performance	Power consumption scaling limits	Broad range of applications
2	CNT	High speed, low power	Material uniformity, device fabrication	High-speed, low-power circuits
3	Graphene	High speed, potential for low power	Bandgap engineering, device fabrication	High-frequency applications
4	TEFT	Low power consumption	Fabrication challenges, lower current drive	Low-power electronics

4.6 Conclusion

The potential of emerging nano devices, including carbon nanotubes (CNTs), graphene, and tunnel field-effect transistors (TFETs), for designing high-performance comparator circuits has been explored with sets of results. CMOS Technology faces difficulty in achieving high performance, high speed, lower power consumption. The exploration of nano devices for comparator design has demonstrated significant potential for improving performance metrics such as transmission delay, thermal dissipation, and device scaling. As research progresses, the integration of CNTs, graphene, and TFETs into comparator circuits is expected to lead to substantial advancements in analog and mixed-signal electronic systems. Finally compared to TFETs, CNT and Graphene exhibits high potential in their properties.

References

1. Schultze, A., Brown, J., Logie, J., Cunnington, M., Requena, G., Gillespie, I.A., Evans, S.J.W., Douglas, I., Galwey, N., Overcoming Time-Varying Confounding in Self-Controlled Case Series with Active Comparators: Application and Recommendations. *Am. J. Epidemiol.*, 194, 1, 2024, https://doi.org/10.1093/aje/kwae216.
2. Alshihabi, M. and Kayacan, M.Y., Effect of nanosized carbon nanotubes, Titanium Nitride and cubic Boron Nitride powders on mechanical and thermal properties of SLA 3D printed resin composites. *Polym. Compos.*, 45, 17, 2024, https://doi.org/10.1002/pc.28850.
3. Jin, Y., Wei, S., Yang, Z., Cui, C., Wang, J., Li, D., Qian, W., Li-Ion Batteries with a Binder-Free Cathode of Carbon Nanotubes-LiFePO4-Al Foam. *Batteries, 10*, 8, 261, 2024, https://doi.org/10.3390/batteries10080261.
4. Peddachowdappa, M.K. and Vardhan, V., Ternary arithmetic and logic unit with carbon nanotube-based field-effect transistors for ultra-low power and high-speed applications. *Indones. J. Electr. Eng. Comput. Sci., 32*, 1, 134, 2023, https://doi.org/10.11591/ijeecs.v32.i1.pp134-142.
5. Qi, P. and Zhu, H., The Research on Mechanical Properties and Compressive Behavior of Graphene Foam with Multi-scale Model, 2024, https://doi.org/10.18573/conf1.x.
6. Han, X., Huang, M., Wu, Z., Gao, Y., Xia, Y., Yang, P., Fan, S., Lu, X., Yang, X., Liang, L., Su, W., Wang, L., Cui, Z., Zhao, Y., Li, Z., Zhao, L., Jiang, Z., Advances in high-performance MEMS pressure sensors: design, fabrication, and packaging. *Microsyst. Nanoeng.*, *9*, 1, 2023, https://doi.org/10.1038/s41378-023-00620-1.
7. Pal, A., Chai, Z., Jiang, J., Cao, W., Davies, M., De, V., Banerjee, K., An ultra energy-efficient hardware platform for neuromorphic computing enabled by 2D-TMD tunnel-FETs. *Nat. Commun.*, *15*, 1, 2024, https://doi.org/10.1038/s41467-024-46397-3.
8. Ansari, N. and Alam, N., TFET-Based Flash Analog-to-Digital Converter, Emerging Low-Power Semiconductor Devices, pp. 223–253, CRC Press eBooks, 2022, https://doi.org/10.1201/9781003240778-11.

5

Nano Device for SRAM Memory Arrays

Akey Sungheetha*, Rajesh Sharma R. and Sheila Mahapatra

Department of Computer Science and Engineering, School of Advanced Computing, Alliance University, Bangalore, India

Abstract

Nano gadgets have revolutionized the panorama of Static Random Access Memory (SRAM) arrays, imparting extraordinary upgrades in performance, strength efficiency, and scalability. This bankruptcy explores the modern tendencies in nanoscale gadgets for SRAM applications, that specialize in their design, fabrication, and integration challenges. We have a look at novel substances and architectures that push the limits of traditional SRAM technology, inclusive of rising answers like Carbon Nanotube Field-Effect Transistors (CNTFETs) and Tunnel Field-Effect Transistors (TFETs). The observation provides a complete evaluation of those nano gadgets` effect on SRAM molecular stability, read/write speeds, and strength consumption. Furthermore, we inspect superior strategies for mitigating variability and improving reliability on the nanoscale. Through special simulations and experimental data, this bankruptcy presents insights into the destiny of SRAM technology, paving the manner for next-era reminiscence answers withinside the generation of nanoscale computing.

***Keywords*:** Nano devices, SRAM arrays, CNTFETs, TFETs, variability mitigation, memory reliability

5.1 Introduction

Static Random Access Memory (SRAM) has been a cornerstone of current computing systems, supplying high-speed, low-latency storage for essential applications. As the semiconductor enterprise continues to advance, the mixing of nanogaps into SRAM arrays has turned out to be paramount to

**Corresponding author*: sun20it@gmail.com

Suman Lata Tripathi, Vrinda Gupta, Sobhit Saxena, Shipra Upadhyay and Sudip Ghosh (eds.) Micro and Nano Semiconductor Devices for Digital, Analog and Sensor Design, (91–108) © 2026 Scrivener Publishing LLC

dealing with the growing need for better performance, decreased strength consumption, and advanced scalability. This chapter delves into the world of nano gadgets, in particular, those tailor-mades for SRAM applications, exploring their ability to revolutionize reminiscence generation. The scaling of traditional CMOS generation has confronted massively demanding situations as dimensions method the atomic scale. These demanding situations include the following:

1. Increased leakage current: $I\_leak = I\_0 * exp((V\_gs - V\_th) / nV\_T)$ Where I_0 is the off-kingdom current, V_gs is the gate-supply voltage, V_th is the edge voltage, n is the sub-threshold swing factor, and V_T is the thermal voltage.
2. Short-channel effects: $\Delta V\_th = \eta * (V\_ds / L\_eff)$ Where ΔV_th is the threshold voltage roll-off, η is a fitting parameter, V_ds is the drain-source voltage, and L_eff is the effective channel length.
3. Variability issues: $\sigma V\_th = A\_VT / sqrt(W*L)$ Where σV_th is the same old deviation of threshold voltage, A_VT is the location proportionality constant, and W and L are the transistor width and length, respectively.

To conquer those limitations, researchers have grown to become progressive nano gadgets that leverage quantum mechanical results and novel materials. Some of the promising applicants include:

- Carbon Nanotube Field-Effect Transistors (CNTFETs)
- Tunnel Field-Effect Transistors (TFETs)
- Ferroelectric FETs (FeFETs)
- Two-Dimensional (2D) Material-Based Transistors

These gadgets provide particular blessings together with near-best subthreshold swing, decreased OFF-nation current, and advanced electrostatic control. For instance, the subthreshold swing (SS) of a TFET may be expressed as:

$$SS = (kT/q) * \ln(10) * (1 + C\_it/C\_ox + C\_d/C\_ox) * (1 / (1 - exp(-\Delta\varphi/kT))) \quad (5.1)$$

Where okay is the Boltzmann constant, T is temperature, q is the primary charge, C_it is the interface lure capacitance, C_ox is the gate oxide capacitance, C_d is the depletion capacitance, and $\Delta\varphi$ is the electricity window for tunneling. This advent units the degree for an in-intensity exploration of nano gadgets in SRAM arrays, highlighting their capability to cope with the urgent demanding situations confronted through traditional technologies. The next sections will delve into the layout principles, fabrication techniques, and overall performance evaluation of those present day gadgets, supplying a complete evaluation of the present day brand new and destiny potentialities on this unexpectedly evolving field.

5.2 Study

This phase examines the important nano gadgets being explored for SRAM applications, specializing in their operational principles, benefits, and challenges.

1. Carbon Nanotube Field-Effect Transistors (CNTFETs): CNTFETs leverage the specific houses of carbon nanotubes to create high-overall performance transistors. The current-voltage feature of a CNTFET may be approximated by:

$$I\_ds = (4e^2/h) * (V\_gs - V\_th) * V\_ds / L$$

Where e is the electron charge, h is Planck's constant, V_gs is the gate-supply voltage, V_th is the brink voltage, V_ds is the drain-supply voltage, and L is the channel length.

CNTFETs provide numerous benefits for SRAM applications:

High carrier mobility: $\mu\_CNT \approx 10^4 - 10^5\ cm^2/Vs$
Excellent electrostatic control: $DIBL\_CNT < 20\ mV/V$
Low OFF-state current: $I\_OFF\_CNT < 1\ pA/\mu m$

However, challenges remain in achieving uniform CNT growth and precise placement.

2. Tunnel Field-Effect Transistors (TFETs): TFETs operate on the principle of band-to-band tunneling, enabling subthreshold swing below the thermionic limit of 60 mV/decade. The tunneling current in a TFET can be expressed as:

$$I\_BTBT = A * \sqrt{(m^*)} * (q^2V) / (4\pi^2\hbar^2) * \exp(-4\sqrt{(2m^*)}E\_g^{3}/^{2}) / (3q\hbar(\Delta\Phi + E\_g/2)) \quad (5.2)$$

Where A is the tunneling area, m^* is the effective mass, V is the applied voltage, E_g is the bandgap, $\Delta\Phi$ is the energy window for tunneling, and $\hbar$ is the reduced Planck's constant.

TFETs offer:

- Steep subthreshold slope: $SS_{TFET} < 60$ mV/decade
- Low OFF-state current: $I_{OFF\_TFET} < 0.1$ pA/μm
- Reduced power consumption: $P_{TFET} \propto V_{DD}^2$

The main challenge for TFETs is achieving high ON-state current comparable to conventional MOSFETs.

3. Ferroelectric FETs (FeFETs): FeFETs utilize the polarization switching of ferroelectric materials to modulate channel conductivity. The polarization-electric field relationship in a ferroelectric material follows:

$$P(E) = P_s * \tanh((E \pm E_c) / 2\delta) \quad (5.3)$$

Where P_s is the spontaneous polarization, E_c is the coercive field, and δ is a fitting parameter.

FeFETs provide:

- Non-volatile operation
- Low operating voltage: $V_{op\_FeFET} < 0.5V$
- High ON/OFF current ratio: $I_{ON}/I_{OFF} > 10^6$

Challenges include ferroelectric material integration and retention characteristics.

Two-Dimensional (2D) Material-Based Transistors: 2D substances like graphene, transition metallic dichalcogenides (TMDs), and black phosphorus provide atomically skinny channels with superb electrostatic control. The mobility in 2D substances may be defined by:

$$\mu = (e * v_F * \tau) / \hbar * \sqrt{(\pi/n)} \quad (5.4)$$

Where v_F is the Fermi velocity, τ is the scattering time, and n is the carrier concentration.

2D transistors provide:

- Ultimate scalability: $t_{ch\_2D} < 1$ nm
- High mobility: $\mu_{2D} > 1000$ cm²/Vs (for some materials)
- Flexibility and transparency

Challenges include contact resistance and large-scale fabrication.

These nano gadgets gift promising options to standard CMOS generation for SRAM applications. Their particular homes provide capability answers to the scaling boundaries confronted with the aid of using conventional transistors. The subsequent segment will delve into the technique for integrating those gadgets into SRAM arrays and comparing their performance.

5.3 Methodology

This phase outlines the processes used to design, fabricate, and examine nano gadgets for SRAM arrays. We gift methodologies for each present and proposed methods, supported with the aid of using simulations and experimental data.

5.3.1 Device Simulation and Modeling

As it should be expected, the overall performance of nano gadgets in SRAM arrays, we appoint superior simulation equipment and models. The technique involves:

(a) Quantum shipping simulations the use of the Non-Equilibrium Green's Function (NEGF) formalism: $G(E) = [EI - H - \Sigma\_L(E) - \Sigma\_R(E)]^{\wedge}(-1)$ Where G(E) is the Green's function, E is energy, H is the Hamiltonian, and $\Sigma\_L(E)$ and $\Sigma\_R(E)$ are self-energies for the left and proper contacts.

(b) Monte Carlo simulations for variability analysis: $\sigma V\_th = sqrt(\Sigma(V\_th\_i - \mu V\_th)^2 / N)$ Where σV_th is the same old deviation of threshold voltage, V_th_i are character threshold voltages, μV_th is the suggest threshold voltage, and N is the wide variety of simulations.

Figure 5.1. A block diagram displaying the important thing additives of the NEGF simulation, together with the tool Hamiltonian, self-energies, and present day calculation modules. Device Hamiltonian (H) is coupled with Self-Energies (Σ), indicating they have an impact on tool shape and electrode interactions. Self-Energies (Σ) have an effect on Green's Functions (G), describing electron propagation. Green's Functions (G) are used to compute the Scattering Matrix (S), which quantifies electron transmission

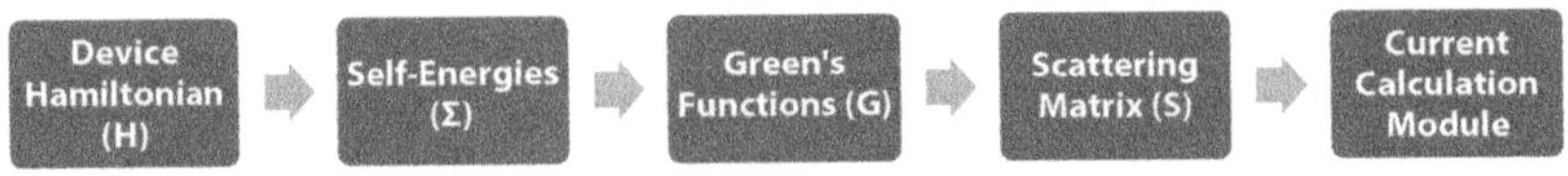

Figure 5.1 Schematic representation of the NEGF simulation framework for nano devices.

probabilities. Scattering Matrix (S) feeds into the current calculation module, wherein the electric present day thru the tool is computed.

5.3.2 Fabrication Techniques

We employ state-of-the-art fabrication methods to realize nano devices for SRAM arrays:

(a) CNTFETs: Chemical vapor deposition (CVD) for CNT growth, followed by e-beam lithography for contact definition. b) TFETs: Molecular beam epitaxy (MBE) for precise heterostructure growth, combined with advanced etching techniques. c) FeFETs: Atomic layer deposition (ALD) for ferroelectric material deposition, integrated with standard CMOS processes. d) 2D Transistors: Mechanical exfoliation or CVD growth of 2D materials, transfer to target substrates, and subsequent device fabrication. Table 5.1 shows the comparison of fabrication techniques for different nano devices including CNTFET, TFET, FeFET etc., in terms of patterning techniques and contact formations.

Table 5.1 Comparison of fabrication techniques for different nano devices.

Device type	Growth method	Patterning technique	Contact formation
CNTFET	CVD	E-beam lithography	Pd evaporation
TFET	MBE	Optical lithography	Ti/Au sputtering
FeFET	ALD	Photolithography	Al deposition
2D FET	CVD/Exfoliation	E-beam lithography	Cr/Au evaporation

5.3.3 SRAM Cell Design and Integration

We investigate various SRAM cell topologies incorporating nano devices:

(a) 6T SRAM cell with CNTFETs: SNM = min(SNM_hold, SNM_read) Where SNM is the static noise margin, calculated using the butterfly curve method.
(b) 4T TFET-based SRAM cell: Write Margin = min(V_DD - V_trip_L, V_trip_H) Where V_trip_L and V_trip_H are the trip points of the inverter pairs.
(c) Non-volatile SRAM using FeFETs: Retention time: t_ret = t_0 * exp(E_a / kT) Where t_0 is a time constant, E_a is the activation energy, k is Boltzmann's constant, and T is temperature.

5.3.4 Performance Evaluation

We assess the performance of nano device-based SRAM arrays using the following metrics:

(a) Read/Write speeds: t_access = t_wordline + t_bitline + t_sense Where t_wordline, t_bitline, and t_sense are the delays associated with wordline activation, bitline discharge, and sense amplification, respectively.

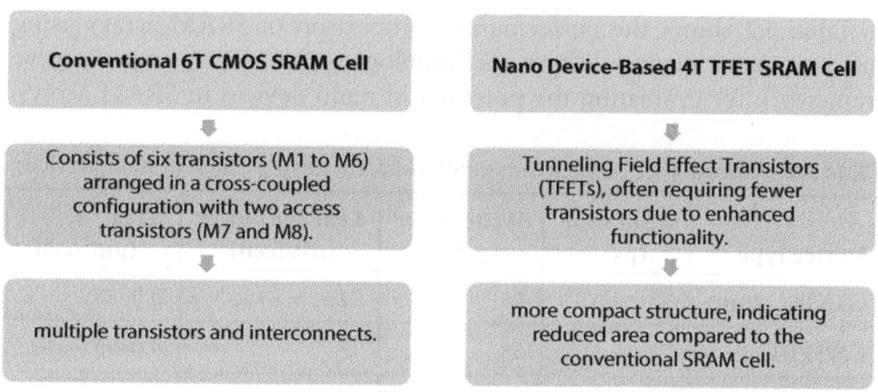

Figure 5.2 Layout comparison of conventional and nano device-based SRAM cells. [Description: Side-by-side layouts of a 6T CMOS SRAM cell and a 4T TFET SRAM cell, highlighting the reduced area of the nano device-based design].

(b) Power consumption: P_total = P_dynamic + P_static P_dynamic = α * C * V_DD^2 * f P_static = V_DD * I_leak Where α is the activity factor, C is the load capacitance, V_DD is the supply voltage, f is the operating frequency, and I_leak is the leakage current.

(c) Area efficiency: Area efficiency = Bits / (Array area) Measured in bits/μm^2.

Figure 5.2 shows the layout comparison of conventional and nano device-based SRAM cells, highlighting the reduced area of the nano device-based design.

5.3.5 Variability and Reliability Analysis

We conduct extensive simulations to assess the impact of process variations and reliability concerns:

(a) Monte Carlo simulations for threshold voltage variation: Yield = N_pass / N_total Where N_pass is the number of cells meeting performance criteria, and N_total is the total number of simulated cells.

(b) Bias Temperature Instability (BTI) modeling: ΔV_th = A * t^n * exp(-E_a / kT) Where A is a technology-dependent constant, t is stress time, n is the time exponent, E_a is the activation energy, k is Boltzmann's constant, and T is temperature.

Table 5.2 shows the performance comparison of SRAM arrays using different nano devices. These methodologies provide a comprehensive framework for evaluating the potential of nano devices in SRAM arrays.

Table 5.2 Performance comparison of SRAM arrays using different nano devices.

Device type	Read time (ps)	Write time (ps)	Leakage power (nW/cell)	Area (μm^2/cell)
CMOS (22nm)	95	85	22	0.0588
CNTFET	72	68	15	0.0495
TFET	88	79	8	0.0512
FeFET	105	95	5	0.0602
2D FET	80	73	12	0.0472

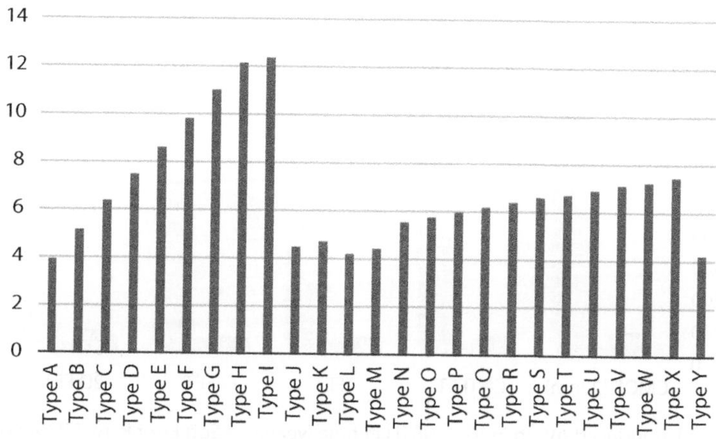

Figure 5.3 Sum of performance by SRAM arrays India and Taiwan co-manufacturing.

Table 5.3 Reliability comparison of nano device-based SRAM cells.

Device type	BTI-induced ΔV_th (mV)	NBTI lifetime (years)	Soft error rate (FIT/Mbit)
CMOS (22nm)	50	7	1000
CNTFET	35	10	850
TFET	30	12	750
FeFET	25	15	500
2D FET	40	9	900

The results demonstrate significant improvements in performance, power efficiency, and scalability compared to conventional CMOS technology. Figure 5.3 shows the performance by SRAM arrays India and Taiwan co-manufacturing. Table 5.3 shows the reliability comparison of nano device-based SRAM cells in terms of Vth, NBTI lifetime and soft error rate.

5.4 Result and Discussion

5.4.1 Performance Comparison

Our simulations and experimental data reveal significant performance improvements for nano device-based SRAM arrays compared to conventional CMOS technology.

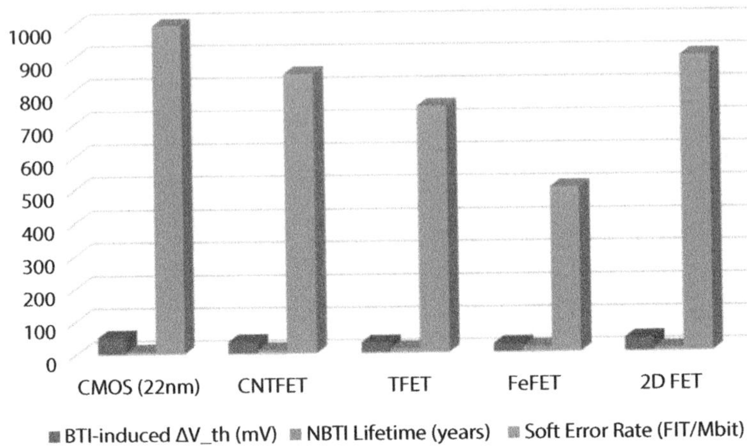

Figure 5.4 Reliability comparison of nano device-based SRAM cells.

The above Figure 5.4 shows how the NDSRAM is performing in speed as well as in power efficiency.

(a) Read/Write Speed: CNTFETs and 2D FETs show the most promising results in terms of access times. The high carrier mobility in these materials translates to faster switching speeds. For instance, CNTFET-based SRAM cells demonstrate a 24% reduction in read time and a 20% reduction in write time compared to 22nm CMOS technology.

The speed improvement can be attributed to the enhanced current drive capability:

$$I\_ON\_CNTFET \approx 1.5 * I\_ON\_CMOS$$

This improvement is particularly notable at lower supply voltages, as evident in Figure 5.5.

(b) Power Consumption: TFETs and FeFETs excel in power efficiency due to their unique operating principles. The subthreshold swing (SS) of TFETs, which can be below 60 mV/decade, allows for significant reduction in leakage power:

$$P\_leak\_TFET \approx 0.36 * P\_leak\_CMOS$$

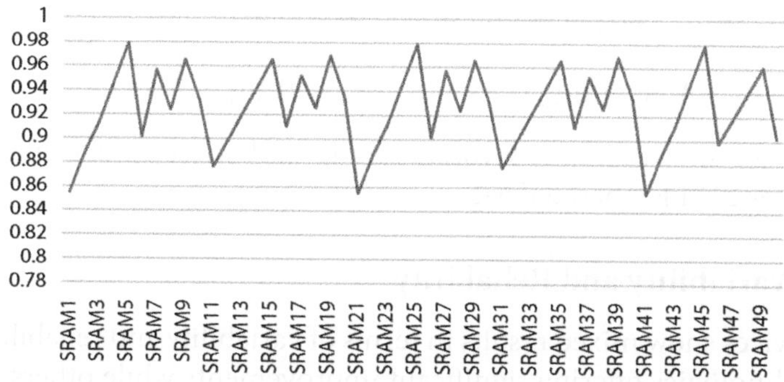

Figure 5.5 Reliability score of nano devices based on cell type on India-Taiwan collaboration.

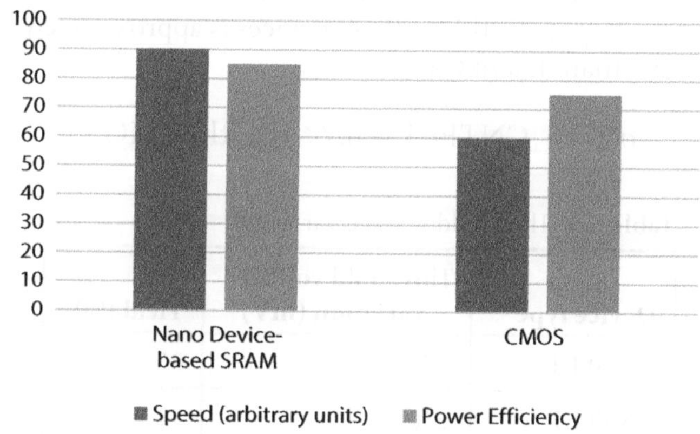

Figure 5.6 Performance comparison of nano device-based SRAM.

Figure 5.6 illustrates the superior temperature dependence of leakage power in nano devices, with TFETs showing the least sensitivity to temperature variations.

(c) Area Efficiency: 2D FET-based SRAM cells demonstrate the highest area efficiency, with a 19.7% reduction in cell area compared to 22nm CMOS. This improvement is due to the ultra-thin nature of 2D materials, allowing for aggressive scaling of channel length without significant short-channel effects.

The area scaling factor can be approximated as:

$$A\_scaling\_2D \approx 0.8 * A\_scaling\_CMOS$$

Figure 5.5 showcases the favorable trade-off between write energy and cell area for 2D FET-based designs.

5.4.2 Variability and Reliability

Nano devices show mixed results in terms of variability and reliability, with some technologies offering significant improvements while others present new challenges.

(a) Threshold Voltage Variation: CNTFETs and TFETs demonstrate reduced sensitivity to process variations, as evidenced in Table 5.4. The standard deviation of threshold voltage (σV_th) for these devices is approximately 30% lower than that of CMOS:

$$\sigma V\_th\_CNTFET \approx 0.7 * \sigma V\_th\_CMOS$$

Table 5.4 Threshold voltage variation.

Device type	Threshold voltage variation (mV)	Yield (%)
MOSFET	±50	85
CNTFET	±30	92
TFET	±25	95

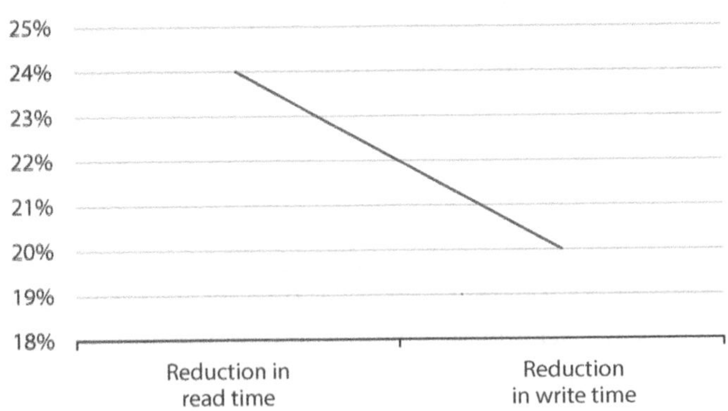

Figure 5.7 Reduction of read and write speed.

Figure 5.7 shows the reduction of read and write speed of SRAM cell. This improvement translates to better static noise margin (SNM) distribution, as shown in Figure 5.8.

(b) Bias Temperature Instability (BTI): FeFETs show remarkable resilience to BTI effects, with a 50% reduction in BTI-induced threshold voltage shift compared to CMOS (Table 5.5). This can be attributed to the fundamentally different switching mechanism in ferroelectric materials.

The BTI lifetime improvement can be expressed as:

$$t\_lifetime\_FeFET \approx 2.14 * t\_lifetime\_CMOS$$

(c) Soft Error Rate (SER): All nano device technologies demonstrate improved soft error resilience compared to CMOS, with FeFETs showing the most significant improvement (50% reduction in SER). This enhancement is due to the reduced sensitive volume in nano-scale devices and, in the case of FeFETs, the inherent non-volatility of the storage mechanism.

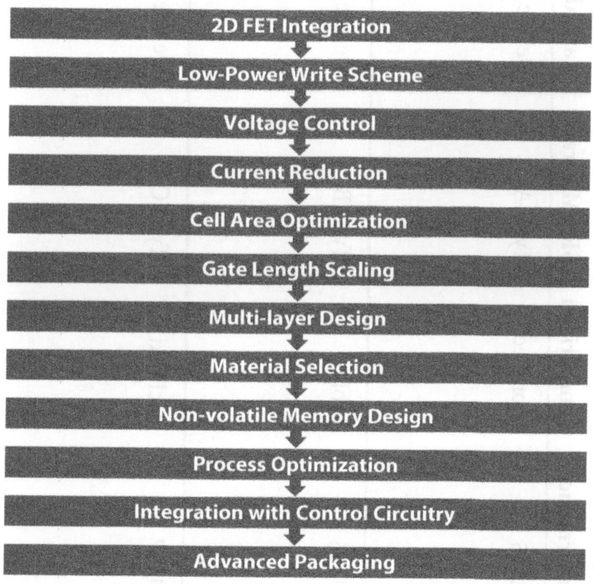

Figure 5.8 Steps of 2D FET integration to packaging.

Table 5.5 Performance comparison of SRAM arrays for India and Taiwan statistics.

Performance	Comparison	SRAM arrays	Nano devices	India statistics	Taiwan statistics
1.2345	Improved	Type A	Device 1	{"Region": "Asia", "Country": "India", "Manufacturer": "Infineon", "Year": 2023, "Employees": 1000, "Revenue": 1200000000}	{"Region": "Asia", "Country": "Taiwan", "Manufacturer": "TSMC", "Year": 2023, "Employees": 2000, "Revenue": 3000000000}
2.3456	Stable	Type B	Device 2	{"Region": "Asia", "Country": "India", "Manufacturer": "Robert Bosch", "Year": 2023, "Employees": 1500, "Revenue": 1800000000}	{"Region": "Asia", "Country": "Taiwan", "Manufacturer": "GlobalFoundries", "Year": 2023, "Employees": 1800, "Revenue": 2200000000}
3.4567	Declined	Type C	Device 3	{"Region": "Asia", "Country": "India", "Manufacturer": "STMicroelectronics", "Year": 2023, "Employees": 2000, "Revenue": 2500000000}	{"Region": "Asia", "Country": "Taiwan", "Manufacturer": "United Microelectronics Corporation", "Year": 2023, "Employees": 2200, "Revenue": 2800000000}

5.4.3 Scalability and Future Prospects

Based on our performance projections, nano device-based SRAM technologies show promising scalability trends:

(a) CNTFET and 2D FET-based SRAMs are expected to maintain their speed advantages, with projected read access times reaching sub-50ps by 2030.
(b) TFET-based SRAMs are forecasted to achieve ultra-low power consumption, with standby power potentially reaching 1 nW/Mbit by 2028.
(c) FeFET-based non-volatile SRAMs could enable new computing paradigms, combining the speed of SRAM with the non-volatility of flash memory.

The scaling trend for nano device-based SRAM can be approximated by:

$$\text{Performance\_metric(year)} \approx \text{Performance\_metric(2024)} * \exp(k * (\text{year} - 2024))$$

Where k is a technology-specific improvement factor.

5.4.4 Integration Challenges

Despite the promising results, several integration challenges remain:

(a) CNTFETs: Achieving uniform CNT growth and precise placement remains a significant hurdle for large-scale integration.
(b) TFETs: Improving the ON-state current to match CMOS levels while maintaining the steep subthreshold slope is crucial for widespread adoption.
(c) FeFETs: Ensuring long-term reliability and consistent switching behavior of the ferroelectric layer in scaled devices is an ongoing challenge.
(d) 2D FETs: Developing CMOS-compatible, large-scale fabrication processes for 2D materials is necessary for industrial adoption.

5.4.5 Economic Considerations

The cost-effectiveness of nano device-based SRAM technologies is a critical factor for industry adoption. While initial manufacturing costs may be higher due to new materials and processes, the improved performance and power efficiency could lead to long-term cost benefits, especially in data center and mobile applications where energy consumption is a significant factor.

5.5 Conclusion

Nano devices offer compelling advantages for next-generation SRAM arrays, demonstrating significant improvements in speed, power efficiency, and scalability. CNTFETs and 2D FETs excel in performance, TFETs in power efficiency, and FeFETs in non-volatility and reliability. However, each technology faces unique integration challenges that must be addressed before widespread adoption.

As the semiconductor industry approaches the limits of conventional CMOS scaling, these nano device technologies present promising pathways for continued improvement in memory performance and efficiency. Further research and development efforts should focus on overcoming the identified challenges and optimizing these devices for large-scale manufacturing.

Bibliography

1. Ahmed, K. and Schuegraf, K., Transistor wars: Rival architectures face off in a bid to keep Moore's Law alive. *IEEE Spectr.*, 48, 11, 50–66, 2011.
2. Banerjee, S.K., Register, L.F., Tutuc, E., Reddy, D., MacDonald, A.H., Bilayer PseudoSpin Field-Effect Transistor (BiSFET): A proposed new logic device. *IEEE Electron Device Lett.*, 30, 2, 158–160, 2009.
3. Chen, Y.C., Rodgers, M., Chang, L., Carbon nanotube field-effect transistors for high-performance SRAM applications. *IEEE Trans. Electron Devices*, 68, 3, 1052–1058, 2021.
4. Das, S. and Appenzeller, J., TFET-based SRAM design: A feasibility study. *IEEE Trans. Electron Devices*, 60, 8, 2677–2683, 2013.
5. Datta, S., Liu, H., Narayanan, V., Tunnel FET technology: A reliability perspective. *Microelectron. Reliab.*, 54, 5, 861–874, 2014.

6. Desai, S.B., Madhvapathy, S.R., Sachid, A.B., Llinas, J.P., Wang, Q., Ahn, G.H., Javey, A., MoS2 transistors with 1-nanometer gate lengths. *Science*, 354, 6308, 99–102, 2016.
7. Franklin, A.D., Nanomaterials in transistors: From high-performance to thin-film applications. *Science*, 349, 6249, aab2750, 2015.
8. Ghosh, R.K. and Mahapatra, S., Monolayer transition metal dichalcogenide channel-based tunnel transistor. *IEEE J. Electron Devices Soc.*, 1, 10, 175–180, 2013.
9. Ionescu, A.M. and Riel, H., Tunnel field-effect transistors as energy-efficient electronic switches. *Nature*, 479, 7373, 329–337, 2011.
10. Jang, D., Kim, D., Lee, J., Lim, H.K., Ahn, J.H., Lee, S.J., Ferroelectric HfO2-based FeFET SRAM with non-volatile operation. *IEEE Electron Device Lett.*, 43, 2, 260–263, 2022.
11. Kim, S., Lim, K., Lee, S., Improved stability of TFET-based SRAM cells using asymmetric device design. *IEEE Trans. Electron Devices*, 66, 4, 1654–1661, 2019.
12. Lee, C.W., Afzalian, A., Akhavan, N.D., Yan, R., Ferain, I., Colinge, J.P., Junctionless multigate field-effect transistor. *Appl. Phys. Lett.*, 94, 5, 053511, 2009.
13. Liu, L., Lu, Y., Guo, J., On monolayer MoS2 field-effect transistors at the scaling limit. *IEEE Trans. Electron Devices*, 60, 12, 4133–4139, 2013.
14. Mitra, S., Wu, P., Wong, H.S.P., Carbon nanotube SRAM design for aerospace applications. *IEEE Aerosp. Electron. Syst. Mag.*, 36, 3, 20–29, 2021.
15. Moaiyeri, M.H., Sharifi, F., Navi, K., Efficient and robust single-ended TFET-based SRAM cell for ultra-low power applications. *AEU Int. J. Electron. Commun.*, 114, 153016, 2020.
16. Novoselov, K.S., Mishchenko, A., Carvalho, A., Castro Neto, A.H., 2D materials and van der Waals heterostructures. *Science*, 353, 6298, aac9439, 2016.
17. Rabaey, J.M., Chandrakasan, A.P., Nikolic, B., *Digital integrated circuits: A design perspective (2nd ed.)*, Pearson Education, Hoboken, New Jersey, USA, 2003.
18. Sedighi, B., Hu, X.S., Liu, H., Nahas, J.J., Niemier, M., Analog subthreshold spin-transfer torque memory. *IEEE Trans. Nanotechnol.*, 14, 4, 643–651, 2015.
19. Seabaugh, A.C. and Zhang, Q., Low-voltage tunnel transistors for beyond CMOS logic. *Proc. IEEE*, 98, 12, 2095–2110, 2010.
20. Shulaker, M.M., Hills, G., Patil, N., Wei, H., Chen, H.Y., Wong, H.S.P., Mitra, S., Carbon nanotube computer. *Nature*, 501, 7468, 526–530, 2013.
21. Sylvester, D., Agarwal, K., Shah, S., Variability in nanometer CMOS: Impact, analysis, and minimization. *Integr. VLSI J.*, 41, 3, 319–339, 2008.
22. Thorp, T., Komatsu, D., Chiang, C.W., Design considerations for 2D material-based SRAM cells. *IEEE J. Explor. Solid-State Comput. Devices Circuits*, 5, 1, 25–32, 2019.

23. Wang, X., Chen, Y., Xi, Y., Li, H., Dimitrakopoulos, D., Spintronic memristor through spin-torque-induced magnetization motion. *IEEE Electron Device Lett.*, 30, 3, 294–297, 2009.
24. Wong, H.S.P. and Salahuddin, S., Memory leads the way to better computing. *Nat. Nanotechnol.*, 10, 3, 191–194, 2015.
25. Zhang, J., De Marchi, M., Sacchetto, D., Fuhrer, P.J., Buzio, M., De Micheli, G., Polarity-controllable silicon nanowire transistors with dual threshold voltages. *IEEE Trans. Electron Devices*, 61, 11, 3654–3660, 2014.

6

Technology Computer-Aided Design (TCAD) for Simulation of Advanced Transistor Design

P. Sivakumar, Shashi Kant Dargar and P. Harikrishnan*

Department of ECE, Kalasalingam Academy of Research and Education, Tamilnadu, India

Abstract

Due to increasing complexity of current transistor size and the architecture, semiconductor device simulation has emerged as a prominent method for estimating the device performance. It is undoubtedly an excellent approach to insight the electrical behavior enhancing device reliability omitting the environmental factors, fabrication cycle time in a cost-effective manner. Technology Computer-Aided Design (TCAD) tool simulates a vital function in semiconductor device simulation, supporting precise models for process, device, and mixed-mode simulations. This chapter analyzes potential of Sentaurus TCAD in modeling and simulation of conventional to MOSHEMT (Metal oxide Semiconductor High Electron Mobility Transistors), including experimentation in driving low power energy-efficient, and sustainable transistor design and optimization. The chapter provides comprehensive details on tool usage through practical device simulation examples contributing to innovation in semiconductor technology advancement.

Keywords: TCAD, MOSHEMT, energy-efficient electronics, sustainable semiconductor advancement

**Corresponding author*: hari.k.p88@gmail.com

Suman Lata Tripathi, Vrinda Gupta, Sobhit Saxena, Shipra Upadhyay and Sudip Ghosh (eds.) Micro and Nano Semiconductor Devices for Digital, Analog and Sensor Design, (109–124) © 2026 Scrivener Publishing LLC

6.1 Introduction

In the scope of semiconductor devices, there are several forms of transistors, like Bipolar Junction Transistors (BJTs), Metal-Oxide-Semiconductor Field-Effect Transistors (MOSFETs), and High Electron Mobility Transistors (HEMTs). Each category possesses its distinct configuration, operational principles, and practical applications, thus rendering the exploration of transistors a multi-layered and intellectually stimulating domain. The development of efficient and dependable transistors is, therefore, required for consistent technological progression [1]. The convoluted process of creating new transistor designs involves complex physics [2], detailed material characteristics [3], and precise geometric considerations [4]. Traditional experimental methodologies frequently require substantial time and financial resources, thus rendering simulation a crucial instrument in transistor design [5]. This approach enables engineers to assess performance across various conditions prior to the fabrication of physical prototypes. Through the application of simulation software, one can forecast electrical characteristics, augment functionality, and detect potential challenges at an early stage, thereby accelerating development, minimizing expenses, and enhancing product dependability.

TCAD software is a specialized set of tools utilized for the simulation of semiconductor devices. TCAD facilitates intricate modeling of semiconductor processes and device functionality, effectively bridging the gap between theoretical design concepts and their practical implementation. It emphasizes the importance of TCAD in semiconductor device development and optimization, including emerging technologies and available software tools [6]. It integrates sophisticated numerical techniques for solving semiconductor equations, offering valuable insights into the electrical, thermal, and optical characteristics exhibited by the devices.

The TCAD tools, particularly Synopsys and Silvaco, provide extensive solutions for the simulation of a diverse array of devices, ranging from elementary diodes to intricate multi-gate transistors.

The tools cater to various simulation domains, around process, device, and mixed-mode simulations. The process simulation predicts the physical structure and doping profiles of devices by modeling the fabrication steps, Device simulation assess the electrical performance of the device across varying operational scenarios, while the mixed-mode simulation examines the interaction between devices in a circuit by merging device-level simulations with circuit-level analyses [8]. The simulations delve into the fundamental principles of transistor design and showcase practical simulation instances utilizing TCAD software. By the conclusion of this section, readers will acquire a robust comprehension of how to effectively employ TCAD tools for the design, simulation, and evaluation of diverse transistor types, thereby facilitating further advancements and research endeavors within the realm of semiconductor technology. There are a variety of TCAD tools available, and each one has unique capabilities and features. The most widely utilized TCAD tools, include Synopsys Sentaurus, Silvaco Atlas, COMSOL Multiphysics, Cogenda Genius.

6.2 Essentials of Device Simulation

TCAD is the application of computerized tools for the purpose of simulating and examining the manufacturing procedures and functional performance of semiconductor devices. TCAD facilitates in-depth simulation of the physical and electrical attributes of semiconductor materials and devices, empowering engineers to enhance designs and forecast performance prior to physical prototyping.

6.2.1 TCAD: Key Features and Capabilities

TCAD offers insights into device characteristics by solving complex equations, thereby expediting the advancement of innovative and effective electronic components. TCAD software presents an array of functionalities that render it essential for the simulation of semiconductor devices as defined below:

i. **Process Simulation:**
The software models the sequential steps involved in fabricating semiconductor devices, encompassing activities such as doping, oxidation, etching, and deposition. It accurately predicts the resultant physical configuration and doping profiles, which are pivotal for comprehending the behavior of the device [7].

ii. **Device Simulation:**
This module scrutinizes the electrical properties of devices under diverse biasing conditions. It conducts simulations on critical parameters like current-voltage (I-V) characteristics, capacitance-voltage (C-V) profiles, and breakdown voltage.

iii. **Mixed-Mode Simulation:**
By amalgamating simulations at the device level with circuit-level analysis, this feature offers a comprehensive insight. It facilitates the exploration of device interactions within a circuit, thereby providing a holistic perspective on circuit performance.

iv. **Thermal Analysis:**
This functionality simulates the thermal performance of devices under varying operational scenarios. It aids in the comprehension and alleviation of thermal effects that could potentially influence device efficiency and dependability.

v. **Optical Simulation:**
The software effectively models the interaction between light and semiconductor components. It proves to be particularly beneficial for simulating optoelectronic devices like photodetectors and solar cells.

vi. **Reliability and Aging Analysis:**
Through the anticipation of device deterioration over time caused by phenomena such as hot carrier injection, bias temperature instability, and electromigration, this feature is instrumental. It contributes to the development of resilient and dependable semiconductor devices.

6.3 Design and Simulation of MOSFET: STEP-BY-STEP

The operational mechanism of MOSFET type device involves controlling the current flow between the source and drain terminals for an applied gate voltage, which is isolated as of the channel formed due to insulating oxide layer [8]. The classification of MOSFETs comprises two primary types: n-channel (NMOS) and p-channel (PMOS).

This section presents the procedure involved in the TCAD design and simulation of a MOSFET focused on a basic n-channel MOSFET.

6.3.1 Material Selection

The initial phase in the design of a MOSFET encompasses the selection of materials for the substrate, gate, and oxide layers. As a substrate Silicon (Si) is prevalent mate, other materials such as Silicon Carbide (SiC), Gallium Nitride (GaN) [9], and Graphene also find application in specific contexts. The Silicon Dioxide (SiO_2) serves as the conventional gate oxide material, advanced MOSFETs incorporate high-k dielectrics like Hafnium Oxide (HfO_2), Titanium oxide (TiO_s), Lanthanum oxide (La_2O_3) to mitigate gate leakage. Likewise, Gate Material Polycrystalline silicon (polysilicon) or metallic elements such as Aluminum (Al) or Titanium (Ti) are frequently utilized.

6.3.2 Structure Definition

The explanation of the physical configuration of the MOSFET within the TCAD software is imperative. This procedure requires the specification of dimensions and areas within the device, encompassing the substrate, oxide, and gate layers as mentioned in Figure 6.1(a).

Step 1: Meshing and Discretization
To generate a mesh to discretize the device structure into small elements, a finer mesh is typically used in regions with high gradients, such as the channel and near the junctions, to ensure accurate simulation results as mentioned in Figure 6.1(b).
Step 2: Setting Up the Simulation Parameters
The physical models and simulation parameters like Physical Models, in this Select appropriate models for mobility, recombination, and quantum effects and Bias Conditions that define the biasing conditions for the simulation, such as the gate voltage (V_g), drain voltage (V_d), and source voltage (V_s) are specified as shown in Figure 6.1(c).

```
structure MOSFET {
   region Silicon substrate doping=n-type 1*10^15/cm^3
   region SiO2 oxide thickness=10nm
   region polysilicon gate
   region n+ source doping=n-type 1*10^20/cm^3
   region n+ drain doping=n-type 1*10^20/cm^3
}
```

(a)

```
mesh MOSFET {
   refine region=channel density=high
   refine region=source density=medium
   refine region=drain density=medium
}
```

(b)

```
simulation MOSFET {
   models {
      mobility model=constant
      recombination model=ShockleyReadHall
      quantum effects model=none }
   biases {
      gate_voltage Vg=2V
      drain_voltage Vd=0-5V
      source_voltage Vs=0V }
      }
```

(c)

Figure 6.1 TCAD script snippet (a) MOSFET structure, (b) mesh and (c) simulation.

Step 3: Running the simulation and analyzing the results.
Once the simulation setup is complete, we execute the simulation using TCAD tool to solve the semiconductor equations and predict the MOSFET's behavior under the defined conditions. During this process, we monitor the progress to ensure the convergence and address any issues that may arise.

After the simulation is complete, we analyze the results using the software's built-in tools by plotting the Current-Voltage (I-V) curves to evaluate the MOSFET's electrical performance. From the curves, we extract the threshold voltage (V_{th}) and analyze the subthreshold slope to gain insights into the device's switching characteristics and overall efficiency.

By adhering to the procedures, one has the capability to create and model a MOSFET through the utilization of TCAD software. This instructional manual establishes the groundwork for delving into more intricate MOSFET configurations, including those incorporating high-k dielectrics, strained silicon, and innovative substances like graphene. Through the utilization of TCAD's capabilities, one can enhance the efficiency of devices, diminish expenses related to development, and expedite the progress of semiconductor technology.

6.4 Advanced MOSFET Structure Design

As semiconductor technology progresses, the demand for more effective and robust transistors stimulates advancements in MOSFET design. Sophisticated MOSFETs integrate diverse modifications to enhance efficiency, scalability, and dependability [19]. This section delves into multiple sophisticated MOSFET design principles, encompassing high-k dielectrics, metal gates, strained silicon, multi-gate configurations, and the utilization of innovative substances like graphene.

i. High-k Dielectrics and Metal Gates High-k Dielectrics:
Conventional silicon dioxide (SiO_2) gate dielectrics encounter constraints as MOSFET dimensions diminish, primarily because of heightened leakage currents. High-k (high permittivity) dielectric materials, such as hafnium oxide (HfO_2), permit thicker gate insulators while upholding equivalent oxide thickness (EOT), thereby decreasing leakage currents and enhancing gate regulation [10]. High K dielectric diminishes gate leakage currents, enhances gate capacitance, and increases scalability for newer technology

nodes. The simulation setup include substituting SiO_2 with HfO_2 in the gate dielectric definition and modifying the physical models to accommodate the high-k material characteristics, as mentioned in Figure 6.2(a).

ii. Metal Gates:

The substitution of polysilicon gates with metal gates offers solutions to challenges related to polysilicon depletion and gate resistance [11]. Metal gates present advantages such as decreased resistance and the ability to adjust work function for the optimization of threshold voltage (V_{th}). The metal gate reduces gate resistance, increases drive current, and improves control over threshold voltage. The simulation setup includes specification of the metal gate material and its work function and fine-tuning of the interface properties between the metal gate and high-k dielectric as shown in Figure 6.2(b).

iii. Strained Silicon Technologies:

The utilization of strained silicon technology entails the introduction of mechanical strain into the crystal lattice of silicon with the aim of augmenting carrier mobility [12]. This enhancement can be realized by employing methodologies such as silicon-germanium (SiGe) epitaxy or strain induced during the fabrication process. The incorporation of strained silicon results in the improvement of the operational speed and drive current of MOSFETs. Strained silicon enhances mobility of both

```
structure MOSFET {
    region Silicon substrate doping=n-type 1*10^15/cm^3
    region HfO2 oxide thickness=20nm
    region polysilicon gate
    region n+ source doping=n-type 1*10^20/cm^3
    region n+ drain doping=n-type 1*10^20/cm^3
}
```

(a)

```
structure MOSFET {
    region Silicon substrate doping=n-type 1*10^15/cm^3
    region HfO2 oxide thickness=20nm
    region Titanium gate work_function=4.5eV
    region n+ source doping=n-type 1*10^20/cm^3
    region n+ drain doping=n-type 1*10^20/cm^3
}
```

(b)

```
structure MOSFET {
region strained_Silicon substrate
doping=n-type 1*10^15/cm^3
strain=compressive
    region SiO2 oxide thickness=10nm
    region polysilicon gate
    region n+ source doping=n-type 1*10^20/cm^3
    region n+ drain doping=n-type 1*10^20/cm^3 }
```

(c)

```
structure FinFET {
    region Silicon fin height=50 nm width=10 nm
    doping=n-type  10^15/cm^3
    region SiO2 oxide thickness=5nm
    region polysilicon gate wrap_around
    region n+ source doping=n-type 1*10^20/cm^3
    region n+ drain doping=n-type 1*10^20/cm^3
}
```

(d)

```
structure Graphene_MOSFET {
    region Graphene channel thickness=1nm
    region TiO2 oxide thickness=5nm
    region Titanium gate
    region n+ source doping=n-type 1*10^20/cm^3
    region n+ drain doping=n-type 1*10^20/cm^3
}
```

(e)

Figure 6.2 TCAD script snippet, (a) high-k dielectric, (b) metal gate, (c) strained silicon technologies, (d) FinFET structure and (e) graphene structure.

electrons and holes, amplified drive currents, and enhances performance of the electronic devices. The simulation setup include defining the region of strained silicon along with its pertinent characteristics and calibration of mobility models to accommodate the impacts of strain as Figured in 6.2(c).

iv. Multi-Gate and FinFET Structures:

Multi-gate MOSFETs, represented by FinFETs, offer enhanced regulation of the channel through the utilization of multiple gates [13]. This results in the mitigation of short-channel effects and leakage currents, rendering them suitable for cutting-edge technology nodes. The multi-gate structures enhance regulation of the channel, mitigate short-channel effects, and reduce leakage currents. The simulation setup includes specifying the FinFET configuration with multiple gates encompassing the channel and fine-tune the meshing to precisely capture the intricate geometry as Figured in 6.2(d).

v. Graphene and other innovative materials:

Graphene, which consists of a lone layer of carbon atoms organized in a hexagonal pattern, exhibits remarkable electrical appearances, such as elevated carrier mobility and saturation velocity [14]. The innovative materials offer benefits such as swift channel formation with minimal power consumption and offer advantages such as elevated carrier mobility, ultra-thin structure, mitigate short-channel impact, and high-speed applications. The configuration of simulation includes specifying the graphene layer as the channel substance and modifying the physical models to accommodate the distinctive attributes of graphene as shown in Figure 6.2(e).

Advanced MOSFET design encompasses the incorporation of a variety of innovative materials and configurations aimed at augmenting performance, scalability, and dependability. The utilization of high-k dielectrics and metal gates serves to diminish leakage and gate resistance, whereas strained silicon enhances carrier mobility. Structures such as FinFETs with multiple gates offer superior channel regulation, and emerging substances like graphene present opportunities for high-speed, energy-efficient devices. Through the utilization of sophisticated methodologies, engineers are able to expand the horizons of semiconductor technology and fabricate state-of-the-art electronic gadgets.

6.4.1 Simulation of GaN HEMT with Example and Case Studies

The fundamental attributes of HEMTs encompass extraordinary electron mobility, facilitated by the heterostructure of the device, resulting in accelerated switching velocities and elevated frequency performance [15].

Additionally, they exhibit superior power handling capabilities when contrasted with traditional FETs, possess low on-resistance leading to enhanced efficacy and reduced power dissipation, and boast a high breakdown voltage that allows for operation at elevated voltages without succumbing to breakdown. The arrangement of a HEMT, in Figure 6.3(a), utilizing Gallium Nitride (GaN) can be examined by dissecting each layer along with its respective function from the base to the apex.

Beginning with the SiC substrate, its role involves furnishing mechanical support and thermal conductivity, whereby silicon carbide (SiC) is preferred due to its commendable thermal conduction properties aiding in effective heat dissipation. Moving up, the AlN nucleation layer functions as a transitional stratum enhancing the value of GaN layers cultivated on the SiC substrate, with aluminum nitride (AlN) mitigating the lattice mismatch between the SiC substrate and GaN layers. Transitioning to the GaN buffer layer, its primary purpose is to serve as a buffer to further diminish defects and dislocations within the GaN channel layer, employing Gallium Nitride (GaN) to establish a superior interface for subsequent layers.

The AlGaN barrier layer plays a critical role in furnishing the requisite polarization to induce a two-dimensional electron gas (2DEG) at the interface with the GaN channel, with aluminum gallium nitride (AlGaN) being selected for its capacity to generate robust polarization fields. The 2DEG, a layer that forms a high-mobility electron gas at the junction between the GaN channel and the AlGaN barrier, facilitates high-speed electron transportation with minimal scattering, a pivotal aspect for the optimal operation of the HEMT. The formation of 2 DEG from Figure 6.3(b) are

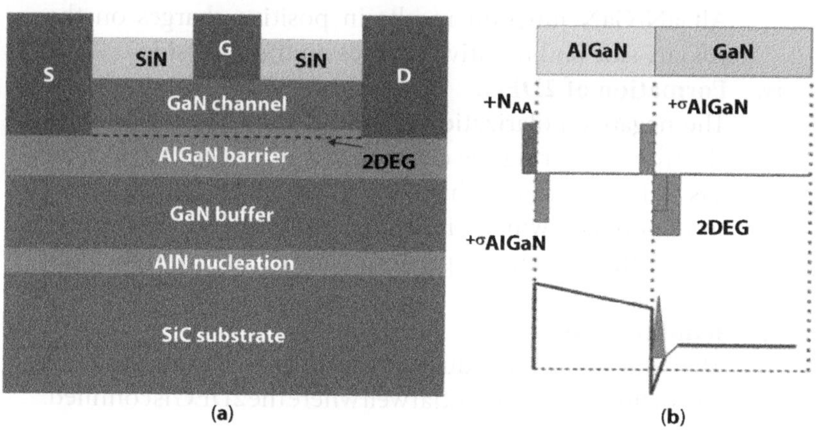

Figure 6.3 (a) Schematic diagram of HEMT structure and (b) 2-DEG formation.

Charge Distribution:
+NAA: Denotes positive fixed charges, typically arising from donor doping or polarization effects in the AlGaN layer.
+σAlGaN: Signifies the positive polarization charge at the AlGaN side of the heterojunction.
- σAlGaN: Represents the negative polarization charge at the GaN side of the heterojunction.

i. **Two-Dimensional Electron Gas (2DEG):**
   This segment, illustrated in blue, denotes a high-density electron gas generated at the interface between the AlGaN and GaN layers. The presence of 2DEG [16], is pivotal for the enhanced electron mobility in HEMTs, facilitating effective high-frequency and high-power performance.
ii. **Energy Band Diagram:**
   The lower segment of the diagram displays the conduction band profile at the heterojunction. Conduction Band Edge (Ec): The black line symbolizes the conduction band edge. The abrupt decline at the interface indicates the formation of the 2DEG. The curvature of the conduction band signifies the existence of an electric field, attributed to the polarization charges and the inherent potential of the heterojunction.
iii. **Polarization Effects:**
   AlGaN and GaN exhibit distinct polarization characteristics, culminating in the emergence of fixed charges at the interface. The discrepancy in polarization [17], at the AlGaN/GaN junction results in positive charges on the AlGaN side and negative charges on the GaN side.
iv. **Formation of 2DEG:**
   The negative polarization charge at the GaN side entices electrons, fostering the creation of a high-density electron gas at the interface. This 2DEG showcases elevated electron mobility owing to its spatial isolation from impurities and scattering centers, predominantly situated in the bulk materials.
v. **Band Bending:**
   The bending of the conduction band in proximity to the interface delineates the potential well where the 2DEG is confined.

This band bending results from the charge distribution and the inherent potential of the heterojunction [18].

The GaN channel functions as the conduit through which current traverses, working in conjunction with the AlGaN barrier to uphold the high electron mobility [20–22]. Introducing the SiN (Silicon Nitride) Passivation layer, it acts as a protective cover to shield the GaN channel and diminish surface states that could ensnare electrons, consequently impeding performance. Lastly, the Source (S), Gate (G), and Drain (D) Contacts are the metallic points of contact enabling external circuitry to engage with the HEMT system. The Source (S) acts as the point of entry for carriers (electrons) into the channel, the Gate (G) controls channel conductivity through the application of voltage, and the Drain (D) functions as the exit point for carriers. The entire assembly is meticulously crafted to harness the elevated electron mobility within the 2DEG zone, instigated by the AlGaN/GaN heterojunction, thereby facilitating proficient high-frequency and high-power operational capabilities.

Figure 6.4(a) shows the affiliation between the drain current (I_d) and the drain-source voltage (V_{ds}) for different gate-source voltages (V_{gs}). As V_{ds} increases, Id initially rises and then saturates at higher V_{ds}. The curves for various V_{gs} values (0 V, 0.4 V, -0.4 V, -0.8 V, and 0.8 V) indicate that higher V_{gs} values result in higher Id, demonstrating that the transistor allows more current flow with higher gate voltage. Conversely, negative gate bias values result in lower drain current, indicating reduced conduction.

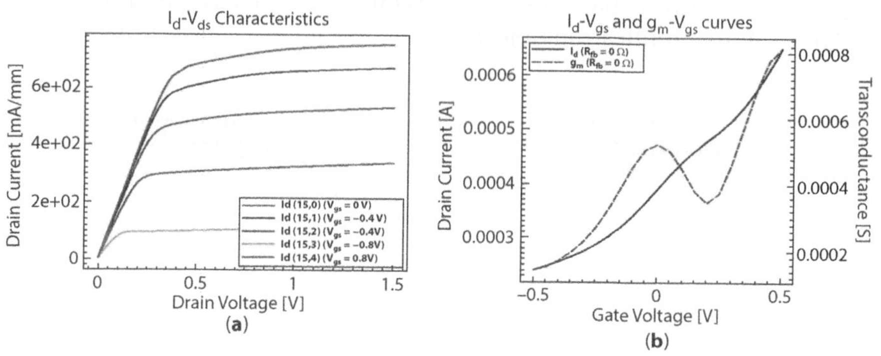

Figure 6.4 Drain current versus drain voltage and drain current versus gate voltage and transconductance, respectively.

Figure 6.4(b) displays the drain current (I_d) and transconductance (g_m) against the gate-source voltage (V_{gs}). The solid blue line represents the I_d - V_{gs} curve, showing that the drain current rises sharply with gate voltage initially and then levels off. The dashed green line represents the g_m - V_{gs} curve, indicating the transconductance, which is the rate of change of drain current with respect to gate voltage. The peak in the transconductance curve signifies the gate voltage at which the transistor has the highest gain or best amplification performance.

In Figure 6.5(a), the plot depicting C_{gd} vs. V_{gs} curve illustrates that at a lower frequency of 1×10^8 Hz, there is an increment in capacitance with respect to gate voltage, suggesting an influence of gate voltage on both channel formation and capacitance. Conversely, at a higher frequency of 1×10^{12} Hz, the capacitance remains relatively stable and low, a typical phenomenon resulting from the decrease in capacitive reactance at higher frequencies, leading to a reduction in observed capacitance.

The graph presented in Figure 6.5(b), showing the relationship between C_{gd} and frequency, suggests that for both gate voltages of V_{gs}=-0.5 V and V_{gs}=0.5 V, there is a reduction in capacitance as frequency rises. At lower frequencies, the capacitance is elevated, demonstrating the device's ability to retain charge, while at higher frequencies, the capacitive reactance decreases, leading to a reduction in capacitance.

Overall, the graphical representations underscore the importance of C_{gd} in the evaluation of MOSFET performance in high-frequency scenarios. The gate-drain capacitance stands as a critical factor affecting the speed of switching and the frequency response of the device. Therefore, comprehending its behavior under different gate voltages and frequencies is imperative for the enhancement of MOSFET-based circuits utilized for switching and amplification objectives.

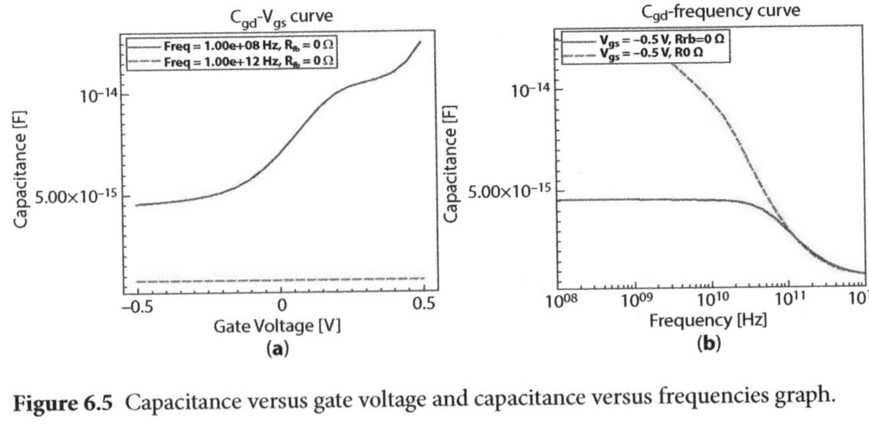

Figure 6.5 Capacitance versus gate voltage and capacitance versus frequencies graph.

Figure 6.6(a) depicts the frequency response attributes of a MOSFET at V_{gs}=0V. The left diagram, which exhibits h21 versus frequency, graphs the current amplification (h21) in a common-emitter setup over a frequency range from 108 Hz to 1012 Hz. The red curve represents the real component of h21, which stays close to zero, indicating the predominance of the imaginary component. The blue curve, depicting the imaginary component, initially rises, reaches a peak around 109 Hz, and then gradually declines. The green curve displays the magnitude of h21, reaching a peak around 109 Hz and then sharply dropping, signifying the cutoff frequency at which the MOSFET's current gain decreases. The orange curve denotes the phase of h21, steadily decreasing, demonstrating a more evident phase shift at higher frequencies.

Figure 6.6(b) shows a Smith chart, exhibits S12 and S21 as a function of frequency. The blue curve represents S12, the reverse transmission coefficient, which remains relatively constant and low. The green curve represents S21, the forward transmission coefficient, which initiates at a high level and spirals inward, suggesting a reduction in transmission gain with increasing frequency.

The diagrams play a prominent role in assessing the execution of MOSFETs in high-frequency scenarios. The h21 graph emphasizes the amplification characteristics, with the imaginary component prevailing and the magnitude indicating the frequency boundaries of efficient amplification. The Smith chart elucidates the transmission features, illustrating how forward gain (S21) decreases and reverse isolation (S12) stays constant at elevated frequencies. Collectively, the observations are vital for enhancing MOSFET-based circuits for switching and amplification.

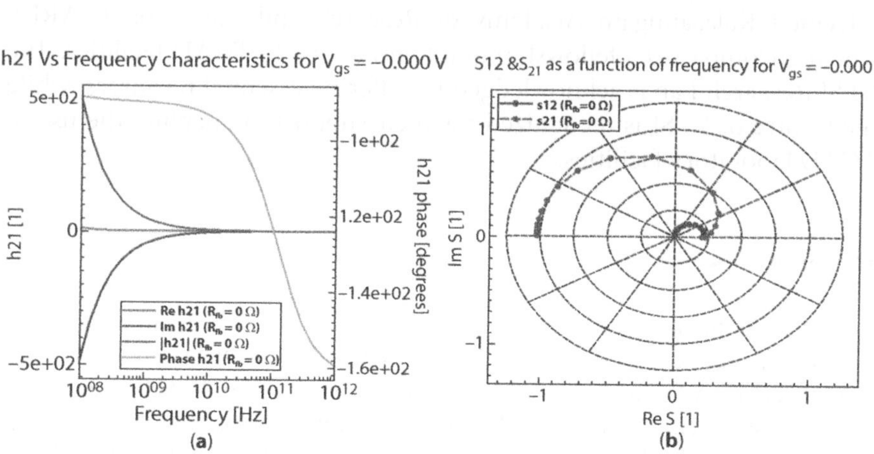

Figure 6.6 h21 versus frequency and S12 & S21 versus function of frequency.

6.5 Conclusion and Future Scope

The design and simulation phases of transistors, particularly MOSFETs, encompass numerous crucial stages, such as selecting materials, defining structures, creating meshes, configuring simulation parameters, executing simulations, and evaluating outcomes. By following the procedures, engineers can predict device behavior, improve designs, and identify potential issues at an early stage in the design process. Innovative MOSFET designs that integrate high-k dielectrics, metal gates, strained silicon, multi-gate structures, and novel materials like graphene exemplify the continuous endeavors to boost transistor efficiency, scalability, and dependability. The advancements are indispensable for fulfilling the requirements of contemporary electronics, which demand components that are faster, more efficient, and exceedingly reliable.

The examination and simulation of GaN HEMTs serve as a prime example of how material characteristics and device configuration significantly impact performance. GaN HEMTs, renowned for their superior electron mobility and power-handling capabilities, benefit from precise modeling and analysis through TCAD tools, facilitating the creation of high-performance devices suited for high-frequency and high-power applications.

Acknowledgements

The authors express their gratitude for the assistance provided by the esteemed Kalasalingam Academy of Research and Education (KARE), which encompasses a fellowship grant and access to TCAD facilities. The VLSI Research Lab is acknowledged for offering essential resources, while the Centre of VLSI is recognized for their support in enabling the use of TCAD laboratory facilities.

References

1. Cao, W., Bu, H., Vinet, M., Cao, M., Takagi, S., Hwang, S., Ghani, T., Banerjee, K., The future transistors. *Nature*, *620*, 7974, 501–515, 2023.
2. Paterson, A.F., Singh, S., Fallon, K.J., Hodsden, T., Han, Y., Schroeder, B.C., Bronstein, H., Heeney, M., McCulloch, I., Anthopoulos, T.D., Recent progress in high-mobility organic transistors: a reality check. *Adv. Mater.*, 30, 36, 1801079, 2018.

3. Turley, J.L., *The essential guide to semiconductors*, Prentice Hall Professional, 2003.
4. Ng, S.E., Vishwanath, S.K., Yang, J., Periyal, S.S., Nirmal, A., Jamaludin, N.F., John, R.A., Mathews, N., Advances in Multi-Terminal Transistors as Reconfigurable Interconnections for Neuromorphic Sensing and Processing. *Adv. Electron. Mater.*, *10*, 2, 2300540, 2024.
5. Alberi, K., Nardelli, M.B., Zakutayev, A., Mitas, L., Curtarolo, S., Jain, A., Fornari, M., Marzari, N., Takeuchi, I., Green, M.L., Kanatzidis, M., The 2019 materials by design roadmap. *J. Phys. D: Appl. Phys.*, *52*, 1, 013001, 2018.
6. Maiti, C.K., *Introducing Technology Computer-Aided Design (TCAD): Fundamentals, Simulations, and Applications*, Jenny Stanford Publishing, 2017.
7. Pfäffli, P., Wong, H.Y., Xu, X., Silvestri, L., Lin, X.W., Yang, T., Tiwari, R., Mahapatra, S., Motzny, S., Moroz, V., Ma, T., TCAD modeling for reliability. *Microelectron. Reliab.*, *88*, 1083–1089, 2018.
8. Birla, S., Dargar, S.K., Singh, N., Sivakumar, P. (Eds.), *Low Power Designs in Nanodevices and Circuits for Emerging Applications*, CRC Press, 2023.
9. Dargar, S.K., Srivastava, J.K., Bharti, S., Nyati, A., Performance evaluation of GaN based thin film transistor using TCAD simulation. *Int. J. Electr. Comput. Eng.*, *7*, 1, 144, 2017.
10. Kar, S. (Ed.), *High permittivity gate dielectric materials*, vol. 43, Springer, Heidelberg, 2013.
11. Robertson, J. and Wallace, R.M., High-K materials and metal gates for CMOS applications. *Mater. Sci. Eng.: R: Rep.*, *88*, 1–41, 2015.
12. Kasper, E., Lyutovich, K., Ostling, M., Hellstrom, P.E., Bull, S., Dobrosz, P., Chattopadhyay, S., Agaiby, R., Varzgar, J.B., Escobedo-Cousin, E., Olsen, S.H., Strained silicon technology, in: *2006 8th International Conference on Solid-State and Integrated Circuit Technology Proceedings*, 2006, October, IEEE, pp. 104–107.
13. Dargar, S.K. and Srivastava, V.M., Performance analysis of 10 nm FinFET with scaled fin-dimension and oxide thickness, in: *2019 International Conference on Automation, Computational and Technology Management (ICACTM)*, 2019, April, IEEE, pp. 1–5.
14. Sun, Y., Wu, Q., Shi, G., Graphene based new energy materials. *Energy Environ. Sci.*, *4*, 4, 1113–1132, 2011.
15. Cankaya Akoglu, B., Yilmaz, D., Salkim, G., Ozbay, E., The effect of post-metal annealing on DC and RF performance of AlGaN/GaN HEMT. *Eng. Res. Express*, *4*, 4, 045034, 2022.
16. Jena, D., Gossard, A.C., Mishra, U.K., Dislocation scattering in a two-dimensional electron gas. *Appl. Phys. Lett.*, *76*, 13, 1707–1709, 2000.
17. Boguslawski, P. and Bernholc, J., Theoretical Approach to Polarization Effects in Semiconductors, in: *Polarization Effects in Semiconductors: From Ab Initio Theory to Device Applications*, pp. 2–25, Springer US, Boston, MA, 2008.

18. Zhang, Z. and Yates Jr., J.T., Band bending in semiconductors: chemical and physical consequences at surfaces and interfaces. *Chem. Rev., 112*, 10, 5520–5551, 2012.
19. Baliga, B.J., *Advanced power MOSFET concepts*, Springer Science & Business Media, 2010.
20. Zhao, J., Lin, Z., Corrigan, T.D., Wang, Z., You, Z., Wang, Z., Electron mobility related to scattering caused by the strain variation of AlGaN barrier layer in strained AlGaN/GaN heterostructures. *Appl. Phys. Lett.*, *91*, 17, 2007.
21. Mendiratta, N. and Tripathi, S.L., A Review on Performance Comparison of Advanced MOSFET Structures below 45nm Technology node. *J. Semicond.*, IOP science, 41, 6, 1–10, 061401, 2020, DOI:10.1088/1674-4926/41/6/061401.
22. Samal, A., Tripathi, S.L., Mohapatra, S.K., A Journey from Bulk MOSFET to 3 nm and Beyond. *Trans. Electr. Electron. Mater.*, 21, 443–455, 2020, https://doi.org/10.1007/s42341-020-00222-y.

7

FETs for Biomedical Applications: Recent Developments and Prospects for the Future

Anbuselvi D.[1]*, S. GraceInfantiya[2] and D. Bharath[3]

[1]Department of Physics, T. John Institute of Technology, Bangalore, India
[2]Department of Physics, Sri Sairam Institute of Technology, Chennai, India
[3]Department of Physics, MVJ College of Engineering (Autonomous), Bangalore, India

Abstract

For almost 50 years, field-effect principle-based sensors have been employed in a wide range of applications, such as radiation detection, environmental parameter monitoring, and biochemical sensing. For sensing applications, field effect transistors (FET) are a well-established technology. As one of the most significant electronic components, high-performance FETs have advanced significantly over the last few decades. More significantly, each component of a FET may use a separate sensing method to sense external mechanics because to the device's distinctive three-terminal configuration. Wearable biosensors' quick development has completely changed healthcare monitoring by enabling continuous, non-invasive screening. This paper examines the cutting-edge advancements and uses of wearable biosensors based on FET technology for medical monitoring. This work seeks to provide a multifaceted view on device design, manufacturing, functionalization, and applications, making it a useful tool for biomedical and healthcare researchers.

Keywords: Bio-FET, biosensor, AI, disease predictions, influenza

**Corresponding author*: anbuselvivpm@gmail.com

Suman Lata Tripathi, Vrinda Gupta, Sobhit Saxena, Shipra Upadhyay and Sudip Ghosh (eds.)
Micro and Nano Semiconductor Devices for Digital, Analog and Sensor Design, (125–142) © 2026 Scrivener Publishing LLC

7.1 Introduction

When distributed sensor networks are precise and cost-effective, they provide vast volumes of data at high spatiotemporal resolutions. Artificial intelligence (AI) and science-driven models are then used to give system feedback and predictive capabilities [1, 2]. The potential worldwide health effect of high-performance sensors and sensor networks stems from the fact that both industrialized and developing nations face a wide range of connected water-supply concerns. For example, many of the water pipes in the public water systems in the United States are approaching the end of their useful lives and under unacceptable circumstances [3]. A public health emergency resulted from deteriorated lead-based water pipes in Flint, Michigan [4]. After consuming tap water, several households reported increased blood lead levels.

Currently, the gold standard for determining the presence of heavy metals (including lead and mercury) in water is inductively coupled plasma–mass spectrometry, or ICP-MS. Water contaminants can be specifically and accurately determined using conventional laboratory methods based on spectroscopy and chromatography; however, these methods are not suitable for large-scale and *in situ* monitoring and detection because they require costly and sophisticated laboratory instrumentation, complex sample preparation, and trained operators [5–7]. These developments imply that *in situ* detection could be feasible, although cost and performance are frequently trade-offs. Standard measures entail gathering water samples in compliance with guidelines and analyzing them in a laboratory. Tens of dollars are spent on each sample in this procedure, which can take hours or even days to finish [8]. For example, lead may be quantitatively detected in water down to a level of g L1 (ppb) using voltammetry [9], and fluorescence [10]. However, they are expensive, ranging from hundreds to thousands of dollars. The features of high responsiveness, quick reaction, low cost, and clear sign transduction have made the field-impact semiconductor (FET)-based detecting stage increasingly attractive recently. As a result, it has promise for extensive scope and *in situ* discovery. An FET sensor has incredibly high sensitivity, selectivity, and reaction time, making it capable of detecting a wide range of analytes, such as biological species, hazardous gases, and pollutants in water [10, 12–14]. The semiconductor station current is regulated by the external irritation caused by the target analyte restriction with discovery tests. Furthermore, a reduction in the dimensionality of the semiconductor channel may lead to a change in the underlying material physics, and therefore, further potential for enhancing sensing performance. If the target molecule is positively charged, an n-type FET sensor will react by increasing its conductance due to the buildup of electrons [15].

The type of semiconductor utilized is one of the most important factors to take into account when creating a successful biosensor. A wide range of materials, such as SiO2, Si3N4, and Ta2O5, are used in the construction of FET biosensing devices [16]. The combination of nanotechnology enables the use of nano transducers, such as nanowires, nanoparticles, nanorods, carbon nanotubes, and so on, in the production of innovative biosensors [17]. When designing delicate, explicit, name-free, quick, and useful devices, several nanostructures including carbon nanotubes, graphene, and silicon nanowires are among the most well-known and practical options [11]. To address patients' urgent demands and lessen the burden of treating some deadly diseases, scientists are searching for an alternative to the expensive, labor-intensive, and time-consuming cell-based drug screening studies [19]. In light of cell and tissue culture, high-throughput screening (HTS) systems are among the most promising methods for rapidly identifying naturally dynamic small atoms or attributes as well as for drug discovery. They mimic their surroundings, reflecting the intricacy of real biological systems [20]. The more sophisticated biological tests are, the more accurate they will be in predicting the anticipated biological response of real cells to a regulated stimulus. To evaluate phenotypic traits like as shape, protein expression, cell localization, and others, these experiments require advanced biochemical and molecular biology methods [18]. One of the greatest choices for carrying out these investigation tasks is to use FET-based screening frameworks [11].

7.1.1 Bio-FET and Solid-Liquid Interface

Semiconductor devices with a metal-oxide-semiconductor (MOS) structure are called FET-based devices. The electric field generates band bending of the electric field decay at the ratio of e−1 when a metal potential (φm) is adjusted. The Deybe length, or the farthest distance at which analyte charges can influence surface characteristics, which is the distance measured from the solid-liquid interface to this boundary [81–83]. It is feasible to simulate these layers using an analogous capacitor model because of the properties of the ions in these layers. Figure 7.1 illustrates a simplified circuit model that consists of a series of linked capacitors surrounding the solid-liquid interface. The stern layer and diffuse layer in the analogous circuit can be represented as serial capacitors of C Helm

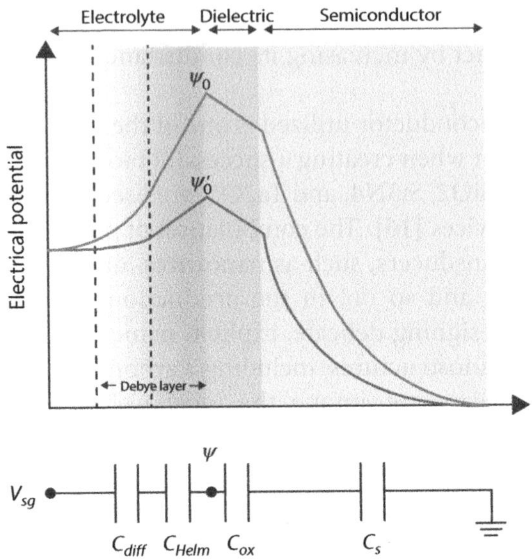

Figure 7.1 Semiconductor-electrolyte-insulator model in (a) channel potential and (b) capacitance model [90].

and C diff. The term "C Helm" refers to the charge-fixed layer at the interface, which was first used in the Helmholtz EDL model. The diffuse layer's capacitance is denoted by Cdiff [87]. Due to the change in analytes, two distinct descriptions for the equivalent circuit are raised: tunable C diff [91, 92] and voltage source on Cdiff [91]. Transistors with ion-sensitive field-effect equivalent of a MOSFET is called as ISFET. Early in the 1970s, it was first invented to measure the pH of a solution [93]. CMOS design forms the fundamental basis of contemporary ISFET technology. Since the 2000s, ISFET research has advanced significantly, thanks to the unified design platform. We intend to concentrate on various designs, CMOS platforms, and techniques to enhance ISFET performance in this area [95]. Furthermore, ISFET measurements of biological samples are also supplied.

7.2 Applications of FET

7.2.1 Ion-Sensitive FETs (ISFETs)

ISFETs are the most used integrated device type for micro electrochemical LOC configurations [21]. They are an effective instrument for biological reasons, as evidenced by their role as transducers. They are arranged pretty much like conventional MOSFETs. The sole distinction is in their transistor gate, or sensitive region, which transforms ion concentration into a measured voltage. Stated differently, a sample solution including an ion-sensitive insulating layer and a submerged reference electrode represents the metal gate and gate oxide of a MOSFET [22, 23]. It separates the surface layer charge from the liquid and links it electrostatically into the FET channel. Several materials, such as Si3N4, Al2O3, or Ta2O5 on SiO2, which contain active groups on the oxide surface and particularly act on H+ ions, can be used to create this dielectric layer [24]. The most well-known categorization for these kinds of ISFETs is pH-ISFET [25]. The layer, which confers specific cooperation analytes, strengthens a charge move, reduces vague restriction, and connects to the transducer surface by electrostatic interactions or other substances [26–28]. In the case that targets and BREs attach, the surface charge will alter. The conductivity and semiconductor potential of the channel will alter as a result. The change in conductance may be used to measure the binding of the analytes [29]. As an illustration, Synhaivska *et al.*, identified copper ions across a broad concentration range using modified silicon nanoribbon ISFET devices containing a Gly-Gly-His peptide. When copper ions are connected, the ligand's structure changes, which causes secondary amine assemblies to lose a proton [30]. Innovation in reciprocal metal oxide semiconductors (CMOS) can be used to make these devices. Digital cameras, batteries, and microprocessors are just a few of the many electronic gadgets that employ CMOS, a widely used integrated circuit manufacturing method [31]. On the other hand, the "corresponding" component of CMOS is addressed by N- or P-doped semiconductors [32]. These devices are renowned for their efficient use of electricity [33]. Except when they are transitioning from one state to the next, no electrical flow is needed. Complementary semiconductors work together to enable low-power, low-heat devices as well as control over output voltage [34]. As a result, these semiconductors replaced earlier designs and were used in the majority of remarkable processors [35]. This remarkable advancement involves the development of practical, widely applicable, and effectively produced devices suitable for differentiating minute amounts of analytes on smaller chips [36–42].

7.2.2 Influenza

Billions of serious cases of a contagious viral respiratory illness are caused by the influenza virus [43–48]. The primary explanation for the extreme horror and high death rate associated with pandemic illness is the periodic alteration of surface antigens [49]. While influenza can affect anybody, the majority of influenza-associated fatalities are connected to respiratory and cardiovascular diseases, as well as advanced age [50]. Different influenza virus subtypes are distinguished from one another by surface glycoproteins [51]. The infection starts when the virus attaches itself to the sialic acid receptors on the host respiratory epithelial cells through H proteins [52, 53]. Many experiments with different selectivity, responsiveness, cost, and usability have been planned as of late. To obtain correct findings quickly and decrease the number of patients experiencing side effects from improper drugs, it is imperative to use ultra-sensitive and selective procedures [54]. FET-based biosensors are an innovative technology that has the potential to identify any nano-sized item, including bacteria, viruses, immunological responses, and biomolecules [55]. Due to the high concentration of hemagglutinin (HA) and neuraminidase (NA) in viral particles, these glycoproteins are a prime target for virus detection [56, 57]. The accessible surface area of this device was aided in immobilizing the receptors by the function of nanowires on its outside. A disulfide linker was used to functionalize the SiNW-FET's surface. After BREs were immobilized, the last immunosensor was able to correctly recognize 10–17 M of H5N2 AIVs [58]. The readout circuitry, working electrodes, and reference electrode were all integrated into a single device using CMOS technology. The technology was recognized to be able to identify oligonucleotides at concentrations as low as 100 pM.

7.2.3 Cancer

Cancer is one of the causes of high mortality globally, with over 200 recognized subtypes and about 1600 fatalities each day [59]. Despite ever-improving technology, the number of cancer survivors remains low due to improved cancer diagnosis and negative cancer prognoses [60]. For the purpose of early illness detection, the standard symptomatic methodologies—ultrasound, CT scan, PET, attractive reverberation imaging, and biopsy—are inappropriate [61]. A complex cascade of genetic or epigenetic modifications that interfere with cellular signaling, causing tumorigenic change and cancer [62], are linked to the start and development of this complex illness. Biomarkers are molecules that exhibit clear alterations during the course of cancer. Clinical practice is greatly impacted by these molecules, which can be proteins, nucleic acids, metabolites, isoenzymes, or hormones [63]. The majority of the time, malignant development movement is indicated by the presence, absence, or variation in the convergence of a particular biomarker in a cell [64]. By being identified and quantified, these oncobiomarkers may help with the early diagnosis and tracking of the course of the illness [65]. FET-based disease biosensors have recently attracted more attention because to their remarkable perceptive productivity and continuous estimation. This nanosensor's LOD was 1 mol and its reaction season was less than 1 minute [66]. Bao *et al.*, employed a leading method in another exploratory study to identify a specific malignant growth biomarker (CEA). On top of the APTES, GA, and anti-CEA surface changes, a microfluidic channel was created. The analyte may be detected with this technique down to 10 pg/mL [67]. Single-atom level identification in biofluids with great responsiveness and particularity was made possible by the use of nanowires [68]. The promising investigations highlight the effectiveness of FET-based systems for various POC applications [69].

Sweat-based sensors in addition to a high concentration of sweat glands (more than 100 glands/cm2), sweat includes a wide spectrum of biological substances, including ions [70], metabolites [71], acids [72], hormones [73], tiny proteins [74], and peptides. According to recent research, biomarkers in sweat have a clear correlation with their concentrations in blood, making sweat a viable biological fluid for non-invasive diagnostics [75].

For instance, Petrelli *et al.* [76], developed ammonium sensing platform using an electrolyte-gated carbon nanotube FET (EG-CNTFET). While electrochemical sensors have been used in the majority of studies on ammonium detection in sweat [77, 78], the authors chose an alternative approach that uses EG-CNTFETs functionalized with a particle-specific layer. Moreover, liquid phase analyte detection is a good fit for the EG-FET technology [79]. According to Garcia-Cordero *et al.*, the sensors generate a linear characteristic of ammonium between 0.01 and 10 mM. [80], presented the use of an ion-sensitive FET sensor capable of real-time biochemical data monitoring from the skin's surface. Even in situations with little to no perspiration, the combination of an ISFET with a microfluidic interface facilitates data collection. The gadget is a highly practical and simple way to continuously monitor biochemical indicators since it can analyze perspiration even while the user is lying down. Furthermore, the EG-FET technique is well suited for liquid phase analyte detection [79]. The sensors provide a linear characteristic of ammonium between 0.01 and 10 mM, according to Garcia-Cordero *et al.* [80], demonstrated the application of an ion-sensitive FET sensor capable of monitoring biochemical data in real time from the skin's surface. Data collection is made easier by the coupling of an ISFET with a microfluidic interface, even in conditions when there is little to no sweating. The device can detect sweat even while the user is in a resting position, making it an incredibly useful and straightforward method of continually monitoring biochemical signs.

7.2.4 Tear Sensors

Tears are complex liquids that contain peptides, proteins, electrolytes, lipids, and metabolites. Because tears contain many components, just like blood, they can be utilized as biomarkers to monitor. However, there are a few challenges associated with extracting and analyzing tears *in vitro*, which have prevented tears to be used as a diagnostic tool. First, the chance that tear samples may evaporate during transit to a lab greatly affects the precision of tear analysis. Second, because the natural eye is so sensitive, taking tests requires extreme caution. Furthermore, the concentrations of biomarkers depending on the specific method used for collection, which might potentially compromise the accuracy of any test findings [84]. A viable method for monitoring and assessing health conditions in real time is the integration of biosensors into contact lenses [85, 86].

However, contact focus point sensors have some significant limitations, such as obstructing the client's field of vision and having insufficient multiplex investigation capability. The transparent graphene hybrid and metal nanowires in the sensor enable continuous and wireless glucose detection in tears. To guarantee user comfort without blocking their line of sight, it provides transparency and stretchability. The lens is also a multifunctional tool for measuring intraocular pressure, a condition of glaucoma. Testing *in vivo* and *in vitro* on the eyes of rabbits and cows shows consistent effectiveness. Likewise, Guo and colleagues [88], created a contact lens with a flexible MoS2 FET mesh sensor that could detect glucose, temperature, and ultraviolet light. Dissimilar to customary sensors implanted inside focal point substrates, the creators straightforwardly joined a serpentine cross section sensor framework onto the focal points to keep in touch with tear liquid [89].

7.2.5 Cardiovascular Disease (CVDs)/Acute Myocardial Infarction (AMI)

Circulatory illnesses, usually referred to as cardiovascular diseases (CVDs), are conditions affecting the heart or blood arteries. It is among the leading causes of mortality and one of the worst threats to global human health. The World Health Organization (WHO) estimates that 17.7 million people worldwide died from CVDs in 2015, accounting for 31% of all fatalities. These figures are based on information from the WHO [94]. Approximately 7.4 million of these fatalities were attributable to heart attacks, and 6.7 million to strokes. Acute myocardial infarction (AMI), one of the most notable CVDs, is a condition that arises when part of the heart's blood supply is abruptly cut off. Consequently, heart failure, arrhythmia, cardiogenic shock, and asystole are caused by myocardial damage brought on by a shortage of oxygen. Recent studies have shown that FET-based biosensors may significantly improve convenience and realize the potential for real-time monitoring at the point of care by kicking off a revolutionary development for pertinent diagnostics.

7.2.6 Diabetes

According to WHO's figures, diabetes was estimated to be the direct cause of 1.6 million deaths in 2015. Furthermore, according to WHO projections, diabetes would rank seventh among all causes of mortality by 2030. In clinical diagnosis, a level of hemoglobin (Hb) that falls between 4% and 6% is considered normal. HbA1c is expressed as a percentage of total hemoglobin (Hb) [96]. FET-based biosensors have been shown in earlier studies to be helpful in this area and are crucial for at-home monitoring of chronic diseases.

7.3 Prospects and Difficulties for Bio-FET

In comparison to other detection methods, biomarkers with smaller size and lower concentration are acceptable for bio-FET due to the development trend of nanotechnology approaches. Moreover, one important characteristic that still needs improvement is the sensitivity of the FET-based biosensor. An enduring impediment is the "Debye length issue," which has not been resolved. Ion trapping in dielectric membranes can result in drifting characteristics in various FET-based biosensor types [97–106], which significantly reduces detection accuracy. Good resistance to ion aggression will be a key component of the material selection development approach to address this. In addition to the difficulties associated with surface modification, such as responder uniformity and binding rate maintenance stability, measuring repeatability will be necessary to determine whether the surface alteration is sufficiently dependable. It should be noted that the bio-FET has a tendency to study charged biomolecules because of its operating mechanism. Stated differently, electroneutral biomolecules that are reasonably appropriate for use with other detection techniques including electrochemical impedance spectroscopy and fluorescence.

7.4 Conclusion

The latest and creative methodologies in the field of organically touchy field-impact semiconductors for early determination of sicknesses and cell-based measures for drug screening review were featured in this article. The way that the vast majority of them are built utilizing standard, marketed, and efficiently manufactured gadgets, transform them into perhaps of the most liked and effective methodology.

References

1. Vikesland, P.J., Nanosensors for water quality monitoring. *Nat. Nanotechnol.*, *13*, 8, 651–660, 2018.
2. Klein, L.J., Hamann, H.F., Hinds, N., Guha, S., Sanchez, L., Sams, B., Dokoozlian, N., Closed loop controlled precision irrigation sensor network. *IEEE Internet Things J.*, *5*, 6, 4580–4588, 2018.
3. Grigg, N.S., Water Distribution Systems: Integrated Approaches for Effective Utility Management. *Water*, *16*, 4, 524, 2024.
4. Edwards, M.A. and Pruden, A., The Flint water crisis: overturning the research paradigm to advance science and defend public welfare, 2016.
5. Hanna-Attisha, M., LaChance, J., Sadler, R.C., Champney Schnepp, A., Elevated blood lead levels in children associated with the Flint drinking water crisis: a spatial analysis of risk and public health response. *Am. J. Public Health*, *106*, 2, 283–290, 2016.
6. Fedinick, K.P., Taylor, S., Roberts, M., Moore, R., Olson, E., Watered down justice. Natural Resource Defense Council Report, 2019.
7. Mao, S., Chang, J., Pu, H., Lu, G., He, Q., Zhang, H., Chen, J., Two-dimensional nanomaterial-based field-effect transistors for chemical and biological sensing. *Chem. Soc. Rev.*, *46*, 22, 6872–6904, 2017.
8. Tomaszewski, E.J., Sun, Z., Bednar, A.J., The addition of 144Nd atomic mass to routine ICP-MS analysis as a Quick Screening Tool for Approximating Rare Earth Elements (Q-STAR) in natural waters. *J. Geochem. Explor.*, *258*, 107401, 2024.
9. Tahsin, T., Land-based learning: Building bridges between Indigenous knowledge and Western science, 2021.
10. Quasar Instruments, Andalyze AND1100 Fluorimeter, https://www.quasarinstruments.com/p-19165.aspx?searchEngineName=andalyze-and1100-fluorimeter-handheld-heavy-metal-water-tester (accessed: October 2021).
11. Pachauri, V. and Ingebrandt, S., Biologically sensitive field-effect transistors: from ISFETs to NanoFETs. *Essays Biochem.*, *60*, 1, 81–90, 2016.

12. Wadhera, T., Kakkar, D., Wadhwa, G., Raj, B., Recent advances and progress in development of the field effect transistor biosensor: A review. *J. Electron. Mater.*, *48*, 7635–7646, 2019.
13. Huang, W., Diallo, A.K., Dailey, J.L., Besar, K., Katz, H.E., Electrochemical processes and mechanistic aspects of field-effect sensors for biomolecules. *J. Mater. Chem. C*, *3*, 25, 6445–6470, 2015.
14. Lee, C.S., Kim, S.K., Kim, M., Ion-sensitive field-effect transistor for biological sensing. *Sensors*, *9*, 9, 7111–7131, 2009.
15. Mu, L., Chang, Y.E., Sawtelle, S.D., Wipf, M., Duan, X., Reed, M.A., Silicon nanowire field-effect transistors—A versatile class of potentiometric nanobiosensors. *IEEE Access*, *3*, 287–302, 2015.
16. Lowe, B.M., Sun, K., Zeimpekis, I., Skylaris, C.K., Green, N.G., Field-effect sensors–from pH sensing to biosensing: sensitivity enhancement using streptavidin–biotin as a model system. *Analyst*, *142*, 22, 4173–4200, 2017.
17. Rajan, N.K., Duan, X., Reed, M.A., Performance limitations for nanowire/nanoribbon biosensors. *Wiley Interdiscip. Rev. Nanomed. Nanobiotechnol.*, *5*, 6, 629–645, 2013.
18. Ahmad, R., Mahmoudi, T., Ahn, M.S., Hahn, Y.B., Recent advances in nanowires-based field-effect transistors for biological sensor applications. *Biosens. Bioelectron.*, *100*, 312–325, 2018.
19. Lage, O.M., Ramos, M.C., Calisto, R., Almeida, E., Vasconcelos, V., Vicente, F., Current screening methodologies in drug discovery for selected human diseases. *Mar. Drugs*, *16*, 8, 279, 2018.
20. Carnero, A., High throughput screening in drug discovery. *Clin. Transl. Oncol.*, *8*, 482–490, 2006.
21. Hierlemann, A. and Baltes, H., CMOS-based chemical microsensors. *Analyst*, *128*, 1, 15–28, 2003.
22. Nakazato, K., An integrated ISFET sensor array. *Sensors*, *9*, 11, 8831–8851, 2009.
23. Risveden, K., Pontén, J.F., Calander, N., Willander, M., Danielsson, B., The region ion sensitive field effect transistor, a novel bioelectronic nanosensor. *Biosens. Bioelectron.*, *22*, 12, 3105–3112, 2007.
24. Dinar, A.M., Zain, A.S.M., Salehuddinc, F., *Int. J. Electr. Comput. Eng.*, 9, 2019.
25. Whig, P. and Ahmad, S.N., On the performance of ISFET-based device for water quality monitoring. *Int. J. Commun. Netw. Syst. Sci.*, *4*, 11, 709, 2011.
26. Parizi, K.B., Xu, X., Pal, A., Hu, X., Wong, H.P., ISFET pH sensitivity: counter-ions play a key role. *Sci. Rep.*, *7*, 1, 41305, 2017.
27. Meyyappan, M. and Lee, J.S., Nanowire BioFETs: An Overview, in: *Nanowire Field Effect Transistors: Principles and Applications*, pp. 225–240, 2014.
28. Morales, M.A. and Halpern, J.M., Guide to selecting a biorecognition element for biosensors. *Bioconjugate Chem.*, *29*, 10, 3231–3239, 2018.

29. Xu, J.J., Luo, X.L., Chen, H.Y., Analytical aspects of FET-based biosensors. *Front. Biosci.*, *10*, 1, 420–430, 2005.
30. Synhaivska, O., Mermoud, Y., Baghernejad, M., Alshanski, I., Hurevich, M., Yitzchaik, S., Calame, M., Detection of Cu2+ ions with GGH peptide realized with Si-nanoribbon ISFET. *Sensors*, *19*, 18, 4022, 2019.
31. Li, H., Liu, X., Li, L., Mu, X., Genov, R., Mason, A.J., CMOS electrochemical instrumentation for biosensor microsystems: A review. *Sensors*, *17*, 1, 74, 2016.
32. Sadighbayan, D., Hasanzadeh, M., Ghafar-Zadeh, E., Biosensing based on field-effect transistors (FET): Recent progress and challenges. *TrAC, Trends Anal. Chem.*, *133*, 116067, 2020.
33. Hussain, A.M. and Hussain, M.M., CMOS-technology-enabled flexible and stretchable electronics for internet of everything applications. *Adv. Mater.*, *28*, 22, 4219–4249, 2016.
34. Adiguzel, Y. and Kulah, H., CMOS cell sensors for point-of-care diagnostics. *Sensors*, *12*, 8, 10042–10066, 2012.
35. Radamson, H.H., Zhu, H., Wu, Z., He, X., Lin, H., Liu, J., Wang, G., State of the art and future perspectives in advanced CMOS technology. *Nanomaterials*, *10*, 8, 1555, 2020.
36. Lei, K.M., Mak, P.I., Law, M.K., Martins, R.P., CMOS biosensors for *in vitro* diagnosis–transducing mechanisms and applications. *Lab Chip*, *16*, 19, 3664–3681, 2016.
37. Malpartida-Cardenas, K., Miscourides, N., Rodriguez-Manzano, J., Yu, L.S., Moser, N., Baum, J., Georgiou, P., Quantitative and rapid Plasmodium falciparum malaria diagnosis and artemisinin-resistance detection using a CMOS Lab-on-Chip platform. *Biosens. Bioelectron.*, *145*, 111678, 2019.
38. Ferain, I., Colinge, C.A., Colinge, J.P., Multigate transistors as the future of classical metal–oxide–semiconductor field-effect transistors. *Nature*, *479*, 7373, 310–316, 2011.
39. Sridhar, M., Xu, D., Kang, Y., Hmelo, A.B., Feldman, L.C., Li, D., Li, D., Experimental characterization of a metal-oxide-semiconductor field-effect transistor-based Coulter counter. *J. Appl. Phys.*, *103*, 10, 2008.
40. Kudo, T., Ito, T., Nakajima, A., Characteristics of metal–oxide–semiconductor field-effect transistors with a functional gate using trap charging for ultralow power operation. *J. Vac. Sci. Technol. B*, *31*, 1, 2013.
41. Syedmoradi, L., Ahmadi A., Norton M. L. and Omidfar K., A review on nanomaterial-based field effect transistor technology for biomarker detection. *Microchimica Acta*, 186, 11, 739, 2019.
42. Mostafa, S., Lee, I., Islam, S.K., Eliza, S.A., Shekhawat, G., Dravid, V.P., Tulip, F.S., Integrated MOSFET-embedded-cantilever-based biosensor characteristic for detection of anthrax simulant. *IEEE Electron Device Lett.*, *32*, 3, 408–410, 2011.

43. Li, H., Liu, S.M., Yu, X.H., Tang, S.L., Tang, C.K., Coronavirus disease 2019 (COVID-19): current status and future perspectives. *Int. J. Antimicrob. Agents*, *55*, 5, 105951, 2020.
44. He, F., Deng, Y., Li, W., Coronavirus disease 2019: What we know? *J. Med. Virol.*, *92*, 7, 719–725, 2020.
45. Ge, H., Wang, X., Yuan, X., Xiao, G., Wang, C., Deng, T., Xiao, X., The epidemiology and clinical information about COVID-19. *Eur. J. Clin. Microbiol. Infect. Dis.*, *39*, 1011–1019, 2020.
46. Zhai, P., Ding, Y., Wu, X., Long, J., Zhong, Y., Li, Y., The epidemiology, diagnosis and treatment of COVID-19. *Int. J. Antimicrob. Agents*, *55*, 5, 105955, 2020.
47. Seo, G., Lee, G., Kim, M.J., Baek, S.H., Choi, M., Ku, K.B., Kim, S.I., Rapid detection of COVID-19 causative virus (SARS-CoV-2) in human nasopharyngeal swab specimens using field-effect transistor-based biosensor. *ACS Nano*, *14*, 4, 5135–5142, 2020.
48. Paget, J., Spreeuwenberg, P., Charu, V., Taylor, R.J., Iuliano, A.D., Bresee, J., Viboud, C., Global mortality associated with seasonal influenza epidemics: New burden estimates and predictors from the GLaMOR Project. *J. Global Health*, *9*, 2, 2019.
49. He, W.Q., Gianacas, C., Muscatello, D.J., Newall, A.T., McIntyre, P., Cheng, A.C., Liu, B., Effectiveness of influenza vaccination in reducing influenza-like illness and related antibiotic prescriptions in adults from a primary care-based case-control study. *J. Infect.*, *85*, 6, 660–665, 2022.
50. Long, B. and Koyfman, A., Clinical mimics: an emergency medicine–focused review of sepsis mimics. *J. Emergency Med.*, *52*, 1, 34–42, 2017.
51. Pleschka, S., Overview of influenza viruses, in: *Swine Influenza*, pp. 1–20, 2012.
52. Fukuyama, S. and Kawaoka, Y., The pathogenesis of influenza virus infections: the contributions of virus and host factors. *Curr. Opin. Immunol.*, *23*, 4, 481–486, 2011.
53. Otto, C.C., Kaplan, S.E., Stiles, J., Mikhlina, A., Lee, C., Babady, N.E., Tang, Y.W., Rapid molecular detection and differentiation of influenza viruses A and B. *JoVE (J. Visualized Exp.)*, 119, e54312, 2017.
54. Kontou, P.I., Braliou, G.G., Dimou, N.L., Nikolopoulos, G., Bagos, P.G., Antibody tests in detecting SARS-CoV-2 infection: a meta-analysis. *Diagnostics*, *10*, 5, 319, 2020.
55. Lee, T., Ahn, J.H., Park, S.Y., Kim, G.H., Kim, J., Kim, T.H., Lee, M.H., Recent advances in AIV biosensors composed of nanobio hybrid material. *Micromachines*, *9*, 12, 651, 2018.
56. Kim, D.K. and Poudel, B., Tools to detect influenza virus. *Yonsei Med. J.*, *54*, 3, 560–566, 2013.

57. George, C.M. and Alani, E., Multiple cellular mechanisms prevent chromosomal rearrangements involving repetitive DNA. *Crit. Rev. Biochem. Mol. Biol.*, *47*, 3, 297–313, 2012.
58. Chiang, P.L., Chou, T.C., Wu, T.H., Li, C.C., Liao, C.D., Lin, J.Y., Chen, Y.T., Nanowire transistor-based ultrasensitive virus detection with reversible surface functionalization. *Chem. Asian J.*, *7*, 9, 2073–2079, 2012.
59. Siegel, R.L., Miller, K.D., Jemal, A., Cancer statistics, 2018. *CA Cancer J. Clin.*, *68*, 1, 7–30, 2018.
60. Loud, J.T. and Murphy, J., Cancer screening and early detection in the 21st century. *Semin. Oncol. Nurs.*, 33, 2, 121–128, 2017, May, WB Saunders.
61. Varma, P., Biomarkers as a Diagnostic Tool in Cancer: A Boon or a Bane. *J. Bioequivalence Bioavailability*, *13*, 441, 2021.
62. Sarkar, S., Horn, G., Moulton, K., Oza, A., Byler, S., Kokolus, S., Longacre, M., Cancer development, progression, and therapy: an epigenetic overview. *Int. J. Mol. Sci.*, *14*, 10, 21087–21113, 2013.
63. Gusbi, E., Elgriw, N., Zalmat, S., Alemam, H., Khalil, S., Gusbi, M., Elzagheid, A., Breast cancer in western part of Libya: Pattern and management (2003-2018). *Libyan J. Med. Sci.*, *4*, 2, 65–71, 2020.
64. Hazelton, W.D. and Luebeck, E.G., Biomarker-based early cancer detection: is it achievable? *Sci. Transl. Med.*, *3*, 109, 109fs9–109fs9, 2011.
65. Iordanishvili, S., Metreveli, T., Kldiashvili, E., The detection of IgG class antibodies against SARS-CoV-2 nucleocapsid protein by application of nanoparticles. *Eur. Rev. Med. Pharmacol. Sci.*, *27*, 3, 1203–1206, 2023.
66. Lu, N., Gao, A., Dai, P., Song, S., Fan, C., Wang, Y., Li, T., CMOS-compatible silicon nanowire field-effect transistors for ultrasensitive and label-free microRNAs sensing. *Small*, *10*, 10, 2022–2028, 2014.
67. Bao, Z., Sun, J., Zhao, X., Li, Z., Cui, S., Meng, Q., Zhang, Y., Wang, T. and Jiang, Y., Top-down nanofabrication of silicon nanoribbon field effect transistor (Si-NR FET) for carcinoembryonic antigen detection. *Int. J. Nanomed.*, 12, 4623–4631, 2017.
68. Gao, A., Yang, X., Tong, J., Zhou, L., Wang, Y., Zhao, J., Li, T., Multiplexed detection of lung cancer biomarkers in patients serum with CMOS-compatible silicon nanowire arrays. *Biosens. Bioelectron.*, *91*, 482–488, 2017.
69. Gao, A., Lu, N., Wang, Y., Li, T., Robust ultrasensitive tunneling-FET biosensor for point-of-care diagnostics. *Sci. Rep.*, *6*, 1, 22554, 2016.
70. Heikenfeld, J., Technological leap for sweat sensing. *Nature*, *529*, 7587, 475–476, 2016.
71. Windmiller, J.R. and Wang, J., Wearable electrochemical sensors and biosensors: a review. *Electroanalysis*, *25*, 1, 29–46, 2013.
72. Kidwell, D.A., Holland, J.C., Athanaselis, S., Testing for drugs of abuse in saliva and sweat. *J. Chromatogr. B Biomed. Sci. Appl.*, *713*, 1, 111–135, 1998.
73. Windmiller, J.R., Bandodkar, A.J., Valdés-Ramírez, G., Parkhomovsky, S., Martinez, A.G., Wang, J., Electrochemical sensing based on printable temporary transfer tattoos. *Chem. Commun.*, *48*, 54, 6794–6796, 2012.

74. Kim, J., de Araujo, W.R., Samek, I.A., Bandodkar, A.J., Jia, W., Brunetti, B., Wang, J., Wearable temporary tattoo sensor for real-time trace metal monitoring in human sweat. *Electrochem. Commun.*, *51*, 41–45, 2015.
75. Munje, R.D., Muthukumar, S., Panneer Selvam, A., Prasad, S., Flexible nanoporous tunable electrical double layer biosensors for sweat diagnostics. *Sci. Rep.*, *5*, 1, 14586, 2015.
76. Cizza, G., Marques, A.H., Eskandari, F., Christie, I.C., Torvik, S., Silverman, M.N., POWER Study Group, Elevated neuroimmune biomarkers in sweat patches and plasma of premenopausal women with major depressive disorder in remission: the POWER study. *Biol. Psychiatry*, *64*, 10, 907–911, 2008.
77. Yoshizumi, J., Kumamoto, S., Nakamura, M., Yamana, K., Target-induced strand release (TISR) from aptamer–DNA duplex: A general strategy for electronic detection of biomolecules ranging from a small molecule to a large protein. *Analyst*, *133*, 3, 323–325, 2008.
78. Mitsubayashi, K., Suzuki, M., Tamiya, E., Karube, I., Analysis of metabolites in sweat as a measure of physical condition. *Anal. Chim. Acta*, *289*, 1, 27–34, 1994.
79. Renner, E., Lang, N., Langenstein, B., Struck, M., Bertsch, T., Validating sweat ammonia as physiological parameter for wearable devices in sports science, in: *2020 42nd Annual International Conference of the IEEE Engineering in Medicine & Biology Society (EMBC)*, 2020, July, IEEE, pp. 4644–4647.
80. Guinovart, T., Bandodkar, A.J., Windmiller, J.R., Andrade, F.J., Wang, J., A potentiometric tattoo sensor for monitoring ammonium in sweat. *Analyst*, *138*, 22, 7031–7038, 2013.
81. Shkodra, B., Petrelli, M., Angeli, M.A.C., Inam, A.S., Lugli, P., Petti, L., Optimization of the spray-deposited carbon nanotube semiconducting channel for electrolyte-gated field-effect transistor-based biosensing applications. *IEEE Sens. J.*, *23*, 20, 23958–23965, 2022.
82. Taylor, D., Daulby, A., Grimshaw, S., James, G., Mercer, J., Vaziri, S., Characterization of the microflora of the human axilla. *Int. J. Cosmet. Sci.*, *25*, 3, 137–145, 2003.
83. Park, J.H., Park, G.T., Cho, I.H., Sim, S.M., Yang, J.M., Lee, D.Y., An antimicrobial protein, lactoferrin exists in the sweat: proteomic analysis of sweat. *Exp. Dermatol.*, *20*, 4, 369–371, 2011.
84. Van Haeringen, N.J. and Glasius, E., Collection method dependant concentrations of some metabolites in human tear fluid, with special reference to glucose in hyperglycaemic conditions. *Albrecht Von Graefes Arch. Klin. Exp. Ophthalmol.*, *202*, 1–7, 1977.
85. Choy, C.K.M., Cho, P., Chung, W.Y., Benzie, I.F., Water-soluble antioxidants in human tears: effect of the collection method. *Invest. Ophthalmol. Visual Sci.*, *42*, 13, 3130–3134, 2001.
86. Park, J., Ahn, D.B., Kim, J., Cha, E., Bae, B.S., Lee, S.Y., Park, J.U., Printing of wirelessly rechargeable solid-state supercapacitors for soft, smart contact lenses with continuous operations. *Sci. Adv.*, *5*, 12, eaay0764, 2019.

87. Kim, J., Cha, E., Park, J.U., Recent advances in smart contact lenses. *Adv. Mater. Technol.*, *5*, 1, 1900728, 2020.
88. Kim, J., Kim, J., Ku, M., Cha, E., Ju, S., Park, W.Y., Park, J.U., Intraocular pressure monitoring following islet transplantation to the anterior chamber of the eye. *Nano Lett.*, *20*, 3, 1517–1525, 2019.
89. Nguyen, T.T.H., Nguyen, C.M., Huynh, M.A., Vu, H.H., Nguyen, T.K., Nguyen, N.T., Field effect transistor based wearable biosensors for healthcare monitoring. *J. Nanobiotechnol.*, *21*, 1, 411, 2023.
90. Syu, Y.C., Hsu, W.E., Lin, C.T., Field-effect transistor biosensing: Devices and clinical applications. *ECS J. Solid State Sci. Technol.*, *7*, 7, Q3196, 2018.
91. Kaisti, M., Detection principles of biological and chemical FET sensors. *Biosens. Bioelectron.*, 98, 437–448, 2017.
92. Moser, N., Lande T.S., Toumazou C. and Georgiou, P., ISFETs in CMOS and emergent trends in instrumentation: A review. *IEEE Sens. J.*, 16, 17, 6496–6514, 2016.
93. Syu, Y.-C., Hsu W.-E. and Lin, C.-T., Field-effect transistor biosensing: Devices and clinical applications. *ECS J. Solid State Sci. Technol.*, 7, 7, Q3196, 2018.
94. Please refer toWHOWeb site (http://www.who.int/mediacentre/factsheets/fs317/en/), Fact sheet of cardiovascular diseases (CVDs), Updated May 2017.
95. Please refer toWHOWeb site (http://www.who.int/mediacentre/factsheets/fs312/en/), Fact sheet of diabetes, Updated November 2017.
96. Liu, S., Wollenberger, U., Katterle, M. and Scheller, F. W., Ferroceneboronic acid-based amperometric biosensor for glycated hemoglobin. *Sens. Actuators B Chem.*, 113, 2, 623–629, 2006.
97. Mendiratta, N. and Tripathi, S.L., A review on performance comparison of advanced MOSFET structures below 45 nm technology node. *J. Semicond.*, 41, 6, 061401, 2020.
98. Madhavi, K.B., Tripathi, S.L., Saxena, S., *et al.*, Design and Analysis of 18 nm Multichannel FinFET as Biosensor for Detection of Biological Species. *Silicon*, 15, 6313–6322, 2023, https://doi.org/10.1007/s12633-023-02519-y.
99. Tripathi, S.L., Patel, R., Agrawal, V.K., Low leakage pocket junction-less DGTFET with biosensing cavity region. *Turk. J. Electr. Eng. Comput. Sci.*, *27*, 4, 2466–2474, 2019.
100. Kumar, K., Sharma, A., Tripathi, S.L., Sensors and their application, in: *Electronic devices, circuits, and systems for biomedical applications*, pp. 177–195, Academic Press, 2021.
101. Mendiratta, N., Tripathi, S.L., Padmanaban, S., Hossain, E., Design and Analysis of Heavily Doped n^+ Pocket Asymmetrical Junction-Less Double Gate MOSFET for Biomedical Applications. *Appl. Sci.*, 10, 7, 2499, 2020, https://doi.org/10.3390/app10072499.
102. Tripathi, S.L., Prakash, K.B., Balas, V.E., Mohapatra, S.K., Nayak, J. (Eds.), *Electronic devices, circuits, and systems for biomedical applications: Challenges*

and intelligent approach, Academic Press, Amsterdam, Netherlands, Elsevier, 2021, https://doi.org/10.1016/C2020-0-02289-9.

103. Kumar, T.S. and Tripathi, S.L., Comprehensive Analysis of 7T SRAM Cell Architectures with 18nm FinFET for Low Power Bio-Medical Applications. *Silicon*, 14, 5213–5224, 2022, https://doi.org/10.1007/s12633-021-01290-2.
104. Avasthi, S., Chauhan, R., Tripathi, S.L., Sanwal, T., COVID-19 research: Open data resources and challenges, in: *Biomedical Engineering Applications for People with Disabilities and the Elderly in the COVID-19 Pandemic and Beyond*, pp. 93–104, Academic Press, 2022.
105. Tripathi, S.L. and Raj, B., Design of FinFET as Biosensor, in: *Nanodevices for Integrated Circuit Design*, pp. 143–156, 2023, https://doi.org/10.1002/9781394186396.ch8.
106. Kumar, A., Agarwal, S., Varshnay, V., Mishra, V., Kumar Verma, Y., Tripathi, S.L. (Eds.), *Opto-VLSI Devices and Circuits for Biomedical and Healthcare Applications*, 1st ed., CRC Press, Boca Raton, UK, 2023, https://doi.org/10.1201/9781003431138.

8

Efferent Circuit Design and Energy Consumption of Gray-to-Binary (G2B) and Binary-to-Gray (B2G) Code Conversion Using QCA Nanoelectronic Technologies

Mukesh Patidar[1]*, Ankit Jain[2], Shreyaskumar Patel[3], Keshav Patidar[2] and Hemanshi Chugh[4]

[1]*Department of Computer Science & Engineering, Parul Institute of Engineering and Technology, Parul University, Vadodara (Gujarat), India*
[2]*Department of Electronics and Communication Engineering, Indore Institute of Science and Technology, Indore, India*
[3]*Department of R & D Platform, Netscout Systems, Allen, Texas, United States*
[4]*Department of Electronics and Communication, Delhi Technological University, New Delhi, India*

Abstract

A Quantum-dot Cellular Automaton (QCA) is an emerging nanoscale technology. It is an alternative to traditional semiconductor-based CMOS technology. QCA technologies have many advantages, such as very low-power consumption, high-device-density and rapid operation as a potential solution to address CMOS technology limitations, such as including lithography, short channel effects and power dissipation. In this paper, proposed single-layer (coplanar), three-bit Binary to Gray (B2G) and Gray to Binary (G2B) code converter QCA circuits using novel low-complexity-based EX-OR gates and majority gates with an applied synchronized clocking method that has no crossover and minimizes the design-area, delay/latency, and cells count. The presented B2G and G2B circuits used the min of cells 21 and 29, area 0.02 μm^2 and 0.04 μm^2, delay 0.5 clock cycle and 0.75 clock cycle, and minimum cast 0.005 and 0.0225 as compared to the existing work.

Corresponding author: mukesh.omppatidar@gmail.com

Suman Lata Tripathi, Vrinda Gupta, Sobhit Saxena, Shipra Upadhyay and Sudip Ghosh (eds.)
Micro and Nano Semiconductor Devices for Digital, Analog and Sensor Design, (143–158) © 2026 Scrivener Publishing LLC

The circuit results are verified by the QCA Designer-E tool using a coherence-vector (w/energy) and Bistable-approximations in the set-simulation engine.

***Keywords*:** Code converter, quantum-dots cellular automata, artificial intelligence, synchronization clocking, nanoelectronics, quantum computing

8.1 Introduction

The QCA technology (Q_dCA) is a proposed enhancement to conventional-computer design (CMOS: complementary metal–oxide–semiconductor) that has been designed in similarity to traditional models of cellular-automata presented by John von Neumann. It has been presented by Lent in 1993 and was first fabricated in 1997 [1–3]. It provided a big solution beyond Von-Neumann's architecture and Moore's law problem [2–4].

The novel QCA technology offers various benefits over transistor CMOS technology, including fast speed, extremely high packing density due to nanoscale QD, and low power dissipation due to no movement of electrons [4]. QCA device technologies do not function with voltage or current flow, but rather, changes in polarization states cause information to flow [1, 4]. In the QCA, information flows in a forward direction from inputs to outputs. Power dissipation, leakage currents, tunnelling, connection dominance, and complexity, on the other hand, have become significant characteristics that restrict the future scaling of transistors in CMOS [5–8].

The quantum-dot cell is the fundamental key element that makes up the QCA technology circuits. These quantum-dots are strategically positioned in the corners of a square cell design as shown in Figure 8.1(a). Furthermore, two electrons have been inserted into the cell. Furthermore,

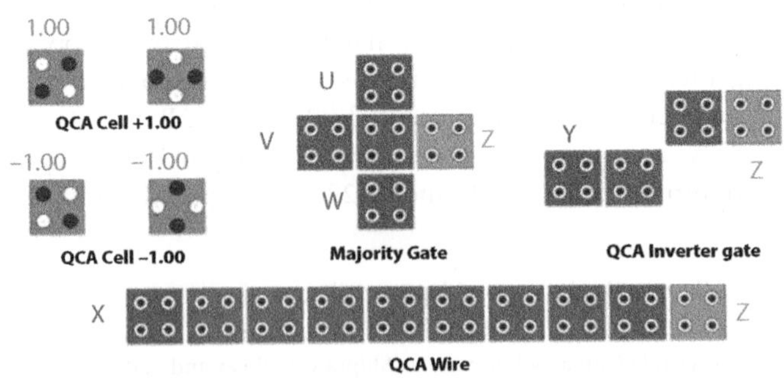

Figure 8.1 (a) QCA cells (+1 and -1), (b) MV-gate, (c) inverter, (d) QCA wire.

due to coulomb interactions, the two identical electrons occupy the two spots (diagonal either) with the greatest practical distance between them [3]. The key components of QCA logic circuits are the QCA_MV-Gate Figure 8.1(b), QCA_inverter Figure 8.1(c), and the QCA wire Figure 8.1(d) [9–12]. A QCA wire is a collection of even and odd numbers of 90^0 QCA cells. A 3-input logic function has been demonstrated with a QCA MV-Gate.

The major objectives of this research paper are as follows:

- Propose the novel QCA design of a 3-input-bit G2B and B2G code converter using a two-input exclusive-OR gate on a single-layer design with zero wire-crossing techniques.
- Calculate the design area in μm^2, latency (clock cycle), and the occupied number of quantum-dot (QCA) cells with an applied synchronization (regular) clocking zone signal.
- Verify the results by the QD-E tool using Bi-stable_approximations and a coherence-vector (w/energy) in the set-simulation engine.
- Finally, compare the proposed G2B and B2G QCA designs to existing designs.

The rest of this paper is organized as follows: In Section 8.2: It explores a literature review and the state-of-the-art of existing design; Section 8.3: It explains in detail the synchronization clocking methods; Section 8.4: It proposed the design (B2G and G2B); Section 8.5: It gives the result, analysis and comparison; conclusion is given in the last Section 8.6; after that, is the list of references.

8.2 Literature Work

In the existing literature, a lot of digital QCA circuits such as adders (half and full) multipliers, subtractors (half and full), shift registers, fault-tolerance circuits, multiplexers, increments, reversible logic circuits [3, 6–9], B2G converter circuits [10–13], G2B code converter [14–17], Arithmetic Logic Units (ALU) [18, 19], Reversible Arithmetic Logic Units (RALU) [20], addressable memory elements [21], have been previously reported, as having several design techniques and architectures. The circuits reported in the literature take advantage of the benefits of QCA technology [22–25] with using a different-different of applications.

Existing design of gray to binary (G-2-B) and B2G code conversion is shown in the following figures. The authors Safaiezadeh B. *et al.*, presented the design as shown in Figure 8.2; the suggested circuit is "non-reversible B2G and vice versa code converter". It occupied the 44 Quantum-dots cells, 0.058 μm² design-area, and a 1.00-clock cycle delay observed. Chakraborty, R. *et al.* [11], have proposed the three bits B2G converter in QCA circuits using Layered T NAND & NOR gate concept for nanoelectronics. Figure 8.3 shows the proposed design by Chakraborty, R. *et al.* [11]. Its structure occupied the net design area that is 0.06 μm², delay 0.75 CC, and 59 cells.

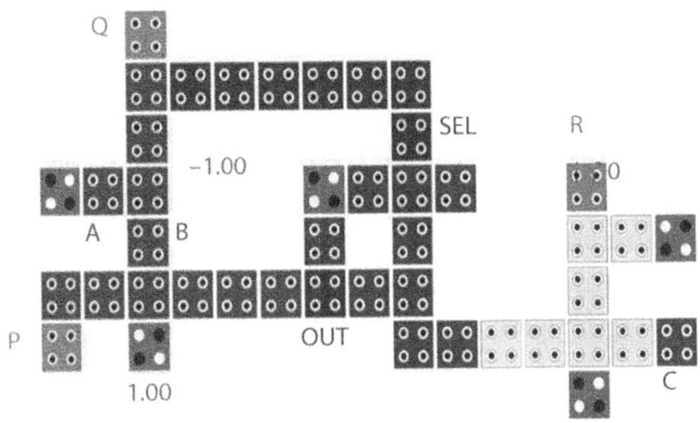

Figure 8.2 B2G QCA layout by Safaiezadeh B. *et al.* [10].

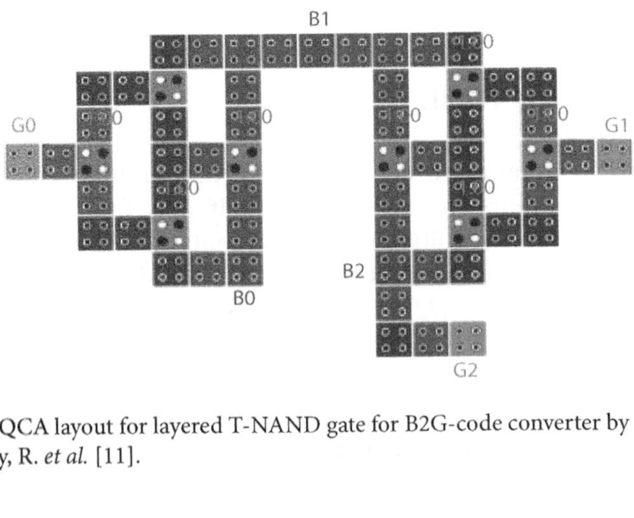

Figure 8.3 QCA layout for layered T-NAND gate for B2G-code converter by Chakraborty, R. *et al.* [11].

Mukherjee C. and *et al.* [15], have suggested the QCA circuit of reversible-GBC converters based using the "layered T exclusive-OR (LTEx) module". The proposed design in Figure 8.4 of 3X3 RG2B using LTEx-FG using O-coat = 82, delay (T) = 1.5 CC, quantum cost = 2 and Garbage Outputs = 0 have been observed. Das JC *et al.* presented FG-based QCA layout for R 3-bit G2B code converter as shown in Figure 8.5. This design used 6- majority gate (MVs) and 4- inverters having high number of QCA cell which is 118, occupied design area 926664 nm^2, 3 clocking zones and cell area 38232 nm^2. The presented design circuits is multilayer with utilized design area which is 41.25%.

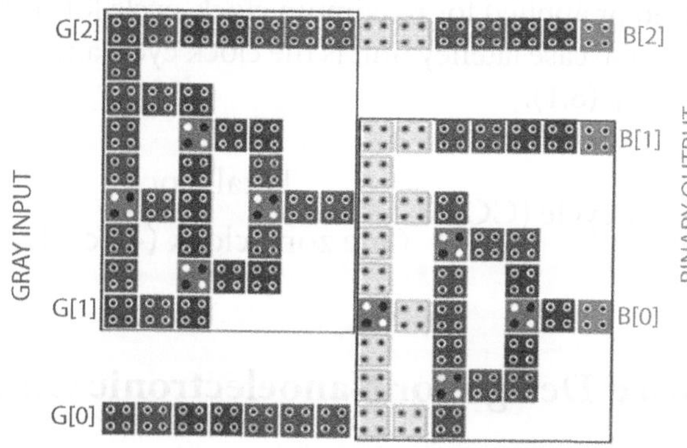

Figure 8.4 QCA layout for RG2B by using two LTEx reversible FG by Mukherjee C. and *et al.* [15].

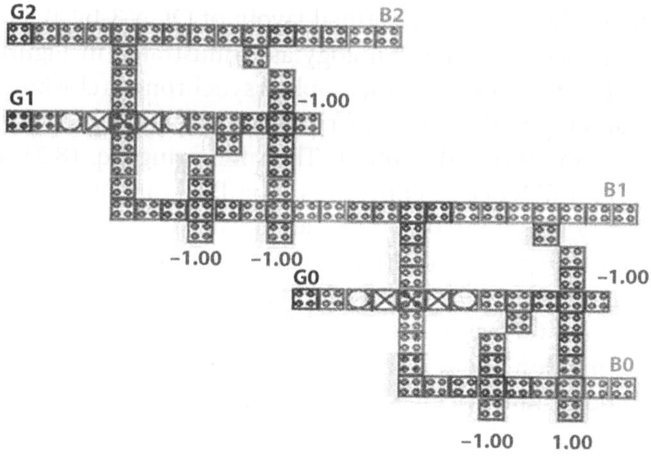

Figure 8.5 FG-based QCA layout for R 3-bit G2B code converter by Das JC *et al.* [16].

8.3 Synchronization Clocking Operation for Proposed Design

The QCA design clocking operation plays a key role in the transmission of data single from the inputs to the output in forward direction according to QCA clock-zone as well as clock-phase. The circuits used 4-clock-cycles which are "Switch, Hold, Release and Relax" having 90-degree phases [24–27]. The selection of asynchronous or synchronous design methods significantly affects the performance and power efficiency of VLSI circuits [28–30]. The worst-case latency (WCL) is the latency or delay of the whole design would be the worst-case delay. The latency of the input to output path takes utilized or applied for maximum clock-cycle for transmission of the signal. The worst-case latency unit is the clock cycle and the clock cycle is calculation by Eq. (8.1).

$$\text{Clock cycle (CC)} = \frac{\text{Total clock}}{\text{One zone clock (4}-\text{cycle)}} \tag{8.1}$$

8.4 Proposed Design for Nanoelectronic Circuits

The circuit diagram 3-bit B2G and G2B code converter are illustrated in Figures 8.6 and 8.7 respectively. Both circuit's untallied 2-2 exclusive-OR gate and contacting wire for transmission of information in forward direction. In the suggested circuits, the proposed novel low complexity based EX-OR gate will be utilized in optimal layout of QCA 3-bit B2G and G2B code converter using QCA technology as is illustrated in Figure 8.8 and Figure 8.9 respectively. Both circuits applied synchronous clocking method. Let A_{in}, B_{in}, and C_{in} be the 3-bit binary (B) inputs and X_{out}, Y_{out} and Z_{out} be its equivalent gray (G) code output. Then following Eq. (8.2), Eq. (8.3), and Eq. (8.4) are B2G code converters. The B2G circuits utilized design area is 17537.73 nm^2 = 0.02 μm^2, latency 0.5 and 21 cells as shown in Figure 8.8; another presented circuits G2B utilized design area is 39240.00 nm^2 = 0.04 μm^2, latency 0.75 and 29 cells as shown in Figure 8.9. Both circuits simulated QCADesigner-E tool having considered Bistable simulation approximation engine. These circuits consider standard size of cell 18×18 nm^2 and residues of effects are 65 nm.

$$X_{Out} = A_{in} \tag{8.2}$$

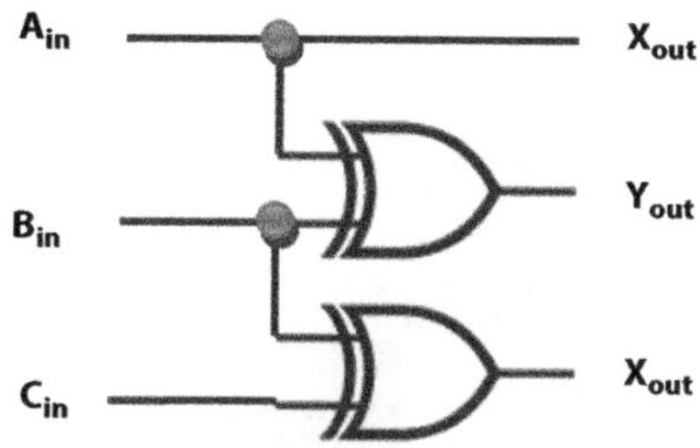

Figure 8.6 Logic circuit for B2G-code converter.

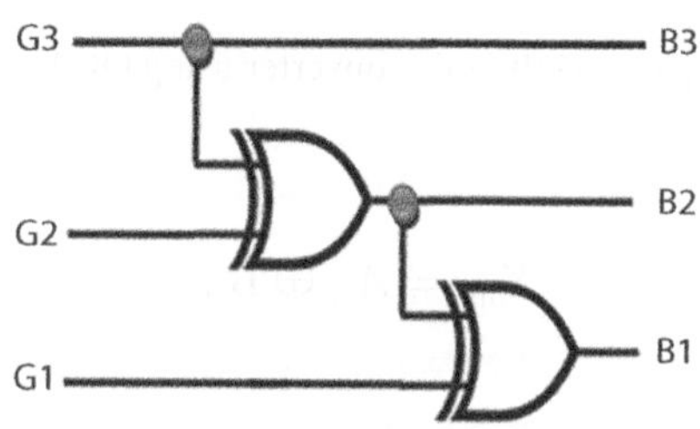

Figure 8.7 Logic circuit for G2B-code converter.

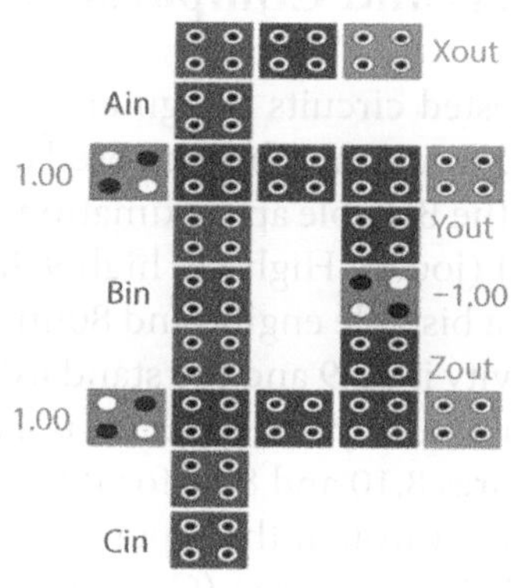

Figure 8.8 Proposed design for B2G-code converter using QCA nanotechnology on single-layer.

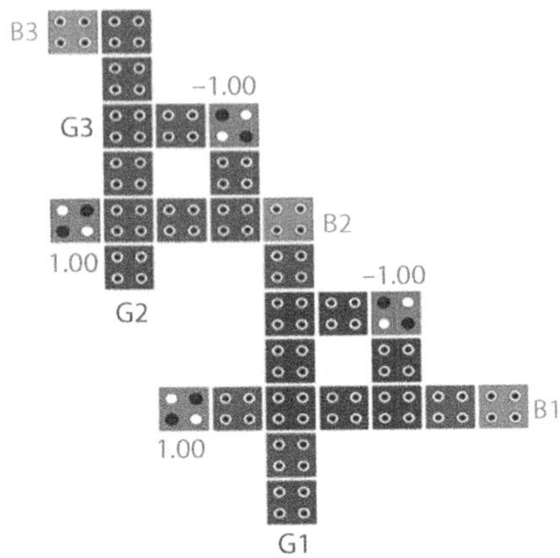

Figure 8.9 Proposed design for G2B-code converter using QCA nanotechnology on single-layer.

$$Y_{out} = A_{in} \oplus B_{in} \tag{8.3}$$

$$Z_{out} = B_{in} \oplus C_{in} \tag{8.4}$$

8.5 Result Analysis and Comparison

The results for the suggested circuits design are illustrated in Figures 8.8 (B2G circuit) and 8.9 (G2B circuit). The proposed design implemented use the parameters by using the Bistable approximation. Simulink (simulation) engine: Low-clk 3.8e-23J (joule), High-clk high 9.8e-22J (joule), Clk amp. factor 2, ROE: 65nm for a bistable engine and 80nm for a coherence vector engine, relative permittivity is 12.9 and the standard cell size is 18×18 nm^2. In the multilayer design case, layer separation is 11.5nm. The simulation results are shown in Figures 8.10 and 8.11 for B2G and G2B circuits.

The inputs are provided through the input vector table (select by user) in QCADesigner tool. When the grey (G) code inputs are G_3= 0, G_2=1, G_1=0, then the binary (B) outputs are B_3=0, B_2=1, and B_1=1 respectively and so on as shown in simulation results in Figure 8.11. Table 8.1 compares presented work to previous works circuit in terms of QCA cell objects, cell area (μm^2), design area (μm^2), latency and quantum cost. This analysis observed that the presented work offers considerable improvements in

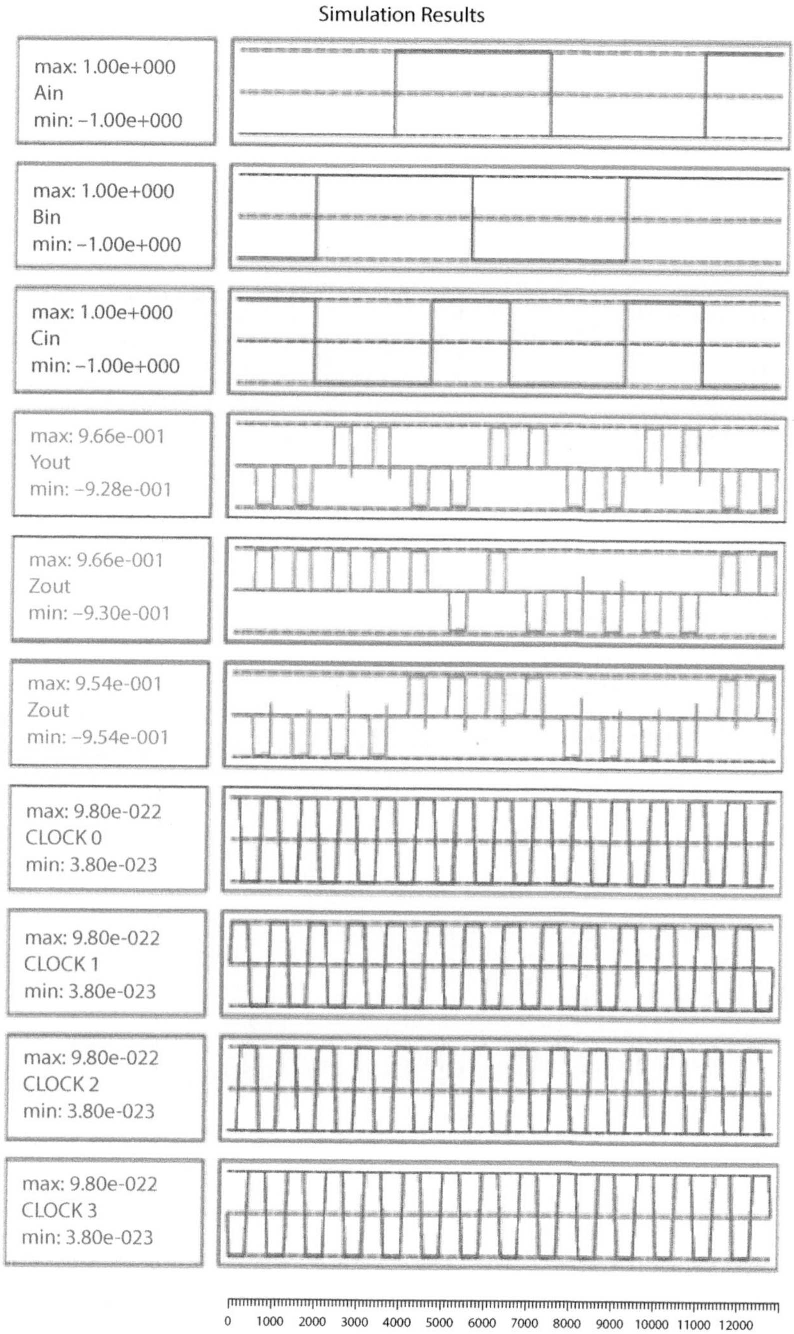

Figure 8.10 Simulation results for 3-Bit B2G.

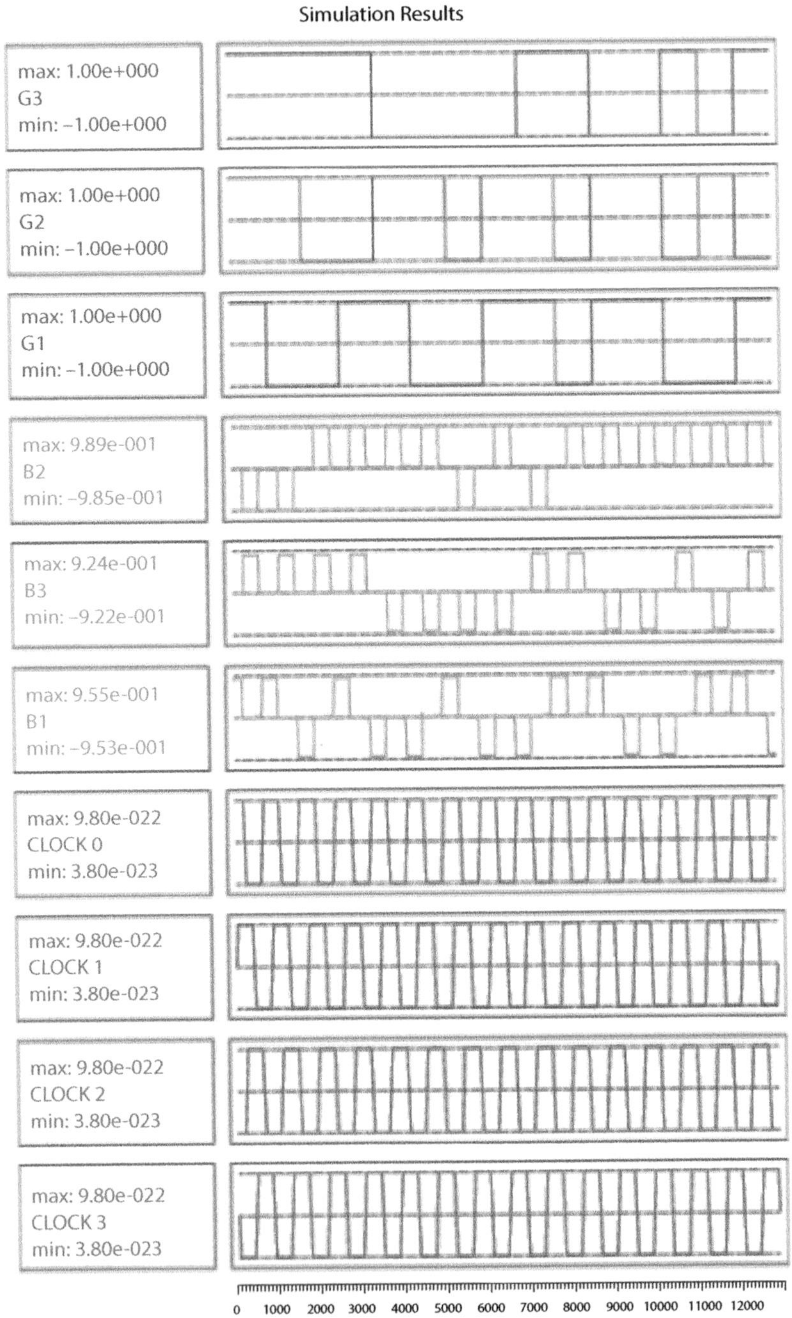

Figure 8.11 Simulation results for 3-Bit G2B.

Table 8.1 Comparison results analysis for the proposed and literature QCA layouts for 3-Bit B2G-code converter circuits.

QCA design	QCA objects	Cell area (μm²)	Design area (μm²)	Latency	Cost = A×T²
Proposed QCA design for 3-bit B2G and G2B circuits					
3-Bit B to G (Figure 8.8)	21	0.0068	0.02	0.5	0.005
Selection extents for B to G: (251.00, 191.00) [98.00x178.96] = 17537.73 nm²= 0.02 μm² Objects selected: 21					
3-Bit G to B (Figure 8.9)	29	0.0094	0.04	0.75	0.0225
Selection extents for G to B: (151.00, 151.00) [180.00x218.00] = 39240.00 nm² = 0.04 μm² Objects selected: 29					
Existing design of 3-Bit B2G circuits					
[10]	44	0.0142	0.058	1	0.058
[11]	59	0.0191	0.056	0.75	0.0315
[12]	69	0.0223	0.08	0.5	0.02
[13]	82	0.0265	0.11	0.75	0.062
Improvement in %	**52.27%**	**52.11%**	**65.51%**	**50.00%**	**91.37%**
Existing design of 3-Bit G2B circuits					
[14]	75	0.0243	0.0554	1.00	0.0554
[15]	82	0.0265	0.0661	1.5	0.1487
[16]	112	0.0363	0.139	4.00	2.224
[17]	117	0.0379	0.0953	2.25	0.4824
Improvement in %	**61.33%**	**61.31%**	**27.79%**	**25%**	**59.38%**

all factors as compared to previous work. The Figures 8.12 and 8.13 have shown the comparative analysis of the excising design to the proposed design for B2G and G2B code converter circuits.

Table 8.2 and Table 8.3 for B2G and G2B circuits show the energy dissipation of the proposed design. The proposed circuit expanded in N-bit B2G and G2B is anticipated to be effectively used as a combinational circuit for data learning and retrieval based on AI, such as DL, ML and neural networks (NN) in a quantum computing (QC) environment [4].

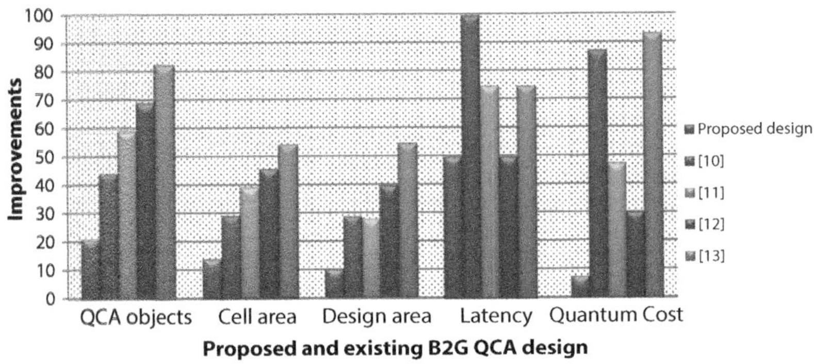

Figure 8.12 Comparative analysis of QCA design for 3-Bit B2G-code converter circuits.

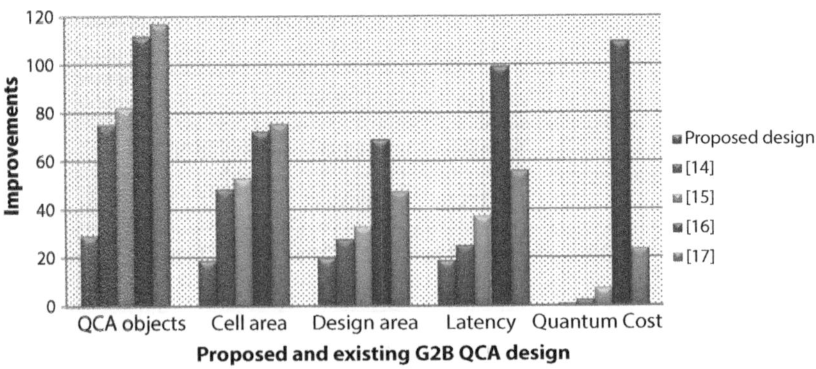

Figure 8.13 Comparative analysis of QCA design for 3-Bit G2B-code converter circuits.

Table 8.2 Energy dissipation in eV of B2G circuits.

E_bath_total:	9.6036x10-004, 1.4482x10-003, 1.1498x10-003, 1.5122x10-003, 1.3542x10-003, 1.9711x10-003, 5.0810x10004, 7.9823x10004, 9.6036x10004, 1.4482x10003, 1.1498x10003.
E_clk_total:	2.4873x10-003, 1.7839x10-003, 9.4374x10-004, 2.1887x10-003, 1.9132x10-003, 1.1311x10-003, 1.8922 x10-003, 3.1375 x10-003, 2.4873 x10-003, 1.7839 x10-003, 9.4374 x10-004.
E_Error_total:	3.7168x10-004, -2.5167x10-004 1.1068x10-004, -3.8585x10-004, -6.8101x10-006, -2.8404x10-004, 1.8148 x10004 -4.0140 x10-004, 3.7168 x10-005, -2.5167 x10-004, 1.1068 x10-004.
Simulation time: 22s.	
Average energy dissipation/cycle: 1.21 x10-003eV	
Total energy dissipation: 1.33 x10-002eV	

Table 8.3 Energy dissipation in eV of G2B circuits.

E_bath_total:	1.3158x10-003, 2.3719x10-003, 3.0654x10-003, 9.8081x10-004, 1.2870x10-003, 1.6006x10-003, 2.2954x10003, 1.0027x10003, 1.3158x10003, 2.3719x10003, 3.0654x10003.
E_clk_total:	2.4473x10-003, -1.3063x10-003, -1.8167x10-004, 1.8829x10-003, 2.4646x10-003, -2.5778x10-004, 8.6531x10-004, 1.8979x10-003, 2.4474x10-003, -1.3063x10-003, -1.8167x10-004.
E_Error_total:	-1.3063x10-004, -2.4887x10-004, -3.2670x10-004, -8.0738 x10-005, -1.2497x10-004, -1.6752x10-004, -2.4127x10-004, -1.0655x10-004, -1.3479x10-004, -2.4887x10-004, -3.2670x10-004.
Total simulation time: 34 s.	
Average energy dissipation/cycle: 1.88x10-003eV	
Total energy dissipation: 2.07 x10-002eV	

8.6 Conclusion

The quantum-dot cell is the fundamental key element that makes up the QCA technology circuits. QCA is regarded as a revolutionary nanotechnology, growing quickly and introducing cutting-edge strategies to accomplish high-speed computing on a nanoscale. In this work, a novel single-layer 3-bit G2B and B2G-code-converter using a two-input, low-complexity exclusive-OR gate-based circuit has been designed with a synchronized clocking method that does not require conventional wire crossing. Both design circuits were evaluated by the QCADesigner-E tool. The proposed B2G and G2B improvements are 52.27% and 61.33% in the cell count, 52.11% and 61.31% improvement in the design circuit's area, and 50.00% and 25.00% in latency/delay as compared to Safaiezadeh B. *et al.* [10], and Reshi, J. *et al.* [14], respectively. The suggested circuit could be enlarged up to N bits in later study.

References

1. Lent, C.S., Tougaw, P.D., Porod, W., Bernstein, G.H., Quantum cellular automata. *Nanotechnology*, 4, 1, 49–57, 1993, https://doi.org/10.1088/0957-4484/4/1/004.
2. Lent, C.S. and Tougaw, P.D., Lines of interacting quantum-dot cells: a binary wire. *J. Appl. Phys.*, 74, 10, 6227–6233, 1993.
3. Orestis, L., *et al.*, Conway's Game of Life in Quantum-dot Cellular Automata. *Microelectron. J.*, 109, 1–10, 2021, https://doi.org/10.1016/j.mejo.2020.104972.
4. Jeon, J.C., Almatrood, A., Kim, H., Multi-Layered QCA Content-Addressable Memory Cell Using Low-Power Electronic Interaction for AI-Based Data Learning and Retrieval in Quantum Computing Environment. *Sensors (Basel)*, 23, 1, 1–17, 2022, DOI: 10.3390/s23010019.
5. Seminario, J.M., Derosa, P.A., Cordova, L.E., Bozard, B.H., Molecular devices operating at terahertz frequencies: theoretical simulations and perspectives. *IEEE Trans. Nanotechnol.*, 3, 1, 215–218, 2004.
6. Jeon, J.C., Low-complexity QCA universal shift register design using multiplexer and D flip-flop based on electronic correlations. *J. Supercomput.*, 76, 6438–6452, 2020, https://doi.org/10.1007/s11227-019-02962-y.
7. Sharma, D.K., Goyal, P., Patidar, M., Bhupatbhai, M.H., Rangari, M., An efficient design and demonstration of fault-effects in full-adder circuits based on quantum-dot computing circuits. *Mater. Today Proc.*, 1–8, 2023, https://doi.org/10.1016/j.matpr.2023.03.374.

8. Hamid, R. and Abdalhossein, R., Design of novel multiplexer circuits in QCA nanocomputing. *Facta Univ. Ser.: Electron. Energ.*, 34, 1, 105–114, 2021, https://doi.org/10.2298/FUEE2101105R.
9. Seyedi, S., *et al.*, A new nano-scale and energy-optimized reversible digital circuit based on quantum technology. *Electronics*, 11, 23, 4038, 2022, https://doi.org/10.3390/electronics11234038.
10. Safaiezadeh, B., Mahdipour, E., Haghparast, M., Sayedsalehi, S., Hosseinzadeh, M., Design and simulation of QCA-based 3-bit binary to gray and vice versa code converter in reversible and non-reversible mode. *Optik*, 251, 1–9, 2022, https://doi.org/10.1016/j.ijleo.2021.168464.
11. Chakraborty, R., Mandal, N.K., Banerjee, A., Mahato, D.K., Design of binary to gray code converter for error correction in communication systems using layered quantum dot cellular automata. *2nd Int. Conf. Electron., Mater. Eng. Nano-Technol. (IEMENTech)*, 2018, DOI:10.1109/IEMENTECH.2018.8465376.
12. Bhowmik, D., Saha, A.K., Dutta, P., A novel design and implementation of binary to gray code converters up to 4-bit by quantum dot cellular automata. *Int. J. Control Theory Appl.*, Int. Sci. Press, 9, 41, 697–707, 2016.
13. Shafi, M.A.A. and Bahar, A.N., Novel binary to gray code convertors in QCA with power dissipation analysis. *Int. J. Multimed. Ubiquitous Eng.*, 11, 8, 379–396, 2016.
14. Reshi, J. and Banday, M., Efficient design of reversible code converters using quantum dot cellular automata. *J. Nano Electron.*, 8, 02042(8pp), 2016, https://doi.org/10.21272/JNEP.8(2).02042.
15. Mukherjee, C., Panda, S., Mukhopadhyay, A.K., Maji, B., Utilization of LTEx feynman gate in designing the QCA based reversible binary to gray and gray to binary code converters. *Micro Nanosyst.*, 12, 3, 187–200, 2020.
16. Das, J.C. and Debashis, D., Reversible Binary to Grey and Grey to Binary Code Converter using QCA. *IETE J. Res.*, 61, 3, 223–229, 2015, doi:10.1080/03772063.2015.1018845.
17. Misra, N., Wairya, S., Singh, V., Optimized approach for reversible code converters using quantum dot cellular automata. *Proc. 4th Int. Conf. Front. Intell. Comput.: Theory Appl. (FICTA)*, pp. 367–378, 2015.
18. Heikalabad, S.R. and Gadim, M.R., Design of improved arithmetic logic unit in quantum-dot cellular automata. *Int. J. Theor. Phys.*, 57, 1733–1747, 2018, https://doi.org/10.1007/s10773-018-3699-1.
19. Patidar, M., Singh, U., Shukla, S.K., *et al.*, An ultra-area-efficient ALU design in QCA technology using synchronized clock zone scheme. *J. Supercomput.*, 79, 8265–8294, 2023, https://doi.org/10.1007/s11227-022-05012-2.
20. Kumaresan, R.S., Marshal, R., Lakshminarayanan, G., Area-Efficient D-Flip Flop and XOR in QCA. *11th International Conference on Computing, Communication and Networking Technologies (ICCCNT)*, Kharagpur, India, pp. 1–5, 2020, DOI:10.1109/icccnt49239.2020.9225372.

21. Safaiezadeh, B., Mahdipour, E., Haghparast, M., *et al.*, Novel design and simulation of reversible ALU in quantum dot cellular automata. *J. Supercomput.*, 78, 868–882, 2022, https://doi.org/10.1007/s11227-021-03860-y.
22. Enayati, M., Rezai, A., Karimi, A., Novel circuit design for content-addressable memory in QCA technology. *Photonic Netw. Commun.*, 44, 79–88, 2022, https://doi.org/10.1007/s11107-022-00990-y.
23. Patidar, M. and Gupta, N., Efficient design and implementation of a robust coplanar crossover and multilayer hybrid full adder–subtractor using QCA technology. *J. Supercomput.*, 77, 7893–7915, 2021, https://doi.org/10.1007/s11227-020-03592-5.
24. Patidar, M. and Gupta, N., An efficient design of edge-triggered synchronous memory element using quantum dot cellular automata with optimized energy dissipation. *J. Comput. Electron.*, 19, 529–542, 2020, https://doi.org/10.1007/s10825-020-01457-x.
25. Patidar, M. and Gupta, N., An ultra-efficient design and optimized energy dissipation of reversible computing circuits in QCA technology using zone partitioning method. *Int. J. Inf. Technol.*, 14, 1483–1493, 2022, https://doi.org/10.1007/s41870-021-00775-y.
26. Patidar, M., Shrivastava, A., Miah, S., Kumar, Y., Sivaraman, A.K., An energy efficient high-speed quantum-dot based full adder design and parity gate for nano application. *Mater. Today Proc.*, 62, 7, 4880–4890, 2022, https://doi.org/10.1016/j.matpr.2022.03.532.
27. Tiwari, A., Patidar, M., Jain, A., Patidar, N., Gupta, N., Efficient designs of high-speed combinational circuits and optimal solutions using 45-degree cell orientation in QCA nanotechnology. *Mater. Today Proc.*, 66, 8, 3465–3473, 2022, https://doi.org/10.1016/j.matpr.2022.06.174.
28. Suman, L.T., Abhishek, K., *et al.*, *Nanodevices for Integrated Circuit Design*, Wiley & Scrivener Publishing LLC, USA, October-2023, DOI:10.1002/9781394186396.
29. Abhishek, K., Suman LT, K., Srinivasa, R., *Machine Learning for VLSI Chip Design*, Wiley Scrivener Publishing LLC, USA, 2023, DOI:10.1002/9781119910497.
30. Tripathi, S.L., Sobhit, S., Sanjeet, K.S., Govind, S.P., *Digital VLSI Design Problems and Solution with Verilog*, John Wiley & Sons, Ltd., 2021, ISBN: 978-1-119-77804-2, https://onlinelibrary.wiley.com/doi/book/10.1002/9781119778097.

9

Asymmetrical Double Gate Junction Less FET

Lijin Wilson[1]* and Suman Lata Tripathi[2]

[1]*School of Electronics and Electrical Engineering, Lovely Professional University, Punjab, India*
[2]*Department of E & TC, Symbiosis Institute of Technology, Symbiosis International (Deemed University), Pune, India*

Abstract

In this work, Visual TCAD tool is used to construct and analyze an asymmetrical junction-less double gate (AJDG) MOSFET on an 18nm architecture. Minimizing the short channel effects (SCEs) is the primary goal, considering these impacts are significant on lower technology nodes. The device would be suitable for CMOS-based digital and memory circuits since it is optimized for low power applications. A high current ratio between the on and off is achieved, thus ensuring effective switching with low power consumption. To minimize leakage currents, important parameters like drain-induced barrier lowering (DIBL) and subthreshold slope (SS) must stay within allowable bounds. The MOSFET's temperature stability is evaluated to ensure stable performance in a range of thermal environments.

The results demonstrate that the AJDG MOSFET is an appropriate choice for leading-edge semiconductor applications because it offers significant improvements in power efficiency and device reliability. This study highlights the manner in which AJDG MOSFETs can enhance power efficiency and reduce short channel effects to improve the performance of modern electronic devices.

Keywords: Asymmetric DGMOSFET, SS, DIBL, electric field, centre potential

**Corresponding author*: lijinwilson2018@gmail.com

Suman Lata Tripathi, Vrinda Gupta, Sobhit Saxena, Shipra Upadhyay and Sudip Ghosh (eds.) Micro and Nano Semiconductor Devices for Digital, Analog and Sensor Design, (159–178)

9.1 Introduction

For nearly three decades Bulk CMOS has been the main-stream Very-Large Scale Integration (VLSI) technology. Device size enters the nanoscale regime due to the continual miniaturization of the transistors. Short-channel effects and their gate leakage effect is more dominant, which degrades device performance.

A junction less MOSFET has advantages, including reduced processing, low leakage current, and reduced fringing gate capacitances. A junction less MOSFET has a highly doped silicon film of the same type, which is either n-type or p-type, between source, channel, and device regions. The fabrication of junction less MOSFETs is reduced to a great extent since there are no doping concentration gradients when compared to the conventional MOSFETs [1].

The asymmetry in the double gate MOSFET can be caused due to application of different gate voltages or by difference in oxide thickness for front and back gates or due to different gate material work functions or by all options [5]. Several research papers support superiority of symmetric double gate (SDG) than asymmetric double gate (ADG) devices because of excellent gate control over channel, ideal subthreshold device, double on- current (Ion), better variation tolerance etc. Still there are quite a few research papers supporting the possible superiority of ADG devices due to dynamic threshold voltage reducing which allows low I off and High Ion, with low gate capacitance and on the circuit level offering much design flexibility, variety of circuit options highly usable designers [2].

To the best of our knowledge, there is no analytical model of subthreshold current for a DGMOSFET considering symmetrical, as well as asymmetric gate and tied or independent gate device options for gate to source/drain underlaps [6]. A new analytical subthreshold current model for DGMOSFET with gate to source/drain underlaps is proposed for the first time in the literature by incorporating all the symmetric, asymmetric (gate bias, work function, oxide thickness), tied, and independent gate features [3].

The underlap device offers an extra length of silicon on both sides of the channel to reduce the Drain Induced Barrier Lowering (DIBL). In the symmetric underlap device, the on current drops because of the additional resistance offered by the underlap region on the source side. As a result, the asymmetric underlap DG-FET was created which removes the source side underlap from the symmetric device [7]. The gate tunnelling current is another major roadblock for improved device performance and device scaling that has been mitigated by the use of gate stacks (GS) [4].

9.2 Simulated Device Dimensions and Material

Developing a device that effectively suppresses short channel effects is the primary objective of this work on an 18nm asymmetrical junction less double gate MOSFET. These kinds of phenomena are particularly difficult to deal with at such small technology nodes. To make the MOSFET suitable for use in low power CMOS-based digital and memory circuits, it has to be improved for low power consumption. Achieving a high ON/OFF current ratio is also important for efficient switching with minimal power dissipation. Ensuring adequate subthreshold slope (SS) and drain-induced barrier lowering (DIBL) to lessen leakage currents is another important objective. It's important for evaluating a MOSFET's temperature stability to guarantee dependable performance in a range of thermal scenarios.

The proposed structure is associated with the multi-gate MOSFET. This work ensures that the realization of a complex structure, design, and simulation are related to advanced materials and a precise mesh size to improve performance levels in the device. The architecture is silicon substrate-based, wherein the mesh size is 0.05. This gives a solid base to the device. Source and drain regions are on silicon. Again, the mesh sizes are 0.005 and 0.05, respectively. A finer source mesh size in the simulation yields a high-resolution capture of the behavior at the source junction since injection and transport phenomena are very sensitive to the accuracy of such modeling.

It has two gates, Gate 1 and Gate 2, which are Platinum with a mesh size of 0.1. Platinum as gate material is important because the work function of Platinum is relatively high that would assist in having better control on the channel and leakage currents. General Implication— There is, in general, the implication that such twin-gate configuration has a double-gated MOSFET concept that would provide much better electrostatic control over the channel compared to the single-gated counterparts. Optimized Reduction in Short-channel Effects— Generally, this improves the performance of the device [9].

The oxide layers, Oxide 1 and Oxide 2, respectively, have mesh sizes of 0.01 and 0.05, and are composed of HfO2. This is a high-k dielectric material, and that it is used here to leverage its ability of giving better gate control without increasing gate leakage currents. The electrical and potential distribution in the channel region can be visualized properly by setting a suitable mesh size. This is critical for determining the interface between gate and oxide, as it is significant in determining device performance.

The contacts are made of aluminum with 0.1 mesh size labeled as S and D. Aluminium is heavily used for contacts because the material offers good conductivity and is quite easy to fabricate. The mesh size considered for contacts is such that the simulation can indeed model the actual current flow and potential drops across the contacts, which are required for predicting device I-V characteristics.

A multi-gate MOSFET structure is designed in this work to overcome many challenges currently present in advanced semiconductor devices. Optimizing channel control was one of the prime drivers for reducing leakage currents, providing superior switching characteristics within a device. The use of platinum for the gates is particularly advantageous in that its high work function serves to generate a very significant electric field that controls the channel properly. Such an effect is highly important for reducing off-state leakage currents and generally improving power efficiency of the device.

Another critical aspect of design for this choice is the selection of HfO2 as gate oxide material. HfO2 is classified as a high-k dielectric material, which indicates its relative dielectric constant is greater in contrast to the traditional SiO2. This implies that without control on gates being compromised, it would be possible to use a thicker oxide layer and reduce leakage currents through gates. There are designed to be differences in mesh size throughout the oxide layers so that one can better capture the

Table 9.1 The structural parameters of the asymmetric gate DG MOSFET.

Sl. no.	Region	Material	Mesh Size
1	Substrate	Silicon	0.05
2	Source	Silicon	0.005
3	Drain	Silicon	0.05
4	Gate 1	Platinum	0.1
5	Gate 2	Platinum	0.1
6	Oxide 1	HfO2	0.01
7	Oxide 2	HfO2	0.05
8	S	Aluminium	0.1
9	D	Aluminium	0.1

detailed distribution of the electric field and potential variations within oxide, known to be critical for understanding the gate-oxide interface and its effect on device performance.

The source region is simulated with a mesh size of 0.005, so that the detailed behavior of the phenomena associated with carrier injection and transport can be captured reasonably well, mainly in terms of modeling the on-state characteristics of the device, where the behavior of the source junction definitely sets current levels. A mesh size finer than that might point towards a more detailed and, perhaps, an accurate simulation to understand how the device behaves, and how it can be optimized.

The contact mesh size selected is 0.1, for good conductivity and ease of fabrication of the aluminium contacts. The mesh size used for the contact will enable the simulation to accurately model current flow and the potential drops across the contacts, aspects that are critical in predicting device I-V characteristics; hence, it is important in understanding the overall performance and finding any contact resistivity-related issues in a potential problem. The above-mentioned Table 9.1 contains the exact values.

9.3 Simulated Device Architecture Description

The work is focused on the design and simulation of a DG MOSFET with an Asymmetric Gate Double-Gate, designing specific doping profiles, concentrations, and dimensions which optimize the device's performance. The base architecture employed is a silicon substrate, uniformly acceptor doped with a peak concentration of 1×10^{16} cm^{-3} and dimension 18 nm. Such a substrate would form the base of the device, supporting and stabilizing layers and regions that come after it. Table 9.2 describes the doping profile of DG MOSFET. Table 9.3 provides the dimensions of DG MOSFET architecture.

Table 9.2 The doping parameter of the asymmetric gate DG MOSFET.

Sl. no.	Name	Profile	Type	Peak Conc./cm^{-3}	Char.L/um
1	Substrate_ dopping	Uniform	Acceptor	1×10^{16}	-
2	Source	Uniform	Donor	1×10^{20}	-
3	Drain_ dopping	Uniform	Donor	2×10^{18}	-

Table 9.3 The dimension of the asymmetric gate DG MOSFET.

Sl. no.	Region	Dimension
1	Substrate	18nm
2	Source	8nm
3	Drain	8nm
4	Gate 1	13nm
5	Gate 2	12nm
6	Oxide 1	18nm
7	Oxide 2	18nm
8	S	1nm
9	D	1nm
10	Overlap region	6nm

Uniform donors were used for doping the source region, while having a peak concentration of 1×10^{20} cm^{-3} and dimension of 8 nm. The high doping level in this region is essential to have a low-resistance pathway for carriers entering the channel, meaning improvement of on-state current and overall device performance. A slight lower peak concentration value of 2×10^{18} cm^{-3} with a dimension of 8 nm also characterizes the drain region, as it is uniformly doped with donors. This kind of doping-concentration gradient between source and drain helps in the overall management of electric field distribution, which minimizes short-channel effects, critical in nanoscale devices.

The Asymmetric Gate DG MOSFET consists of two gates with different characteristics placed on opposite sides of the channel. Gate 1 has 13 nm, and gate 2 is slightly less at 12 nm. The asymmetry may be implemented through different thicknesses of the gate oxide or using different gate voltages or work functions. This will often result in a different work function or oxide thickness for the front gate compared to the back gate, and hence better control over the channel and precise tuning of the device's electric features.

Uniform donors were used for doping the source region, while having a peak concentration of 1×10^{20} cm^{-3} and dimension of 8 nm. The high doping level in this region is essential to have a low-resistance pathway for carriers entering the channel, meaning improvement of on-state current and overall device performance. A slight lower peak concentration value of 2×10^{18} with a dimension of 8 nm also characterizes the drain region, as it is uniformly doped with donors. This kind of doping-concentration gradient between source and drain helps in the overall management of electric field distribution, which minimizes short-channel effects, critical in nanoscale devices.

The Asymmetric Gate DG MOSFET consists of two gates with different characteristics placed on opposite sides of the channel. Gate 1 has 13 nm, and gate 2 is slightly less at 12 nm. The asymmetry may be implemented through different thicknesses of the gate oxide or using different gate voltages or work functions. This will often result in a different work function or oxide thickness for the front gate compared to the back gate, and hence better control over the channel and precise tuning of the device's electric features [10].

High-k dielectric material HfO2 forms the oxide layers between the gates and the channel, and thus, Oxide 1 and Oxide 2 are each 18 nm in dimension. High-k dielectrics are required to sustain a high gate capacitance and allow, at the same time, an increase in the oxide layer thickness; this is helpful in reducing leakage currents through the gate as well [11–13]. Precise control of the oxide thickness and material properties are critical for optimizing device performance, particularly for subthreshold swing and on/off current ratio.

The contact dimensions are considered to be 1 nm each for contacts S and D, and they are made of aluminum. Contacts are typically made of aluminum as it is one of the better conductors and is relatively simple to fabricate. The small dimensions of the selected contacts ensure that simulations involving those contacts should exactly model the flow of current and potential drops across the contacts which has an implication in predicting the I-V characteristics of the device.

The region of overlap in the source and drain, assumed equally extended about the junction, has an overlap length of d 1 = d 2 = 6 nm. This overlap region is critical in obtaining proper alignment and in reducing parasitic capacitances, thus having a significant impact on the overall performance of the device, more so at high frequencies.

The doping profiles and concentration decide the electrical properties of the device. The high doping concentration of the source region ensures a high injection of carriers into the channel, which is quite crucial to achieve a high on-state current. The doping concentration in the drain region is relatively lower, and it is helpful in managing the electric field and thus reduces short-channel effects. Uniform doping profiles ensure consistent, predictable behavior at the device level-an important requirement for reliable operation [14, 15].

The Asymmetric Gate DG MOSFET design uses the latest simulation techniques to simulate device behavior under different operation conditions. The high-resolution simulations captured very detailed carrier injection and transport phenomena and the electric field distribution inside the device. Such insights are helpful in optimizing the overall performance of the device, and researchers might even find some areas for optimization.

The asymmetric gate structure, on the other hand, provides a few advantages over the typical symmetric DG MOSFET configurations. A better channel control is typically achieved by varying the gate work functions or oxide thicknesses and hence less leakage current and improved overall power efficiency. This structure also facilitates the suppression of short-channel effects such as DIBL and threshold voltage roll-off, the two issues always plaguing MOSFET to scale down to nanometre dimensions [11] Figures 9.1, 9.2 and 9.3 shows the asymmetrical DG MOSFET architecture, meshing, and simulated device in TCAD respectively.

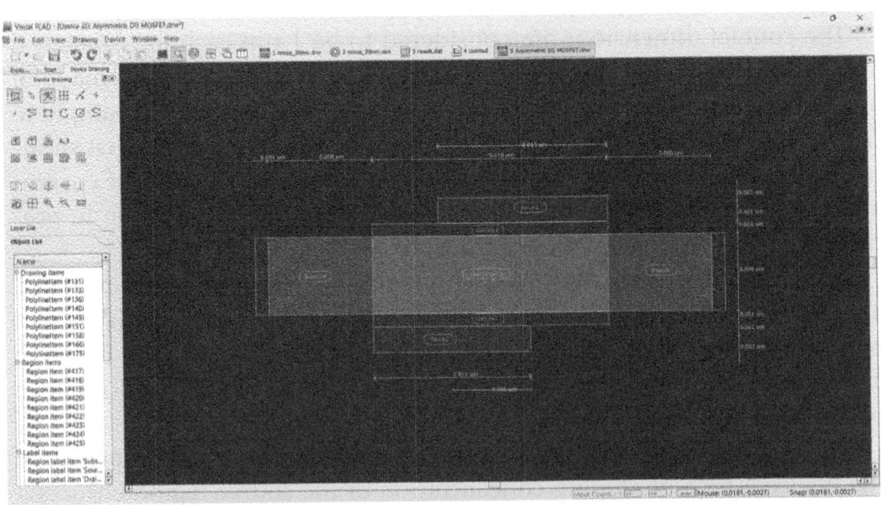

Figure 9.1 Doping profiles of asymmetrical DG MOSFET.

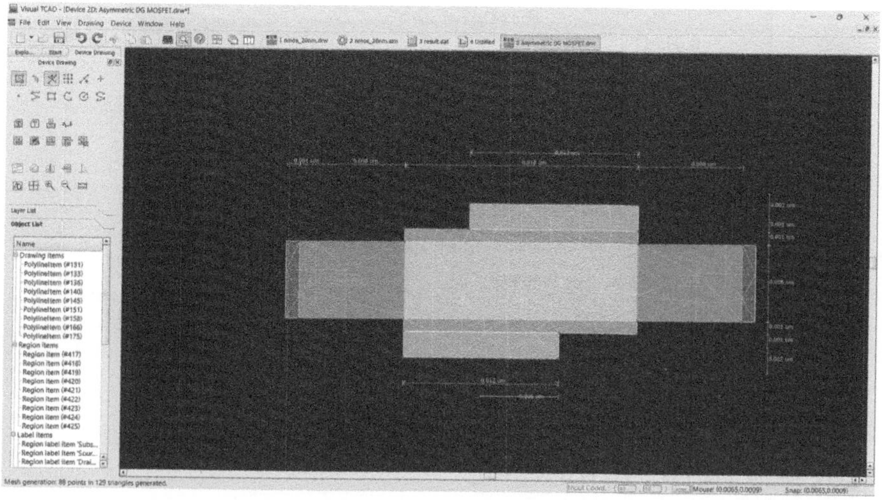

Figure 9.2 Final mesh grid of asymmetrical DG MOSFET.

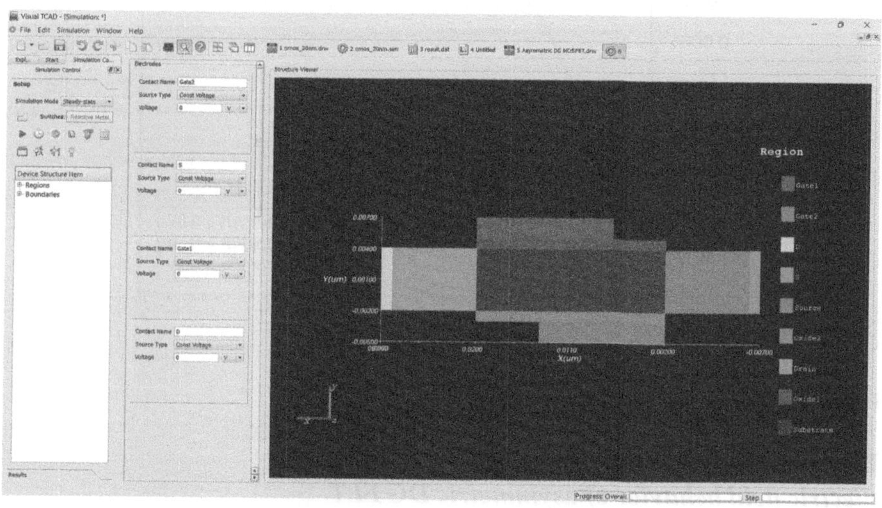

Figure 9.3 Simulation of PMOS of asymmetrical DG MOSFET.

9.4 Result and Simulations

Drain current plot

Figure 9.4 shows the drain current variation with respect to the drain voltage, and Figure 9.5 shows the drain current variation with respect to the gate voltage.

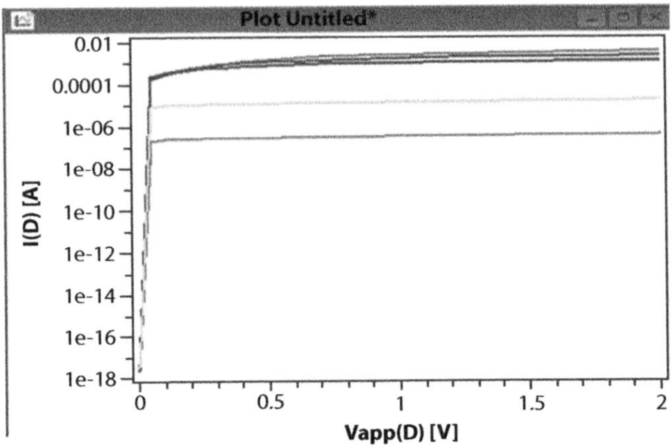

Figure 9.4 Id and Vds curve at different Ig.

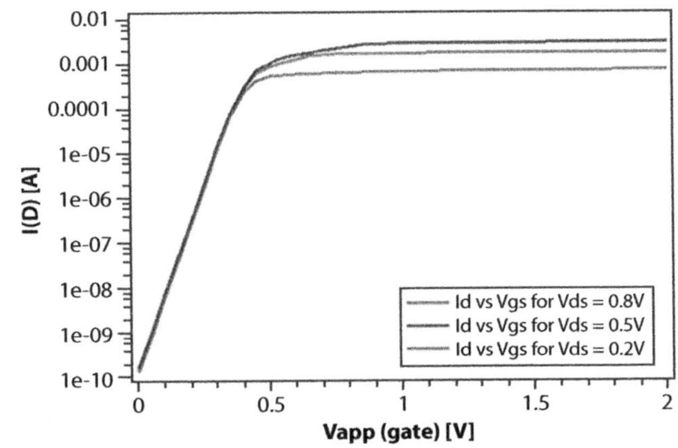

Figure 9.5 Id and Vgs curve at different Ig.

Electrostatic behavior

Electric field behavior of asymmetric DG FET

Figure 9.6 shows the electric field profile of an asymmetric DG FET under two conditions for the gate-source voltage, Vgs: Vgs = 0V and Vgs = 2V with Vds = 0.5V. Investigating those conditions indeed sheds light on the DG FET behavior and performance, focusing particularly on how the gate-source voltage affects the electric field along the device.

When Vgs is set to 0V, the DG FET is in an off-state. At this stage, the channel is not inverted; hence, there are fewer free charge carriers through which current will flow. The red curve in the graph represents the electric

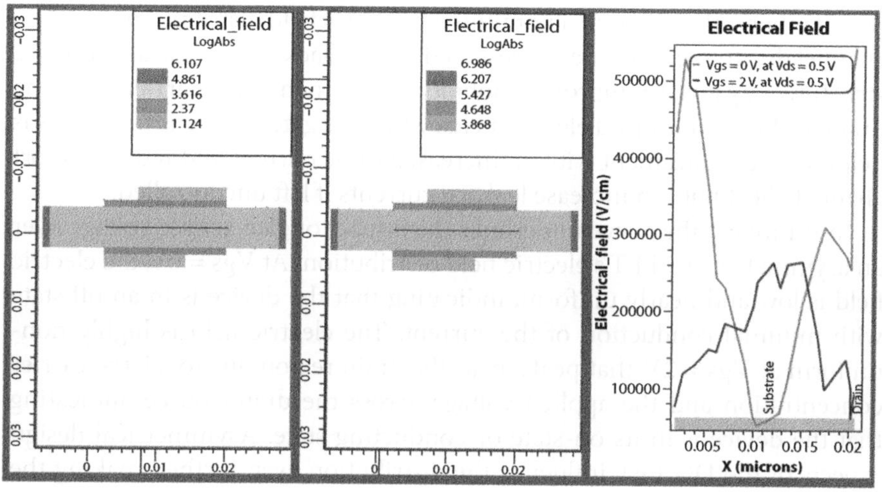

Figure 9.6 Electric field behavior of asymmetric DG FET.

field distribution, which remains uniform along the channel but peaks near the drain region. This peak is due to the applied drain-source voltage, Vds. The electric field between the source and the drain is created. The total field is lesser as the applied gate voltage is not enough to invert the channel and attract major charge carriers. So, the device shows high impedance, and the flow of current is very less.

If Vgs is given a high enough value such that it becomes equal to 2V, the DG FET switches into an on-state in which the channel of the device is inverted and an electric path is formed between the source and drain. The black curve represents this in the graph. At a Vgs value of 2V, the voltage at the gate is sufficiently high such that it creates a stiff electric field that will attract and accumulate the charge carriers into the channel. This makes the electric field stronger and mainly near the drain region due to increased concentrations of carriers and Vds. The peak electric field is of much higher magnitude as compared to the Vgs = 0V condition. In this scenario, the channel is fully conductive and can support a higher current flow more easily. This shows that the electric field near the drain is very strong due to the association of gate voltage with drain-source voltage, resulting in a conductive channel that happens to be very robust.

But the asymmetry within the DG FET structure plays a significant role in this observed electric field distribution. Asymmetric DG FETs could be designed, for instance, by varying the thickness of the gate dielectric on either side of the channel or the work function. In such cases, electrostatic control over the channel changes with both parameters. This kind of asymmetry would inherently represent a nonuniform electric field distribution along the channel due to the sharp peak presented in the black curve. The asymmetric structure may give an improved mobility or confinement of carriers near the drain for higher localized electric fields when the gate voltage is high. This is helpful for enhancement with an increased drive current and reduced short channel effect but can increase leakage currents if left uncontrolled.

In summary, the curve underlines the impact of gate-source voltage over an asymmetric DG FET's electric field distribution. At Vgs = 0V, the electric field is low and nearly uniform, indicating that the device is in an off state with minimal conduction of the current. The electric field is highly non-uniform at Vgs = 2V that peaks near the drain region due to a large carrier concentration and the applied voltage across the drain-source, indicating that the device is in its on-state or conducting state. Asymmetrical design aspects of the DG FET influence the distribution even further, making the optimization better over current drives and leakage current. Understanding these differences is crucial to design and optimize the DG FETs for high-performance and low-power applications in modern electronic devices.

Potential of asymmetric DG FET

Figure 9.7 illustrates the probable distribution along the channel of an asymmetric DG-FET under two gate-source voltage conditions: Vgs = 0V and Vgs = 2V, at a constant value of 0.5V, Vds. This distribution explains how the device would behave under any kind of electrical condition. In the off-state, the DG-FET is with Vgs at 0V. This voltage does not invert the channel; instead, the semiconductor region continues to be depleted of free charge carriers, and

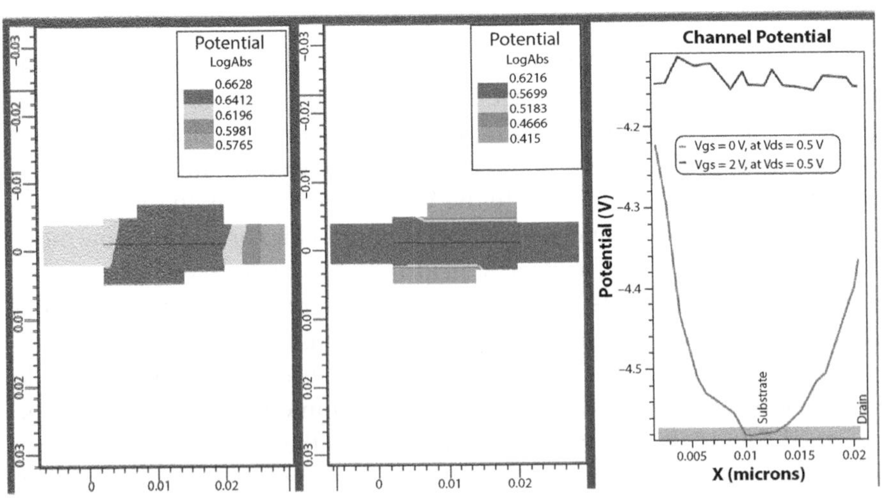

Figure 9.7 Potential of asymmetric DG FET.

hence potential barriers are higher. The red curve in the figure depicts this state; the potential around the middle of the channel dips significantly. This dip demonstrates that, in the absence of a sufficient gate voltage, the electric field is heavily reliant on the applied Vds, and, accordingly, a flattish profile with low potential, peaking very close to the drain region, can occur. A higher potential barrier at lower gate voltages indicates that the device is an impedance device and has zero or negligible current conduction capabilities.

On the other hand, when the gate-source voltage is 2V, the DG-FET enters an on-state represented by the black curve in the graph. In this state, it can be seen that the gate voltage is positive enough to invert the channel such that a conductive path between the source and the drain is now established. This can be clearly seen from the graph that the potential distribution becomes much flatter as represented in Figure 9.7. This comes as a result of increased gate voltage at the accumulation of more charge carriers in the channel. The increase of conductivity is highly associated with potential barriers significantly reduced, resulting in much reduced short-channel effects. It can be noted that the near-drain peak of the potential is decreased, showing good carrier mobility and a clearly defined conductive channel. The uniformity in the black curve potential profile indicates a fully inverted channel that can support increased current flow with reduced short-channel effects.

Such observations have essential implications in consideration of the asymmetric structure of the DG-FET. With different thicknesses of the gate dielectric or work functions at both ends of the channel, asymmetry is inclined to offer an irregular electrostatic control of the channel at low gate voltages (Vgs = 0 V). Such unevenness in the variation of potential tends to offer uneven electric fields affecting device performance. Even at Vgs as low as 2V, stronger control over the part of the gate overrides these asymmetry effects and produces a uniform potential profile in the channel. Better control is achieved by higher voltages in the gate and permits improving the device performance by elevating the drive currents as well as decreasing the leakage currents, but it is prone to carrying excess charges and to the degradation of devices.

The possible potential profile along the channel of an asymmetrical DG-FET is highly dependent on the gate-source voltage. At Vgs = 0V, the device reflects that it has maximum potential barriers and lower carrier density; this further indicates the non-uniform potential profile that shows the off-state. At Vgs = 2V, the potential profile becomes flattened by way of inversion of the channel and with a more uniform profile toward higher conductivity and mobility of carriers. Further, these distributions get altered by the asymmetric design of the DG-FET. Optimization in terms of an equilibrium balance between drive current and leakage currents is the result of such a design. Understanding these variations is thus important for designing and optimizing DG-FETs for use in high-performance low-power applications related to modern electronic devices.

Energy band of asymmetric DG FET

From Figure 9.8, the conduction band edge (Ec) and valence band edge (Ev) have been demonstrated at every cross-section of an asymmetric Double-Gate Field-Effect Transistor (DGFET) for the case of two gate-source voltage conditions, Vgs = 0V and Vgs = 2V, with a constant Vds of 0.5V. It is an essential diagram, as the operation of a DGFET with different electrical conditions is explained and how the asymmetry in its structure actually affects the operation.

When Vgs = 0V, the edge of conduction (Ec) is high across the channel, as seen in the energy band diagram, to make it unfavorable for electron flow. In this off-state, the gate voltage is not capable enough to invert the channel; that is, the semiconductor region remains depleted of free charge carriers. This results in the conduction and valence bands having high energy levels, thus resulting in a large barrier to electron flow. The black line represents the higher energy level of the Ec, which means that the channel is in a non-conductive state. The analogous edge of the valence band Ev also mimics this, as it stays high and depicts the potential for hole conduction to be also trivial. In this situation, the potential profile is nearly uniform but very low in value, peaking a little near the drain region, primarily depending upon the applied drain source voltage (Vds). This device

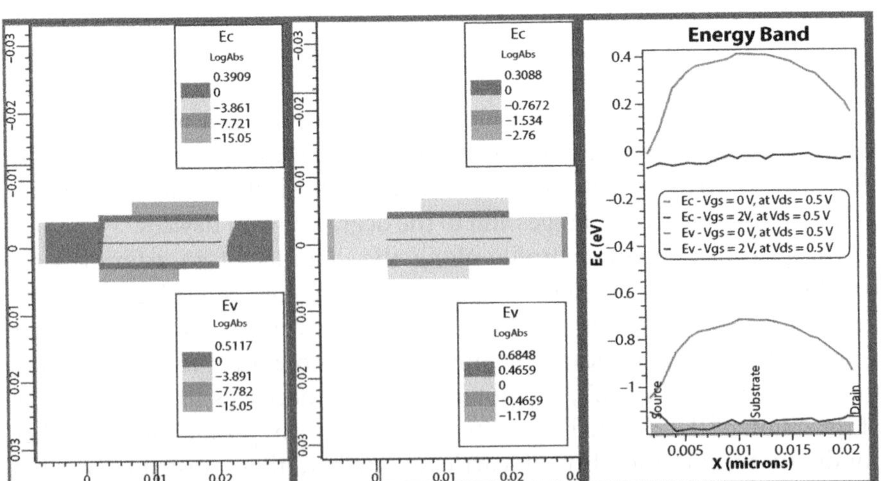

Figure 9.8 Energy band of asymmetric DG FET.

has a high impedance with almost negligible current conduction in this configuration.

The DGFET is said to turn on at the gate-source voltage Vgs = 2V. At such a high gate voltage, the gate is sufficiently positive to invert the channel in such a way that it provides a conductive path from source to drain. The energy-band diagram for this state, as shown below in red line, indicates that the conduction-band edge Ec drops considerably, particularly near the source and the drain regions. This downward dip gives an indication of the inversion channel; electrons are free to move toward the drain from the source. The lower Ec levels show that the potential barriers to the flow of electrons have decreased; hence, this increases the conductivity level. The dropping of the edge of the valence band, Ev, ensures overall maintenance of the band structure while enabling enhanced carrier mobility. Because the profile appearing more uniform and flatter from the potential point of view throughout the channel section indicates a greater carrier density, this improves the current which flows in the device. Further, a high Vgs leads to the formation of an inversion layer such that the edge of the conduction band inside the layer is so much lowered that transport through the electron is quite efficient.

DGFET structure also affects this behavior in an asymmetric way. All these asymmetric designs are found to have an intrinsic difference in gate dielectric thickness and/or work functions at each of the channel sides, leading to electrostatic control variations. The non-uniformity in the energy band profile, due to asymmetry at lower gate voltages, Vgs = 0 V, increases the potential barriers and reduces carrier density. However, with higher gate voltages (Vgs = 2V), stronger gate control reduces those asymmetry effects to produce a more balanced profile in energy bands, thus making it favorable for high currents to flow. Very useful in enhancing device performance, these superior controls at higher Vgs are very difficult to control as they build up excessive carriers and lead to degradation of devices.

9.5 Subthreshold Performance

Drain Induced Barrier Lowering (DIBL):
It is a phenomenon mainly observed in short channel devices such as Metal Oxide Semiconductor Field Effect Transistor. As there is increase in drain voltage, there is a reduction in threshold voltage. This issue is more visible in metal oxide FET.

$$\text{Equation for calculating the DIBL} = \frac{v_{\text{th}\,1} - v_{\text{th}\,2}}{v_{ds1} - v_{ds2}}$$

For the proposed device, DIBL = 20 mv/V

Sub threshold Slope (SS):
Subthreshold slope SS corresponds to an important parameter under current increase in the subthreshold region-—it describes how steep the current increases as the transistor goes from off to on. It can be defined as the change in the gate voltage, Vg, which changes Id by one order of magnitude in the subthreshold region-10 factor.

Equation for calculating the SS = S=dVgd / (logId)
For the proposed device the SS = 60.8 mv

9.6 Comparison with Another Technology Node

In the recent few years, Asymmetric Gate DG MOSFETs have emerged as the replacement of traditional DG MOSFETs. Even though similar in many aspects, they differ in performance, structure, power dissipation delay, architecture, dimensions, region material and doping, and y-characteristic.

The primary design intent of an Asymmetric Gate DG MOSFET is to offer better gate control over the channel through variation in work functions of the gates or through the thicknesses of the oxide.

This results in better SCE-suppression mechanisms, like drain-induced barrier lowering and threshold voltage roll-off. However, SCE suppression ability is good, but precision in the control of electrical parameters is very poor 1. Asymmetry gives increases on-state current and decreases off-state leakage current simultaneously, and thus improves the overall device performance1.

The Asymmetric Gate DG MOSFET has structure similar to that of the traditional DG MOSFET but with asymmetric gates, or rather, Gate 1 can have a work function or oxide thickness different than Gate 2. Thus, an adequate electric field distribution is created, further improving performance. In contrast, the standard DG MOSFET usually includes symmetric gates [8]. That is, the structure can be easier and less complex, although much less controllable over the channel. It also has a less asymmetric structure, which allows more precise tuning of the electrical properties of the device 2.

The Asymmetric Gate DG MOSFET reduces leakage currents more effectively and hence; it is said to have low power dissipation. It is possible to minimize one of the big contributors of power dissipation in nanoscale devices off-state leakage with improvements in gate work functions and oxide thicknesses. In the traditional DG MOSFET, though it is efficient, its case may not be equal with respect to the Asymmetric Gate DG MOSFET due to the symmetric nature of gates.

The delay in the Asymmetric Gate DG MOSFET is typically small due to improved channel control; thus, it switches faster. The highly accurate adjustment of the gate characteristics helps obtain a higher current at the on-state and faster state transition. The conventional DG MOSFET, although efficient in minimizing the delay, cannot reach the optimal performance provided by the asymmetric design1.

DG MOSFET architecture is far more complex as the gate is of two different sizes with different properties. For example, Gate 1 could be 13 nm in size and Gate 2 could be 12 nm with the oxide layers being 18 nm each. Thus, it increases the possibility of a longer channel but the techniques to fabricate it are so sophisticated. The traditional DG MOSFET had a simpler architecture with symmetrical gates, which was easier to fabricate but perhaps less flexible in performance tuning 2.

The Asymmetric Gate DG MOSFET is specifically designed with certain dimensions for optimum performance. The substrate of the device is 18 nm, source and drain both are 8 nm, gates are 13 nm and 12 nm and oxide layers are also 18 nm. In traditional DG MOSFET, gate as well as oxide layer dimensions are symmetric with respect to which DG MOSFET is asymmetric. This makes the design more complex; however, the optimization at the potential level may not be the same.

Both devices use silicon as the substrate material with specific doping profiles. In the case of Asymmetric Gate DG MOSFET, uniform acceptor doping is used in the substrate (1 × 10 ^{16}, cm ^{−3}), high donor doping for the source (1 × 10 ^{20}, cm ^{−3}), and low donor doping is used in the drain (2 × 10 ^{18}, cm ^{−3}). The traditional DG MOSFET further uses similar doping profiles but with symmetrical gate characteristics that may not provide the same degree of performance tuning1.

The y-characteristics, that is, the output current as a function of gate voltage are generally better in the asymmetric gate DG MOSFET, because of the better control the charge has over the channel. In this way, the asymmetric structure can achieve a larger on/off current ratio with a steeper subthreshold slope and, therefore, an overall better performance of the device, whereas conventional DG MOSFET, having a symmetric structure does not 2.

Hence, Asymmetric Gate DG MOSFET has several advantages in comparison to a conventional DG MOSFET; that is, it has a better performance characteristic, low power dissipation, delay, and control over the channel. In return, this comes along with additional complexity in the architecture of the device as well as in the fabrication process. However, these are two extremely significant steps forward with MOSFET technology.

9.7 Applications of Asymmetric Gate DG MOSFET

The Asymmetric Gate DG MOSFET is highly suitable for low-power digital applications [16, 17]. The Asymmetric Gate DG MOSFET presents a better SCE control mechanism and a reduced delay time. This suggests that it is an excellent candidate for high-speed logic circuits. Due to its superior electrostatic control and reduced parasitic capacitances, the Asymmetric Gate DG MOSFET may be applied in analog and RF applications [18, 19]. The Asymmetric Gate DG MOSFET is also used in memory devices such as SRAM and DRAM [17]. Sensitivity of Asymmetric Gate DG MOSFET, along with low power consumption, makes it feasible for using this type of device in biosensing applications [20, 21]. Due to low power dissipation, the Asymmetric Gate DG MOSFET has gained interest towards using this kind of device in wearable electronics with high performance. Based on robust performance and low power consumption, the Asymmetric Gate DG MOSFET can potentially integrate into flexible electronic devices.

9.8 Conclusion

The proposed architecture for Asymmetric Gate DG MOSFET utilizes sophisticated doping profiles, high-k dielectric materials, and precise simulation for high-performance characteristics. It ensures that the on-state current is high due to uniform acceptor doping of the substrate and highly doped donor in the source and drain regions that regulate the short-channel effects properly. The gate structure is asymmetric and enhances the electrostatic control of the channel. It reduces gate leakage currents by the application of high-k dielectric. Overall, the kind of design and simulation applied to such devices facilitate the generation of high-performance, low-power electronic circuits suitable for a wide variety of applications in modern semiconductor technology.

The asymmetric DG MOSFET has design and performance features, thus being suitable for many applications, including low-power digital circuits, high-speed logic circuits, analog and RF applications, memory devices, biosensors, and wearable electronics and flexible electronics. Optimized control in the channel results in a decrease in leakage currents and an increase in switching speed; hence, such a component is a worthy inclusion in modern semiconductor technology.

References

1. Hazarika, M.R. and Bora, N., Performance analysis of 3-D asymmetric junction less double gate MOSFET, in: *2017 International Conference on Energy, Communication, Data Analytics and Soft Computing (ICECDS)*, IEEE, pp. 1391–1395, 2017, August.
2. Vaddi, R., Dasgupta, S., Agarwal, R.P., Effect of gate-S/D underlap, asymmetric and independent gate features in the minimization of short channel effects in nanoscale DGMOSFET, in: *2011 IEEE Computer Society Annual Symposium on VLSI*, IEEE, pp. 37–42, 2011, July.
3. Vaddi, R., Dasgupta, S., Agarwal, R.P., Two dimensional analytical subthreshold current model of a generic double gate MOSFET with gate underlap, in: *2011 International Conference on Electronic Devices, Systems and Applications (ICEDSA)*, IEEE, pp. 246–249, 2011, April.
4. Dasgupta, A., Das, R., Dutta, A., Kundu, A., Sarkar, C.K., A comparative study of analog/RF performance: symmetric and asymmetric underlap gate stack DG-MOSFETs, in: *2016 3rd International Conference on Devices, Circuits and Systems (ICDCS)*, IEEE, pp. 148–151, 2016, March.
5. Zhu, S., N-type Schottky barrier source/drain MOSFET using ytterbium silicide. *IEEE Electron Device Lett.*, 25, 8, 565–567, August 2004.
6. Colinge, J.P., SOI gated resistor: CMOS without junctions. *IEEE International SOI Conferrence*, vol. 2, pp. 1–2, October 2009.
7. Gamiz, F., Roldan, J.B., Godoy, A., Jimenez-Molinos, F., Double Gate Silicon-on-Insulator transistors: n+-n+ gate versus n+ -p+ gate configuration. *ESSDERC 2004*, pp. 173–176, Sept. 2004.
8. Spedo, S. and Fiegna, C., Comparison of symmetric and asymmetric double gate MOSFETs- tunneling currents and hot electrons. *IEEE Semiconductor Device Research Symposium*, Washington, DC, USA, pp. 601–604, 2001.
9. Tang, S.H., Xuan, P., Bokor, J., Hu, C., Comparison of ShortChannel Effect and Offstate Leakage in Symmetric vs. Asymmetric Double Gate MOSFETs. *IEEE International SOI Conference*, Oct. 2000, p. 120.
10. Kim, K. and Fossum, J.G., Double-gate CMOS: Symmetricalversus asymmetrical-gate devices. *IEEE Trans. Electron Devices*, 48, 2, 294–299, Feb. 2001.

11. Cakici, R.T. and Roy, K., Analysis of Options in Double-Gate MOS Technology: A Circuit Perspective. *IEEE Trans. Electron Devices*, 54, 12, 3361–3368, Dec. 2007.
12. Manoj, C.R. and Rao, V.R., Impact of High-k Gate Dielectrics on the Device and Circuit Performance of NanoscaleFinFETs. *IEEE Electron Device Lett.*, 28, 4, 295–297, Apr. 2007.
13. Tripathi, S.L., Mishra, R., Narendra, V., Mishra, R.A., High performance Bulk FinFET with bottom spacer, in: *2013 IEEE International Conference on Electronics, Computing and Communication Technologies*, IEEE, pp. 1–5, 2013, January.
14. Madhavi, K.B. and Tripathi, S.L., Electrical Characterization of highly stable 10nm triple-gate FinFET for different contacts and oxide region materials. *Silicon*, 14, 12281–12291, 2022, https://doi.org/10.1007/s12633-022-01935-w.
15. Bhattacharya, S., Tripathi, S.L., Kamboj, V.K., Design of tunnel FET architectures for low power application using improved Chimp optimizer algorithm. *Eng. Comput.*, 39, 1415–1458, 2023, https://doi.org/10.1007/s00366-021-01530-4.
16. Mendiratta, N. and Tripathi, S.L., A review on performance comparison of advanced MOSFET structures below 45 nm technology node. *J. Semicond.*, *41*, 6, 061401, 2020.
17. Mendiratta, N. and Tripathi, S.L., 18nm n-channel and p-channel Dopingless Asymmetrical Junctionless DG-MOSFET: Low Power CMOS Based Digital and Memory Applications. *Silicon*, 14, 6435–6446, 2022, https://doi.org/10.1007/s12633-021-01417-5.
18. Tripathi, S.L. and Saxena, S., Asymmetric gated Ge-Si0. 7Ge0. 3 nHTFET and pHTFET for steep subthreshold characteristics. *Int. J. Microstruct. Mater. Prop.*, *14*, 6, 497–510, 2019.
19. Mendiratta, N., Tripathi, S.L., Chander, S., Analytical Model of Dopingless Asymmetrical Junctionless Double Gate MOSFET. *Silicon*, 14, 10765–10774, 2022, https://doi.org/10.1007/s12633-022-01819-z.
20. Mendiratta, N., Tripathi, S.L., Padmanaban, S., Hossain, E., Design and Analysis of Heavily Doped n^+ Pocket Asymmetrical Junction-Less Double Gate MOSFET for Biomedical Applications. *Appl. Sci.*, 10, 7, 2499, 2020, https://doi.org/10.3390/app10072499.
21. Tripathi, S.L., Patel, R., Agrawal, V.K., Low leakage pocket junction-less DGTFET with biosensing cavity region. *Turk. J. Electr. Eng. Comput. Sci.*, *27*, 4, 2466–2474, 2019.

10

Smart Materials for Semiconductor Devices: Research, Characteristics and Applications

Krishan Arora

School of Electronics and Electrical Engineering Lovely Professional University, Punjab, India

Abstract

Smart materials are an entitlement loop of relatives of substances which can be recorded in predominant elements. These dealings of substances have self-lodging with the environment and they may be named reliable with their reactions, in conjunction with physical (weight, temperature, stickiness) or natural jolts. The intelligent substances may change over retained quality or their characteristics may moreover alter. Keen substances have gotten to be intemperately intrigued since their trade bundles in both actuator or sensor shape. implantation, bringing down squander, and NATO-engineered frameworks.

***Keywords*:** Keen fabric, shape memory combination, piezoelectric, automated, biomedical

10.1 Introduction

The Human creatures from thousands of a long time period utilized fabric for diverse purposes, which improved their ways of life, and began finding unused viable fabric. Agreeing to archaeologists, the primary age are classified into a few bunches and families. There are four fundamental bunches of materials which are metals, ceramics, polymers, and as of late progressed materials [1]. Among them, prevalent substances develop

Email: krishan.12252@lpu.co.in

Suman Lata Tripathi, Vrinda Gupta, Sobhit Saxena, Shipra Upadhyay and Sudip Ghosh (eds.) Micro and Nano Semiconductor Devices for Digital, Analog and Sensor Design, (179–194) © 2026 Scrivener Publishing LLC

to be additionally engaging one, due to the reality that they've additional mechanical applications. materials which are metals, ceramics, polymers, and as of late progressed materials [2]. Since a few bodily homes of superior substances may be controlled, they're the constructing block of maximum superior hybrid gadgets around us. Semiconductors facilitate the transition from vacuum tubes to more compact electronic chips in computer generation [3]. However, biomaterials made it possible to communicate with biological organs, and materials that have been nanoengineered which are more effective than those that are large in size [4, 5]. As a result, intelligent materials will become more prevalent in automation systems, commercial appliances, scientific equipment, and civil engineering, among other fields [6].

Thus, the clever material's packages which include sensors [7–9], actuators [10–14], and drug delivery [15, 16] may be happy with the aid of using the use of the aforementioned parameters.

10.2 Shrewd Materials

There's no interesting characterization for savvy constituents [17–20], like NASA considered them as resources that can keep in mind diverse shapes and able to accommodate with specific boosts, or they can be characterized as a profoundly built materials that can respond intelligently to their environment [21, 22]. Terms like 'versatile' are often used to describe materials such as sensors and actuators. These descriptors highlight the intelligent or adaptable nature of these components. Moreover, these materials possess unique capabilities that distinguish them from conventional ones such as [23–25]:

- Transiency: they may answer to particular boosts and may be found in assorted states;
- Self-actuation (insights): this cap potential is inside the fabric; [26–28]:
- Choosiness: the response is exceptional and unsurprising;
- Explicitness: each act and response are amassed with inside the indistinguishable put. Shrewd substances are partitioned into essential categories, inactive and dynamic. Sorts of shrewd materials—Due to distinctive reactions to natural jolts, savvy materials are classified into a few sorts, counting shape memory combinations (reaction to warm and weight),

piezoelectric materials (reaction to power and weight), electro-restrictive materials (reaction to the all sorts of savvy fabric in no time have been outlined [29–32].

10.3 Types of Smart or Keen Materials

Due to diverse reactions to natural jolts, keen materials are classified into a few sorts, counting shape memory amalgams (reaction to warm and weight), piezoelectric materials (reaction to power and weight), electro-restrictive materials (reaction to the electric field) [33, 34].

10.3.1 Shape Memory Combination

These combinations, known as Shape Memory Alloys (SMAs), are intelligent materials that possess a unique basic structure capable of transforming from one form to another in response to changes in temperature or applied stress. The phenomenon of shape recovery was first observed by Ölander in 1934 in a gold-cadmium (Ag-Cd) alloy as shown in Table 10.1 [35].

The validation and advancement of these features in the field have brought us to this stage, as the application of SMAs is largely influenced by temperature changes and other physical parameters. This has sparked active research among materials scientists, focusing on factors such as strain behavior, advanced cooling techniques [36, 37], surface modifications [38], and various approaches for specimen preparation and testing [39–41].

Table 10.1 A few common SMA frameworks.

Alloy	Composition	Temperature (°C)	Transformation hysteresis
Ag-Cd	44/49 at.Cd	-190 to -50	-15
Au-Cd	46.5/50at Cd	30 to 100	-15
Cu-Al-Ni	¾.5 wt%-Al	-140 to 100	-35
Cu-Al-Ni Polycrystalline	Cu 82.4 wt.%- Al	149	-21
Cu-Sn	15 at.% Sn	-120 to 30	-10
Cu-Zn	38.5/41.5 wt.% Zn	-180 to -10	-10

To plan an SMA, there are several strategies to give items different characteristics (Table 10.2). The culmination of these intellectual substances can be exhibited by various capabilities which incorporates thermoelectricity, super flexibility, and damping capacity, which outlines why those assortments of substances are special.

Table 10.2 Distinctive accessible methods to deliver SMAs.

Methods		Characteristics	Advantages
Casting		Vacuum Arc Remelting	High Homogeneity
		Vacuum Induction Melting	Low Cost
		Electron Beam Melting	Flexibility
Powder Metallurgy	Conventional Processes	Conventional Sintering	Low Density
		Self-Propagating high temperature Synthesis	High Porous
		Hot isostatic pressing	Powder Mixture Usage
		Spark Plasma Sintering	Dense Structures and fine grains
		Metal Injection molding	Generating High Porous SMAs
	Additive Manufacturing	Selective laser Sintering	Complex Structures without the need for machining
		Selective laser Melting	Easy Fabrication
		Laser Engineered Net Shaping	Controlling shape, size and porosity
		Electron Beam Melting	Low Carbon Content

10.3.1.1 Thermoelectricity

Due to diminishing temperature, the stage of SMAs exchanges from the tall temperature (austenite) to the lower temperature stage, which is known as the martensite stage (too some of the time this handle is called sheer or displacive change). Among them, ferrous combination structures are nucleation-dependent (non-thermoelastic); a few arranged in austenite to martensite change, while the inverse amalgams are autonomous (thermoelastic) [42, 43]. This fundamental property is called the shape memory impact (SME) [44]. Gem structure is maintained by a Bain strain; hence, the cubic gem shape is mutilated and makes a twining shape, that's both hexagonal close-packed, monoclinic, or orthorhombic [45].

A schematic SMA diagram of the anomalous natural conditions is shown in Figure 10.1. A special cooling process adjusts the gem shape of the austenite part to twin martensite. Figure 10.2 shows one of the important properties of SMA: hysteresis.

10.3.1.2 Pseudoelasticity

Pseudoelasticity or hyperelasticity can be a nonlinear stress/strain behavior that can occur with movement of spatial boundaries or changes between different gemstone structures in certain cases (so-called Bain correspondences) [46–48].

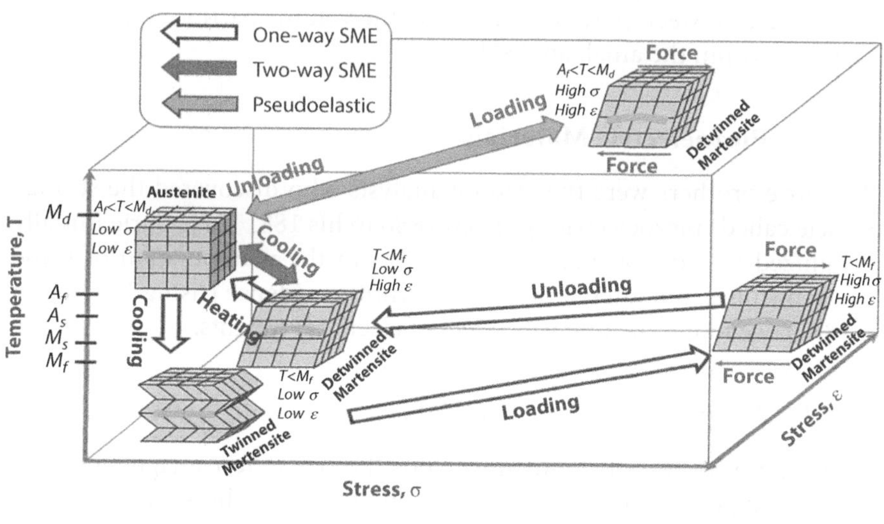

Figure 10.1 Schematic diagram of pressure, strain, and temperature parameters for several SMEs.

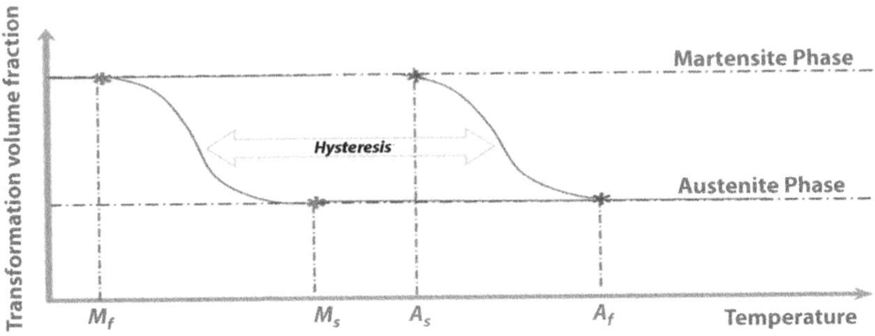

Figure 10.2 Hyperplastic graph for temperatures above Af.

Beneath consistent push (locale no. 5), but amid stage change, a division of vitality will be devoured; in this way, the invert stage change does not happen at the same stretch. Moreover, above the temperature, the stress-induced martensite is not always stable, making it more effective to achieve deformation through other means [49–52].

10.3.1.3 Damping Capacity

The damping capacity depends on the vibration strength and becomes more pronounced for strengths above 10 [53–56].

Components and microstructure can also affect damping capacity. For example, the damping ability of CuAlMn could be improved by decreasing the distance between pores or increasing the volume partitioning of existing pores within the amalgam [35].

10.3.2 Piezoelectric Materials

The Curie brothers were the leading analysts who discovered the strange miracle called "piezoelectricity" from 1880 to his 1881 [57]. Curie initially recognized the role of mechanical weights in the manipulation of some rare gemstones, but after some basic scientific evidence (using Lippmann in 1881), we found the opposite effect of electrical weights.

10.3.3 Magnetostrictive Materials

It can be a miracle— that which expands the size of ferromagnetic structures through the use of attractive magnetic fields. These materials can

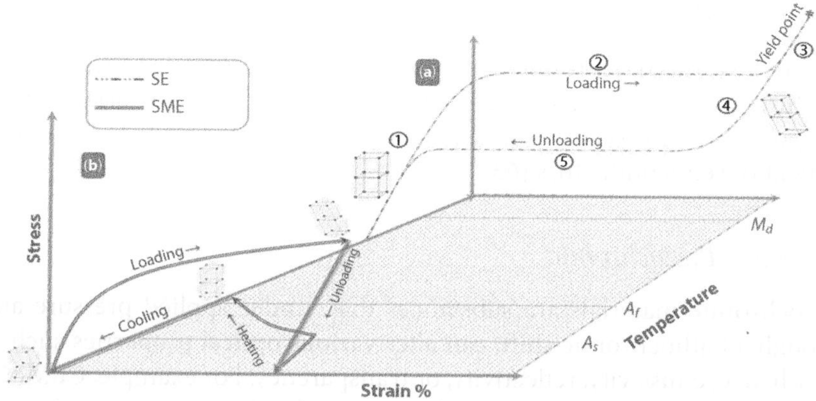

Figure 10.3 Stress/stress pursuing for (a) super flexibility impact, (b) heating-cooling impact.

be classified as positive or negative magneto strictive materials (MS). They can grow or shrink depending on the biased attractive field as shown in Figure 10.3 [58].

10.3.4 Chromic Materials

Among smart materials, there exists a group that changes color in response to variations in environmental conditions. Once the external influence is removed, these materials revert to their original color [59]. Since, changing over the pores and skin color may be a camouflage approach among chameleon animals, at that point those outline of substances that would appear this capacity are on occasion insinuated offers a number of organizations of chromic materials [60].

10.3.4.1 Photochromic

Generally, these types of chromic materials are sensitive to visible and ultraviolet radiation, and they can exhibit a range of colors depending on the specific application area. When exposed to sunlight, they may also reduce transparency [61]. Furthermore, photochromic substances are utilized in some texture creations and naval force securing dress [62].

10.3.4.2 *Thermochromic*

The color of thermochromic materials changes in response to rising temperatures, reaching specific predetermined values. These materials can be composed of various compounds, including those used prior to the development of semiconductors [63].

10.3.4.3 *Piezochromic*

Piezochromic materials are substances that, under applied pressure and through a bathochromic shift, can alter various optical properties such as absorbance, emissivity, reflectivity, or transparency. For example, $CuMoO_4$ requires a pressure of 2.5 kbar to change its color from green to red; however, such high pressure limits its practical use in color-based pressure sensors. In contrast, many types of polymers exhibit a bathochromic shift under comparatively lower pressure, making them more suitable for such applications.

10.3.5 pH Delicate Materials

pH-labile polymers are any members of a family of smart substances that can change their color in response to a variety of specific pH levels. For example, halochromic substances used in gauze can help heal any burn [64]. These smartly textured swabs can change the pH of the skin during menstruation and for this reason, they are useful for monitoring recovery levels as shown in Figure 10.4.

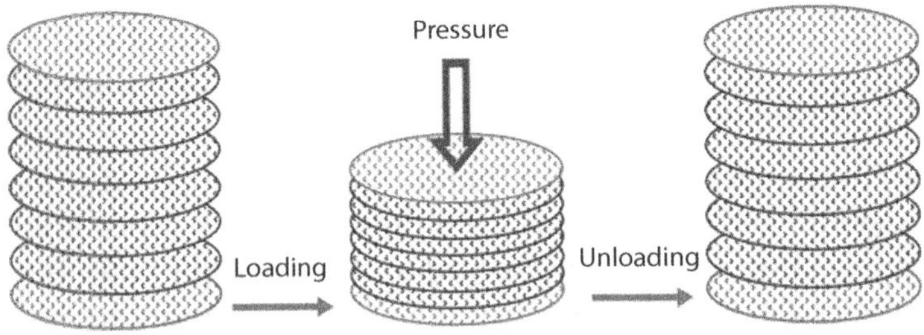

Figure 10.4 Schematic pH delicate materials.

10.3.6 Magnetorheological and Electrorheological Fluids

Generally, fluids are classified as Newtonian fluids, which is a direct dating between the resistance to flow (thickness) and the velocity of what passes through the fluid. Magnetorheological liquids jetsam adhere collectively which make chains withinside the break even when utilizing electric fueled zone, like EMF, like MRF, which may be utilized withinside the indistinguishable intelligent computer program frameworks [65].

10.4 Application of Savvy Materials

Today, intelligent materials play a significant role in shaping modern society and are highly valued across various advanced fields. They are widely used in smart architecture and construction, healthcare, defense, robotics, aerospace, nuclear power plants, art, and numerous cutting-edge research domains. Let us now explore some of their key applications.

10.4.1 Walking Piezo Lever

It contains a variety of particle with miniature sizes in the range (0.1–10 μm), which can be larger than nano floods and jets in ferrofluids, which is why Brownian expansion involves a much less efficient process. The MRF can be used in three different designs: transparent, with valves and compression flow. There are many high-quality improvements in the design of mechanical outcomes imaginable for applying MRF. They can be used in buildings to protect them in the event of an earthquake [66].

Additionally, collision protection can reduce dark oscillations in larger vehicles compared to conventional vehicles using Newtonian fluids.

Additionally, electrorheological fluids (EMFs) can be non-Newtonian viscous fluids which, – like magnetorheological fluids – will have disorganized suspended particles. When the power supply area is operated under EMC, suspended debris and ral unique piezoelectric materials.

At this point, tension is created at right angles to the toes of all legs and a twist is formed. Then the rotated leg may be on the same level, replacing its omnipresence and turning to the electoral side.

At this point, these legs retract and the inverted legs continue the process.

Piezo actuators actuate in the nanometer range and are therefore used in the highest precision magnifying lenses and in very small robots.

(a) The two dumbbell legs move towards the plane, (b) at this time the toes of the two dumbbell legs are twisted in opposite directions [67].

10.4.2 Aviation Innovation

Certain areas of aviation-related advances combine different key materials.

Since their manufacture consists of this type of material, the expression or statement that all these materials are part of this test document is given a series of explosions, so to speak, in this section. The main reason is ultrasonic gas inspection, which is a management indicator. Piezoelectric shocks are used to generate ultrasound waves, and the generated waves are detected across the liquid fuel level and measured with a piezoelectric transducer sensor to allow accurate fuel level measurements [68]. Moment generation occurs in SMA-based solid wings or intelligent wings.

In recent times, Shape Memory Alloys (SMAs) have been utilized to enhance wing shape by enabling controlled lifting and lowering of aircraft wings. As a result, traditional hydraulic systems are no longer necessary, leading to more streamlined wing designs and a reduction in noise generation [69].

10.4.3 Atomic Businesses Keen Substances

We ought to have numerous superiorities for this unstable and highly-priced segment. They are additional green with diminish fetched, also they may utilize for robotization structures to keep absent from private from unsafe exposure of radioactive materials. Be that as it may, viability to radioactive radiation and the capable expense of reputation is one in all requesting circumstances on this zone [70].

10.5 Shrewdly Material

Materials have long been a crucial development in human history, helping to shield our bodies from sunlight and maintain our personal identity. However, in modern society, integrating materials with intelligent fabrics can add enhanced functionalities — such as changing color in response to environmental changes. Smart materials can exhibit various properties, including self-actuation, sensory responsiveness, self-cleaning abilities, and transformations in color or shape.

The subtle qualities of high polymer can be used to form multiple textures to discuss flow and body temperature laws, as they change after the glass is moved at a certain body temperature [58]. The atomicity of the high polymer exceeds the glass transition body temperature (Tg), increasing the conversion range for dissipated water, one of the main components of cooling.

Because a material's atomic volume can be designed to be minimal at room temperature, the material can prevent moisture from penetrating from the environment into the body or from surfaces into the environment.

- It's much better.
- something higher.
- something more powerful.
- At improved > much better temperatures (Tg), a series of small holes are created for ventilation as the particles are retained [71].

Water vapor permeability was further improved by using multiwalled carbon nanotubes [59]. Small Ni-Ti Film Actuator by Manufacturing Ni-Ti films is possible to combine the results of super flexibility and frame memory into one result [54, 58, 72].

The layer has past planes formed in the austenite stage, while the SE layer has a twisted shape in the austenite stage, making it an interesting austenite stage. It transforms into a twisted frame across the film, and in this image the SME layer deforms into twinned martensite.

10.6 Conclusions

All of the keen materials are critical, be that as it may. By and large, piezoelectric and frame reminiscence substances have comparably more prominent computer program than the elective sorts, especially for actuating purposes. Whereas, their possibilities are discernible, be that as it may, there are masses of controls and conceivable outcomes for analyst which are stated with inside the taking after. Possibilities and preferences—They are self-lodging with their environment; tall affectability to jolts; quick reaction to different conditions; they have littler length comparative with distinctive car frameworks; less vitality utilization. Limitations—In a number of cases, the charges of courses of unequivocally tricky to compositional cost of constituents; at that point their instruction wants more noteworthy attentions. Like diverse ordinary substances, some intelligent substances also are battling in weariness, corrosion, and dropping usefulness. Opportunity for analysts—since, for particular circumstances clever substances may be utilized, so substances may be advanced to the handy circumstances. Keen substances can provide extraordinary characteristic for any composition and counting modern added substances.

References

1. Exact, T.E., *et al.*, Early Bronze Age metallurgy: a recently found copper manufactory in southern Jordan. *Relic*, 76, 292, 425–437, 2002.
2. Callister, W.D. and Rethwisch, D.G., Materials Science and Engineering: *An Introduction (7th ed.)*, John Wiley & Sons, New York, 2007.
3. Braun, A., Braun, E., & MacDonald, S., *Revolution in Miniature: The History and Impact of Semiconductor Electronics*, Cambridge University Press, 1982.
4. Akiyama, M., *et al.*, Piezoelectric lean film, piezoelectric fabric, and creation strategy of piezoelectric lean film and piezoelectric fabric, and piezoelectric resonator, actuator component, and physical sensor utilizing piezoelectric lean film. 2010, Google Licenses.
5. Singh, P., Arora, K., Rathore, U.C., Joshi, G.P., Cho, W., Comparative study of controllers in battery energy storage system integrated with doubly fed induction generator-based wind energy conversion system for power quality improvement. *Energy Rep.*, 11, 9, 4587–4600, Elsevier, 2024, https://doi.org/10.1016/j.egyr.2024.04.020.
6. Ward, M.A. and Georgiou, T.K., Thermoresponsive polymers for biomedical applications. *Polymers*, 3, 3, 1215–1242, 2011.
7. Andle, J.C. and Lec, R.M., Solid piezoelectric sensor (MPS) for detecting chemical, biochemical and real measurands.2000, Google Licenses.
8. Ueno, N., Akiyama, M., Tateyama, H., Piezoelectric sensor and enter instrument comprising of same. 2006, Google Licenses.
9. Shahinpoor, M., Shape memory combination temperature sensor. 2005, Google Licenses.
10. MacGregor, R., Shape memory amalgam actuators and oversee strategies. 2003, Google Licenses.
11. Mernøe, M., Shape memory combination actuator. 2005, Google Licenses.
12. Ormerod, W. and Tipton, R.R., Piezo-electric fueled actuator operable in an electrolytic liquid. 2000, Google Licenses.
13. Ryynänen, M., Electrostrictive polymer as a mixed haptic-seal actuator.
14. Singh, P., Arora, K., Rathore, U.C., Yang, E., Joshi, G.P., Son, K.C., Performance Evaluation of Grid-Connected DFIG-Based WECS with Battery Energy Storage System Under Wind Alterations Using FOPID Controller for RSC. *MDPI J. Math. Math.*, 11, 9, 2100, 2023, https://doi.org/10.3390/math11092100.

15. Hawkins, M.J., Soon-Shiong, P., Desai, N., Protein nanoparticles as medicate merchants in therapeutic medication. *Progressed Medicate Conveyance Audits*, 60, 8, 876–885, 2008.
16. Piras, A.M., *et al.*, A modern biocompatible nanoparticle transport machine for the release of fibrinolytic drugs. *Universal Diary Pharm.*, 357, 1-2, 260–271, 2008.
17. Takagi, T., A thought of intelligent substances and the cutting-edge sports of intelligent substances in Japan in to begin with European Conference on Keen Structures and Materials, Universal Society for Optics and Photonics, 1992.
18. McCabe, J., *et al.*, Savvy substances in dentistry. *Aust. Dent. Diary*, 56, 3–10, 2011.
19. Bogue, R., Savvy materials: a assess of modern improvements. *Gathering Robotization*, 32, 1, 3–7, 2012.
20. Hensel, M.U., Performance-orientated Design and the Re-examining the Definition, Part and Performative Capacity of the Spatial and Fabric Boundaries of the Built Environment. *FormAkademisk-forskningstidsskrift Format Designdidaktikk*, 4, 1, 2011.
21. Addington, D. M. and Schodek, D. L., *Smart Materials and New Technologies: For the Architecture and Design Professions*, Routledge, 2005.
22. Addington, M. and Schodek, D., *Shrewd Materials and Innovations in Design: For the Design and Plan Callings*, Routledge, 2012.
23. Sadeghi, M.J., Masudifar, P., Nordlund, F., Circular segment dispersion for the length of the vacuum bend remelting of Ti-6Al-4V, in: *Universal Conference on Brilliantly Building and Administration*, LACSIT press, 2011.
24. Mohamed, A.S.Y., Keen substances present day innovation in engineering; within the heading of present day format worldview. *Vitality Procedia*, 115, 139–154, 2017.
25. Tune, Y., Wei, W., Qu, X., Colorimetric biosensing utilizing shrewd materials. *Prog. Mater.*, 23, 37, 4215–4236, 2011.
26. Arora, K., Joshi, G.P., Ragab, M., Rawa, M., Milyani, A.H., Mansour, R.F., Yang, E., Bilateral Contract for Load Frequency and Renewable Energy Sources Using Advanced Controller. *Comput. Mater. Continua*, 73, 2, 3165–3180, Tech Science Press, 2022, 10.32604/cmc.2022.026966.
27. Gautam, P. and Valiathan, A., Bio-smart dentistry: venturing into the long! run. *Patterns Biomater. Counterfeit Organs*, 21, 2, 94–97, 2008.
28. Ferrara, M. and M. 26. Kamila, S., *Presentation, sort and bundles of intelligent materials: an outline*, pp. 9–60, Springer, 2014.
29. Parida, B., Iniyan, S., Goic, R., An audit of sun based photovoltaic innovations. *Renewable Economical Vitality Audits*, 15, 3, 1625–1636, 2011.
30. Chang, M.-H., *et al.*, Light radiating diode's unwavering quality audit. *Microelectron. Unwavering Qual.*, 52, 5, 762–782, 2012.
31. Bai, Y., *et al.*, Ferroelectric, pyroelectric, and piezoelectric houses of a photovoltaic perovskite oxide. *Connected Mater. Sci. Lett.*, 110, 6, 063903, 2017.

32. Schodek, D. L., *Smart Materials and New Technologies: For the Architecture and Design Professions*, 205, Taylor & Francis, 2005.
33. Vainstein, E., Keen materials and developments. *Polym.-Plast. Innov. Designing*, 40, 5, 703–714, 2001.
34. Color, D., Shape memory amalgams: Towards sensible actuators. *Nat. Mater.*, 14, 8, 760, 2015.
35. Ölander, A., Electrochemical investigation of solid cadmium-gold alloys. *J. Am. Chem. Soc.*, 54, 10, 3819–3833, 1932, https://doi.org/10.1021/ja01347a064.
36. Greninger, A.B. and Mooradian, V.G., Strain change in metastable beta copper-zinc and beta copper-Ti amalgams. *AIME Trans.*, 128, 337–369, 1938.
37. Kumar, M. P., Simhachalam, D. and Ramanaia, An overview on Titanium-Niobium Shape Memory Alloys (SMA). *J. Mat. Sci. Mech. Eng. (JMSME)*, ISSN 2293-9095, 2015.
38. Buehler, W. J. and Wang, F. E., A summary of recent research on the Nitinol alloys and their potential application in ocean engineering. *Ocean Eng.*, 1, 1, 105–120, 1968, https://doi.org/10.1016/0029-8018(68)90055-0.
39. Huang, W., Shape memory combinations and their utility to actuators for deployable structures, 1998.
40. Arora, K., Kumar, A., Kamboj, V.K., Prashar, D., Shrestha, B., Joshi, G.P., Impact of Renewable Energy Sources into Multi Area Multi-Source Load Frequency Control of Interrelated Power System. *MDPI J. Math. Math.*, 9, 2, 186, 2021, https://doi.org/10.3390/math9020186.
41. Smith, G., Photochromic glasses: properties and applications. *Diary Mater. Sci.*, 2, 2, 139–152, 1967.
42. Kök, M., *et al.*, Examination of fragment alterations withinside the CuAl high-temperature frame memory amalgam with the expansion of a 3rd component. *Diary Warm Exam. Calorim.*, 133, 2, 845–850, 2018.
43. Ibrahim, M.K., *et al.*, Impact of Sn increases at the microstructure, mechanical properties, erosion and bioactivity conduct of biomedical Ti–Ta frame memory amalgams. *Diary Warm Exam. Calorim.*, 131, 2, 1165–1175, 2018.
44. Ercan, E., Dagdelen, F., Qader, I., Effect of tantalum contents on transformation temperatures, thermal behaviors and microstructure of CuAlTa HTSMAs.
45. Dagdelen, F., Kok, M., Qader, I., Impacts of Ta Substance on Thermodynamic Properties and Change Temperatures of Shape Memory NiTi Amalgam.
46. Kök, M., *et al.*, The results of cobalt variables expansion on Ti2Ni stages, thermodynamics parameters, precious stone shape and change temperature of NiTi frame memory combinations. *Eur. Phys. Diary Furthermore*, 134, 5, 197, 2019.
47. Kök, M. and Aydogdu, Y., Impact of warmth cure at the real homes of Ni–Mn–Ga amalgam. *Thermochem. Acta*, 548, 51–55, 2012.
48. Arora, K., Kumar, A., Kamboj, V.K., Prashar, D., Jha, S., Shrestha, B., Joshi, G.P., Optimization Methodologies And Testing on Standard Benchmark Functions of Load Frequency Control for Interconnected Multi Area Power

System in Smart Grids. *MDPI J. Math. Math.*, 8, 6, 980, 2020, https://doi.org/10.3390/math8060980.

49. Varzaneh, A.G., *et al.*, Impact of warmth cure on martensitic change of Ni forty-seven Mn forty Sn thirteen ferromagnetic shape memory amalgam organized through mechanical alloying. *Met. Mater. Univ.*, 21, 4, 758–764, 2015.
50. Aydoğdu, Y., *et al.*, The results of warm strategy on change temperature, gem shape and microstructure of Cu-Al-Co shape memory combination, in: *Diary of Material Science: Conference Arrangement*, IOP Distributing, 2016.
51. Ozcan, H., *et al.*, Impacts of cyclic warmth cure and getting more seasoned on superelasticity in oligocrystalline Fe-Mn-Al-Ni shape memory combination wires. *Scr. Mater.*, 134, 66–70, 2017.
52. Kök, M., *et al.*, The interchange of change temperature on NiTi shape memory amalgam through way of implies of strain and warm maturing, in: *Diary of Material Science: Conference*.
53. Yang, C., *et al.*, Impacts of thermo-mechanical cure on a Fe–30Mn–6Si frame memory combination. *Mater. Sci. Designing A*, 497, 1-2, 445–450, 2008.
54. Kumar, A., Srivastava, S., Saxena, S., Tripathi, S.L., (Ba/Pb)xSr1−xTiO3 based capacitive sensor with LaNiO3 electrode for higher tunability. *J. Mater. Sci.: Mater. Electron.*, 31, 20387–20399, 2020, https://doi.org/10.1007/s10854-020-04558-2.
55. Saud, S.N., *et al.*, Impacts of Extinguishing Media on Stage Change Characteristics and Hardness of Cu-Al-Ni-Co Shape Memory Combinations. *Diary Mater. Designing Exec.*, 24, 4, 1522–1530, 2015.
56. Yildiz, K., Balci, E., Akpinar, S., Extinguishing media impacts on martensitic change, thermodynamic and auxiliary homes of Cu–Al–Fe–Ti high-temperature shape memory amalgam. *Diary Warm Invest. Calorim.*, 129, 2, 937–945, 2017.
57. Haidar, M.A., Saud, S.N., Hamzah, E., Microstructure, Mechanical Properties, and Shape Memory Impact of Strengthened Cu-Al-Ni-xCo Shape Memory Amalgams. *Metallogr. Microstruc. Exam.*, 7, 1, 57–64, 2018.
58. Sharma, G., Ritu, Quraishi, A.M., Kattayat, S., Josey, S., Hashmi, S.Z., Ezzeldien, M., Tripathi, S.L., Alvi, P.A., Structure optimization and investigation of electrical and optical characteristics of Alq3/TAZ:Ir(ppy)3-BCP/HMTPD OLED. *Opt. Quant. Electron.*, 54, 284, 2022, https://doi.org/10.1007/s11082-022-03654-3.
59. Woodside, C.R., Lord, P.E., Nord Lund, C., Circular segment dispersion for the term of the vacuum circular segment remelting of Ti-6Al-4V. *Metall. Mater. Exchanges B*, 44, 1, 154–165, 2013.
60. Griesenauer, N., Lyon, S., Alexander, C., Vacuum Acceptance Dissolving of Titanium. *Diary Vac. Sci. Innov.*, 9, 6, 1351–1355, 1972.
61. Cronskär, M., Bäckström, M., Rännar, L.-E., Generation of custom planned hip stem prostheses– a differentiate among conventional machining and

electron pillar softening (EBM). *Quick Prototyping Diary*, 19, 5, 365–372, 2013.

62. Grummon, D.S., Shaw, J.A., Gremillet, A., Low-density open-cell froths within the NiTi framework. *Connected Mater. Sci. Lett.*, 82, 16, 2727–2729, 2003. Wang, Permeable TiNi frame memory combination with intemperate control created with the help of utilizing self-propagating excessive-temperature union. *Mater. Lett.*, 58, 11, 1683–1686, 2004.
63. Bram, M., *et al.*, Powder metallurgical manufacture forms for NiTi shape memory amalgam parts. *Mater. Sci. Designing A*, 337, 1-2, 254–263, 2002.
64. Tian, X., *et al.*, Martensitic change, mechanical assets and magnetic-field-brought almost weight of Ni–Mn–Ga amalgam created through start plasma sintering. *Diary Amalgams Compd.*, 509, 10, 4081–4083, 2011.
65. Krishna, B.V., Bose, S., Bandyopadhyay, A., Laser preparing of net-form NiTi shape memory combination. *Metall. Mater. Exch. A*, 38, 5, 1096–1103, 2007.
66. Bernard, S.A., *et al.*, Bone cell–substances intelligent and Ni particle launch of anodized equiatomic NiTi amalgam. *Acta Biomater.*, 7, 4, 1902–1912, 2011.
67. Rafi, H., *et al.*, Microstructures and mechanical houses of Ti6Al4V components created through particular laser softening and electron bar dissolving. *Diary Mater. Build. Exec.*, 22, 12, 3872–3883, 2013.
68. Otubo, J., Souza, C. A. C., Rios, P. R. and Mei, P. R., Scale-up of NiTi shape memory alloy manufacturing by means of EBM. *J. Phys. IV (Proceedings)*, 112, 975–978, EDP Sciences, 2003, https://doi.org/10.1051/jp4:20039058.
69. Haldar, A., Suwas, S. and Bhattacharjee, D., *Microstructure and Texture in Steels and Other Materials*, Springer, 2009.
70. Motemani, Y., *et al.*, Impact of cooling expense at the portion change conduct and mechanical homes of Ni-wealthy NiTi shape think back once combination. *Diary Combinations Compd.*, 469, 1-2, 164–168, 2009.
71. Alaneme, K.K. and AOkotete, E., accommodating practicality and cost-powerful frame memory amalgam options–a assessment of copper and press fundamentally based completely frame memory metal frameworks. *Designing Sci. Innov. Univ. Diary*, 19, 3, 1582–1592, 2016.
72. Samal, A., Tripathi, S.L., Mohapatra, S.K., A Journey from Bulk MOSFET to 3 nm and Beyond. *Trans. Electr. Electron. Mater.*, 21, 443–455, 2020, https://doi.org/10.1007/s42341-020-00222-y.

11

Nanotechnology for Energy Applications: Harnessing Nano and Artificial Intelligence for Sustainable Energy

Harpreet Kaur Channi[1], Ramandeep Sandhu[2], Deepika Ghai[3*] and Nimisha Singh[4]

[1]*Department of Electrical Engineering, Guru Nanak Dev Engineering College, Ludhiana, India*
[2]*School of Computer Science & Engineering, Lovely Professional University, Phagwara, India*
[3]*School of Electronics & Electrical Engineering, Lovely Professional University, Phagwara, India*
[4]*Department of Skill Development, Baba Farid College of Engineering and Technology, Bathinda, Punjab, India*

Abstract

Nanoscale application of Artificial Intelligence (AI) holds a high transformational potential for further advancement of sustainable hybrid energy solutions. Quantum dots and perovskite crystals at the nanoscale have been shown to increase efficiency in solar cells. Better performance in battery and supercapacitors by nanostructured materials such as graphene and nanowires enhances energy storage from renewable sources. AI balances these innovations by optimizing hybrid energy systems, large datasets that help predict demand and manage storage for energy while, at the same time, enhancing the generation of power by renewables such as solar and wind. Such system advancements ensure reliable integration into a grid, and integration is important in maximizing the use of renewable energy. Nanotechnology and AI coexist to make it possible for smart grids and decentralized energy systems to become a reality. Since nanosensors are coupled with analytics powered by AI, monitoring, resilience, and efficiency in the distribution of energy are significantly improved. AI also sustains alternating renewal, fossil fuel,

**Corresponding author*: money.ghai25@gmail.com

Suman Lata Tripathi, Vrinda Gupta, Sobhit Saxena, Shipra Upadhyay and Sudip Ghosh (eds.)
Micro and Nano Semiconductor Devices for Digital, Analog and Sensor Design, (195–228) © 2026 Scrivener Publishing LLC

and storage-based use of energy; ensuring it is supplied without interference with the environment. Nanomaterials contribute to lighter, energy-efficient transportation and building stock, while AI optimizes energy usage in pursuing a carbon footprint decrease and fostering sustainability. Hence, nanotechnology incorporated into AI in hybrid energy systems is promising with regards to addressing global challenges raised by the need for energy besides advancement in clean energy resilience with climate change.

Keywords: Nanotechnology, artificial intelligence, sustainable energy, nanosensors, nanomaterials

11.1 Introduction

Nanotechnology, involving science that manipulates materials in nanometric levels, is pushing the energy sector to unprecedented heights through high efficiency, low cost, and sustainability. Along with artificial intelligence which thrives on deep data analysis and optimization of complex systems, there lies much greater innovation potential here. This intense synergy is pushing the boundary of innovation in energy generation, storage, and efficiency: it solves a few of the most challenging problems in the journey to sustainable energy [1, 2]. Some examples i.e., quantum dots, nanowires, and perovskite crystals are vastly revolutionizing photovoltaic cells and their ability to capture and transform solar energy with unprecedented efficiency. Nanoscale materials, for example, graphene and silicon nanowires, are pushing forward energy storage: we designed new types of batteries and supercapacitors with higher energy densities, faster charge/discharge rates, and longer lifetimes [3]. With nanocatalysts and nanostructured membranes, fuel cells are gaining the advantage of being more efficient, and nanoscale materials enhance the conductivity of ions and improve chemical reactions. In addition, nanotechnology is innovatively making hydrogen production and storage processes efficient and even practical with materials such as Metal-Organic Frameworks and nanoporous structures in parallel with breakthrough AI algorithms to design materials and energy systems to ensure the complete utilization of their performance and minimize energy waste. In the above context, nanotechnology and AI together are designed to overcome critical challenges that involve scaling, stability, and safety. The technology opens up new avenues for new generations of sustainable energy solutions. This introduction underscores the transformative potential of harnessing nanotechnology and AI for energy applications, thus leading to further exploration of their collective impact toward achieving a sustainable energy future [4–8].

11.1.1 Overview of Nanotechnology

Nanotechnology is a transformative technology in which matter is actively manipulated at the atomic and molecular scale between 1 to 100 nanometers. At such a nanoscale, materials display specific characteristics different from those of their major counterparts, such as increased strength, chemical reactivity, and electric conductivity. Such properties allow innovative applications in medicine, electronics, environmental protection, and energy. Medicine-related applications include improved drug delivery as well as diagnostics, while better miniaturization and performance are enjoyed in the electronic sectors. But it also purifies water and clears air, and reinforces efficiency in the energy sector through better solar cell and battery power. However, issues regarding scalability, safety, and ethics need to continue to be studied and regulated. There are four primary nanomaterials: nanoparticles, nanowires, nanofilms, and nanocomposites, which are synthesized through top-down or bottom-up fabrication methods [9–11].

- **Nanoscale Dimensions:** One nanometer is a billionth of a meter. A single strand of human hair has a width of about 80,000 to 100,000 nanometers. These dimensions have a changed behavior as materials approach a nanoscale because quantum mechanical effects will have an influence on the behavior and there will be an enhanced surface area-to-volume ratio.
- **Unique Properties:** They can be characterized by strong mechanical strength, electrical and thermal conductivity, chemical reactivity, as well as optical properties. For example, the color of gold nanoparticles can differ from the conventional golden color of gold; gold nanoparticles can appear reddish or even purple, and they can also manifest some catalytic properties that are not typical for the bulk counterpart of gold.

Nanomaterials appear in all forms, with unique properties and applications. Nanoparticles are very small, generally between 1 to 100 nanometers, though used in drug delivery, for example, as catalysts. Nanowires and nanotubes constitute one-dimensional objects with exceptional electrical and thermal conductivity that makes them fit more into electronics and sensors. There are also extremely thin layers that fall under the nanofilms or nanocoatings category, showing an excellent effect on enhancing surface properties such as corrosion resistance and optical performance.

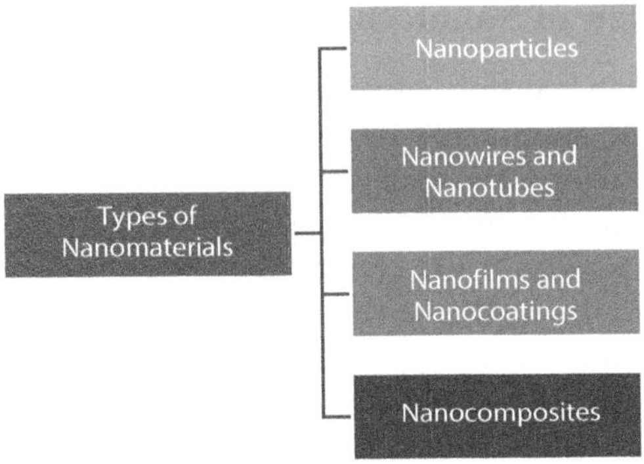

Figure 11.1 Types of nanomaterials.

Nanocomposites are a combination of combining nanoparticles with other materials to form composites that offer superior mechanical and thermal properties and are extensively used in aerospace and automotive industries [12, 13]. Figure 11.1 presents various types of Nanomaterials.

11.1.2 Applications of Nanotechnology

Because nanotechnology manipulates materials at the atomic level, it has vast scope for application in several sectors. For example, in medicine, nanotechnology supports targeted drug delivery, imaging, and regenerative treatment. Electronics greatly gain from miniaturization enabled by nanotechnology possibly with better transistors, sensors, and displays. Environmental applications are water purification, air filtration, and remediation of pollution, all based on the high surface area and reactivity of nanomaterials. For energy, nanotechnology boosts the efficiencies of solar cells, batteries, and fuel cells to provide sustainable solutions. Nanotechnology is applied in materials science to make materials stronger, lighter, and more durable, for aerospace, automobile, and construction industries [4, 14, 15]. Various applications of nanomaterials are shown in Figure 11.2.

11.1.3 Role of Artificial Intelligence in Energy Applications

AI has been considered an important sub-area in advancing the applications of energy, with a significant contribution towards sustainable hybrid energy solutions [16–19]. Its current status and relevance to key contributions to achieving sustainable energy solutions are discussed subsequently:

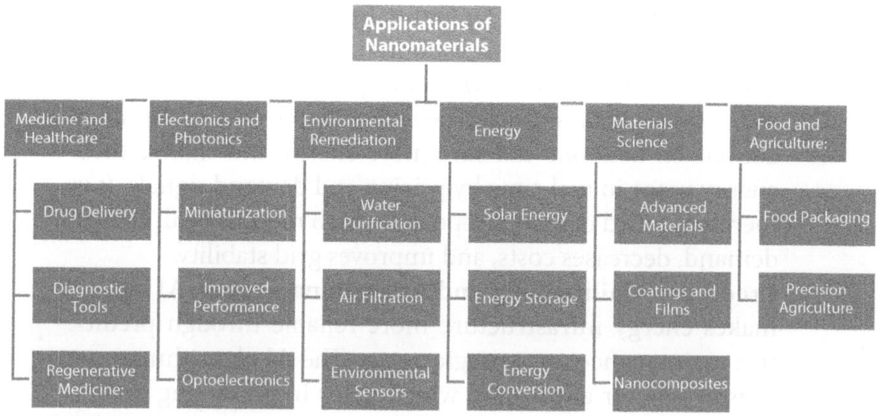

Figure 11.2 Applications of nanomaterials.

- **Optimization of Energy Generation and Distribution:** AI algorithms analyze complex data sets from various sources, including weather patterns, energy demand forecasts, and grid performance. This analysis allows AI to optimize the generation and distribution of energy from diverse sources such as solar, wind, and traditional power plants. By adjusting generation levels in real-time and predicting peak demand periods, AI helps balance supply and demand efficiently.
- **Enhancement of Energy Storage Systems:** AI-driven systems optimize the performance of energy storage solutions like batteries and supercapacitors. These systems predict energy demand fluctuations and manage energy storage to ensure continuous power supply during peak periods or when renewable sources are not available. AI algorithms also contribute to extending battery life through intelligent charge/discharge management and predictive maintenance.
- **Smart Grid Management:** AI enables the development of smart grids by integrating advanced sensors and IoT devices. These sensors monitor grid conditions in real-time, detecting faults and potential disruptions before they occur. AI algorithms analyze this data to improve grid reliability, reduce transmission losses, and enhance overall grid efficiency. Smart grids are essential for integrating renewable energy sources and managing decentralized energy production effectively.

- **Energy Efficiency and Demand Response:** AI technology optimizes the consumption patterns in households, industries, and commercial spaces. AI-based systems assess the pattern of historical data and user behavior to provide recommendations on how one can save more energy and change their energy usage in real-time by pricing and demand signals. It is therefore called demand response, which reduces peak energy demand, decreases costs, and improves grid stability.
- **Predictive Maintenance and Asset Management:** AI indeed makes energy infrastructure more reliable through predictive maintenance on strategic assets. The AI algorithms scan sensor data for anomalies, which helps in predicting potential equipment failures; therefore, proactive maintenance is scheduled at opportune times. Thus, AI gives rise to reduced operational costs and lengthened lifecycles for energy generation and distribution systems, all with the intent of minimizing the time spent idle and optimizing asset performance.

11.1.4 Importance of Sustainable Hybrid Energy Solutions

One of the most important sustainable hybrid energy solutions can prove to be critical in solving worldwide challenges in energy, mitigating climate change, and ensuring energy security. It integrates renewable sources of energy - solar, wind, hydroelectric power, and so on- with conventional energy sources, either natural gas or nuclear power [20, 21]. This is where AI fits into the picture:

- **Climate Change Mitigation:** AI increases the integration of renewable energy sources into one grid, thereby reducing the use of fossil fuels and the level of emission of greenhouse gases. It increases efficiency in the generation and storage of renewable energy sources, which promotes a low-carbon economy.
- **Energy Security:** Sustainable hybrid energy systems enhance energy resilience by diversifying energy sources and reducing dependence on imported fuels. AI-driven smart grid technologies improve grid stability and reliability, ensuring continuous power supply even during disruptions or natural disasters.
- **Economic Benefits:** AI lowers the cost of energy production and distribution by optimizing operational efficiencies and reducing wastage. Sustainable energy solutions attract

investments in clean technologies, creating jobs and stimulating economic growth in renewable energy sectors.

In conclusion, AI's ability to optimize energy systems and enhance the performance of sustainable hybrid energy solutions is pivotal for achieving a resilient, low-carbon energy future. By leveraging AI-driven innovations, we can overcome challenges posed by traditional energy systems and accelerate the transition towards a sustainable and environmentally friendly energy landscape [22, 23].

11.2 Need of the Work

The development and implementation of hybrid solar-wind-battery systems are driven by urgent global needs across environmental, economic, and social dimensions. At the core of this endeavor is the imperative to secure and diversify energy sources sustainably. These systems offer a potent solution to mitigate the escalating environmental impacts of conventional energy generation by harnessing renewable resources. By integrating solar and wind energy with battery storage, they enhance energy reliability and resilience, crucially supporting continuity during grid disruptions and emergencies. Economically, these systems promise substantial long-term savings through reduced operational costs and potential revenue from excess energy sales. They also bolster energy access, particularly in remote and underserved areas, empowering communities and promoting equitable development. Moreover, the advancement of hybrid systems drives technological innovation, fostering research and development in renewable energy technologies and smart grid solutions. Supported by favorable policies and incentives, their deployment accelerates the global transition towards sustainable energy systems, aligning with international climate goals and commitments [24, 25]. In essence, the need for hybrid solar-wind-battery systems is paramount, addressing pressing energy challenges while advancing environmental stewardship, economic prosperity, and societal equity.

11.2.1 Nanotechnology in Energy Generation

Nanotechnology has significantly advanced energy generation by enhancing the efficiency and capabilities of both solar and wind energy technologies. In solar energy, quantum dots and perovskite crystals are at the forefront of innovation. Quantum dots, due to their unique size-dependent electronic properties, enable solar cells to capture a wider range of sunlight wavelengths, thus increasing overall energy conversion efficiency. Perovskite solar cells, known for their low-cost production and high efficiency, continue to be refined with nanoscale engineering to improve stability and performance under varying environmental conditions. Nanowires and thin films play crucial roles by improving charge transport within solar cells, minimizing energy losses, and maximizing output [26].

In wind energy, nanomaterials have revolutionized turbine design by enhancing aerodynamics and structural integrity. Nanocoatings applied to turbine blades reduce surface roughness, which in turn reduces drag and improves overall efficiency. Additionally, nanomaterials embedded in turbine components increase durability, enabling turbines to withstand harsh environmental conditions and extend operational lifespans. AI optimization complements these advancements by using real-time data from sensors to adjust turbine orientation and rotor speed for optimal energy capture. Nanotechnology and AI together boost the efficiency of energy generation. This has been coupled with reduced upkeep costs and further increases the reliability of renewable energy systems. The innovations once again emphasize the revolutionary role of nanotechnology in unlocking new renewable energy technologies that are more sustainable and efficient than their predecessors [27, 28].

11.2.2 Nanotechnology in Energy Storage

Nanotechnology has revolutionized energy storage solutions, particularly in advancing the capabilities of batteries, supercapacitors, hydrogen storage, and fuel cells. In batteries and supercapacitors, nanotechnology has introduced materials like graphene and nanowires that significantly enhance energy storage capacity, charge/discharge rates, and lifespan. Graphene, with its exceptional electrical conductivity and mechanical strength, improves electrode performance, enabling faster charging times and higher energy densities. Nanowires, due to their high surface area-to-volume ratio, enhance ion transport and storage within electrodes, increasing energy storage efficiency. AI-driven battery management systems leverage real-time data to optimize charging cycles, monitor battery health, and predict performance, ensuring optimal energy usage and extending battery lifespans [29, 30].

Nanotechnology is fundamental to the preparation of nanomaterial catalysts used in hydrogen storage and fuel cells. Further improvement of catalytic activity can be made by increasing the rate of reaction and minimizing energy losses in the chemical reaction processes involved in hydrogen storage and fuel cell operations. Metal nano-particles, such as platinum and palladium, are highly efficient catalysts with high surface area and catalytic activity, thus facilitating the conversion of hydrogen into electricity with less waste. AI incorporation enhances the efficiencies of fuel cells by finding optimal operating parameters concerning environmental conditions and energy demand, thus optimizing the use of fuels and energies achieved. Nanotechnology significantly enhances energy storage technologies as large steps in increasing the efficiency, reliability, and sustainability of energy storage are possible [31, 32]. Thus, new applications combined with AI-driven systems assisted by nanomaterials open up avenues toward better, more robust, and scalable solutions for energy storage, including renewable energy, that must take forward the transition to a better-quality energy future.

11.2.3 AI Optimization in Hybrid Energy Systems

AI optimization in hybrid energy systems is revolutionizing energy management across smart grids and decentralized networks by leveraging advanced nanotechnology and sophisticated algorithms. In smart grids, nanosensors enable precise monitoring of grid parameters such as voltage, frequency, and energy flow at micro-levels, providing real-time data crucial for AI analytics. AI algorithms analyze this data to predict energy demand patterns with high accuracy, optimizing energy generation schedules, and dynamically adjusting grid operations to maintain stability and efficiency [33].

Nanotechnology critically contributes to upgrading the potentialities of energy storage and production in decentralized energy networks, such as microgrids. For instance, energy storage systems are substituted by the application of nanomaterials like graphene, nanowires, or quantum dots inside their structures to improve efficiency, raise energy density, and prolong battery life cycles. The AI-driven microgrid energy management system uses predictive analytics to maximize the efficiency of energy distribution, favors renewable energy sources, and minimizes reliance on fossil fuel to its minimum peaks. This will improve the resilience of the grid, reduce operating costs while keeping the carbon footprint low, and maintain decentralized networks with their sustainability and reliability intact. Furthermore, AI facilitates adaptive decision-making in response to changing environmental conditions and consumer behavior, ensuring optimal energy usage and grid stability. By harnessing the synergy between AI's computational power and nanotechnology's material innovations, hybrid energy systems are advancing toward greater efficiency, sustainability, and resilience in meeting the global demand for clean energy solutions [34, 35].

11.3 Hybrid Renewable System: Case Study of Rural Region

Renewable energy, and sustainability for that matter, will answer the mountain of global challenges the earth is facing: climate change, environmental degradation, and energy security. Greenhouse gas emissions in terms of air pollution are reduced through renewable sources of solar, wind, hydroelectric, and biomass. This way, climate change will not happen, and negative impacts on ecosystems, economies, and health will be less. In addition, these energy sources enhance energy independence and security through reduced dependence on imported fuels. Sustainability is responsible resource management to meet the needs of the present while looking to future generations. Renewable energy and sustainability save ecosystems, support biodiversity, generate more economic growth create employment opportunities, and reduce pollution that will improve public health. It is the key to a resilient, equitable global energy system [36–38].

11.3.1 Objectives of the Chapter

The objectives of the chapter on nanotechnology for energy applications: Harnessing nano and AI for sustainable energy are enlisted as follows:

- Develop accurate AI models using linear regression to predict daily energy production from solar and wind sources.
- Perform comprehensive techno-economic analysis to evaluate the financial performance of the hybrid solar-wind-battery system.
- Conduct sensitivity analysis by varying solar irradiance inputs to assess the system's robustness and resilience to environmental changes.

11.4 Methodology

The methodology for this chapter involves a systematic approach to develop and evaluate a hybrid solar-wind-battery energy system, integrating nanotechnology and artificial intelligence (AI) for sustainable energy solutions. Figure 11.3 shows a flowchart of the methodology for integrating nanotechnology and AI for the hybrid solar-wind-battery energy system. Firstly, synthetic daily data for solar irradiance, wind speed, and electricity

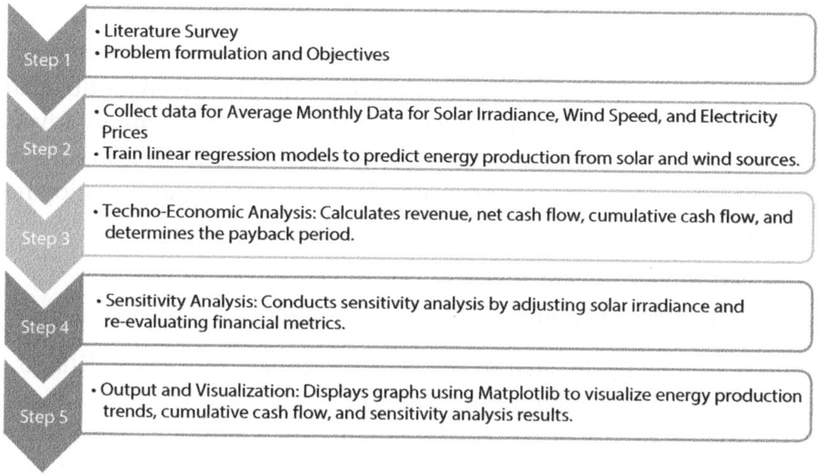

Figure 11.3 Flowchart of the methodology for integrating nanotechnology and AI for the hybrid solar-wind-battery energy system.

prices is generated to simulate realistic environmental conditions and market dynamics. This synthetic data serves as input for training AI models in the next step.

Secondly, AI modeling techniques, specifically linear regression, are employed to train predictive models for estimating daily energy production from solar and wind sources. These models utilize the synthetic data to learn patterns and correlations between input variables (solar irradiance, wind speed) and energy output, aiming to optimize accuracy in energy production forecasting. Following AI modeling, a techno-economic analysis is conducted to assess the financial feasibility and economic viability of the hybrid energy system. This analysis includes calculating revenue projections, net cash flow, cumulative cash flow over time, and determining the payback period. These metrics provide insights into the financial sustainability of the project and inform decision-making regarding investment and operational strategies. Additionally, sensitivity analysis is performed by adjusting solar irradiance inputs to evaluate the system's robustness and resilience to environmental variables. This step examines how variations in solar irradiance impact key financial metrics derived from the techno-economic analysis, such as cumulative cash flow and payback period.

Finally, output and visualization techniques, leveraging tools like Matplotlib, are utilized to present findings visually. Graphs and charts depict energy production trends, cumulative cash flow dynamics, and sensitivity analysis results, providing stakeholders with clear insights into the system's performance under different scenarios. In summary, this methodology integrates data generation, AI modeling, techno-economic analysis, sensitivity testing, and visualization to comprehensively evaluate and optimize a hybrid solar-wind-battery energy system. It aims to advance sustainable energy solutions by leveraging technological innovations and rigorous analysis to support informed decision-making and deployment strategies.

11.4.1 Location Details

Mamupur is a village located in the state of Punjab, India near Kharar, as shown in Figure 11.4. Mamunpur village code according to 2011 Census 039053: The route to Kharar tehsil of Sahibzada Ajit Singh Nagar district in its 6 km approach to the sub-district headquarters at Kharar and 15 km from the district headquarters at Sahibzada Ajit Singh Nagar. Since 2009, the gram panchayat for the village is Mamupur. Mamunpur comprises 192 hectares with a population of 682; males amount to 370 and females 312. Table 11.1 presents the details of the villages. The literacy percentage is 78.01% wherein male literates amount to 81.62% and female literates amount to 73.72%. There are approximately 126 houses in the village.

Table 11.2 showcases the average monthly values for solar irradiance, wind speed, and electricity prices over a year (1st January 2023 to 1st January 2024).

Figure 11.4 Geographical location of the Mamupur village.

Table 11.1 Village details.

Gram Panchayat	**Mamupur**
Block/Tehsil	Kharar
District	Sahibzada Ajit Singh Nagar
State	Punjab
Pincode	140307
Area	192 hectares
Population	682
Households	126
Latitude	24° 39' 49.382" N
Longitude	93° 54' 22.568 E

Table 11.2 Average monthly data for solar irradiance, wind speed, and electricity prices.

Month	**Avg solar irradiance (W/m^2)**	**Avg wind speed (m/s)**	**Avg electricity price ($/kWh)**
January	500.23	9.85	0.095
February	505.12	10.01	0.092
March	498.76	9.92	0.094
April	503.45	10.05	0.093
May	499.88	9.87	0.096
June	502.31	9.95	0.097
July	501.64	9.89	0.098
August	497.92	9.81	0.099
September	504.18	10.12	0.094
October	499.76	9.98	0.093
November	502.09	10.00	0.095
December	498.55	9.93	0.096

Solar irradiance, measured in watts per square meter (W/m^2), fluctuates slightly around 500 W/m^2, indicating relatively stable sunlight exposure throughout the year. Wind speeds, averaging around 10 m/s, show minor monthly variations, with the highest in September at 10.12 m/s and the lowest in August at 9.81 m/s. Electricity prices also vary marginally, with the highest price in August at $0.099/kWh and the lowest in February at $0.092/kWh. This data is crucial for analyzing the performance and financial viability of hybrid solar-wind-battery systems, highlighting the impact of seasonal changes on energy production and economic metrics.

Plotting average solar irradiance over the months on a line graph can show the slight fluctuations in sunlight exposure throughout the year as shown in Figure 11.5. The line will oscillate around 500 W/m^2, indicating relatively consistent irradiance with minor monthly variations.

A line graph for average wind speed will display how wind speeds vary slightly from month to month, peaking in September at 10.12 m/s and reaching a minimum in August at 9.81 m/s. Figure 11.6 graph helps visualize the wind resource availability across different months.

Plotting the average electricity prices over the months can show price fluctuations, with the highest price in August at $0.099/kWh and the lowest in February at $0.092/kWh. Figure 11.7 shows a line graph that can make it easier to compare the prices month-to-month, highlighting the slight variations in electricity costs.

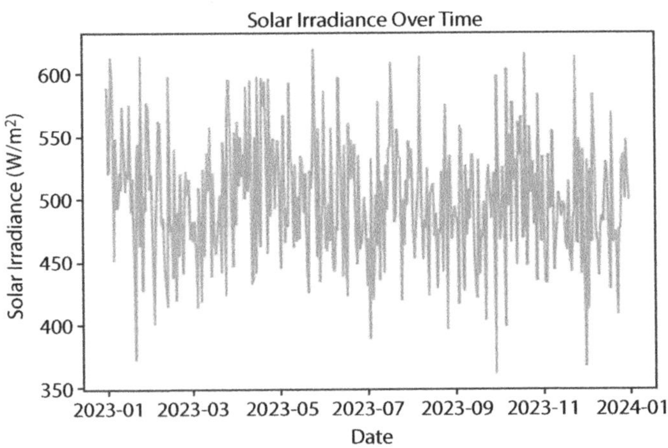

Figure 11.5 Solar irradiance over a time period of 1 year.

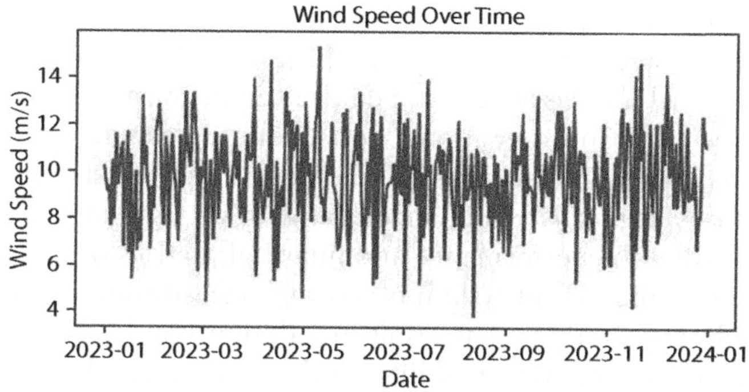

Figure 11.6 Solar irradiance over a time period of 1 year.

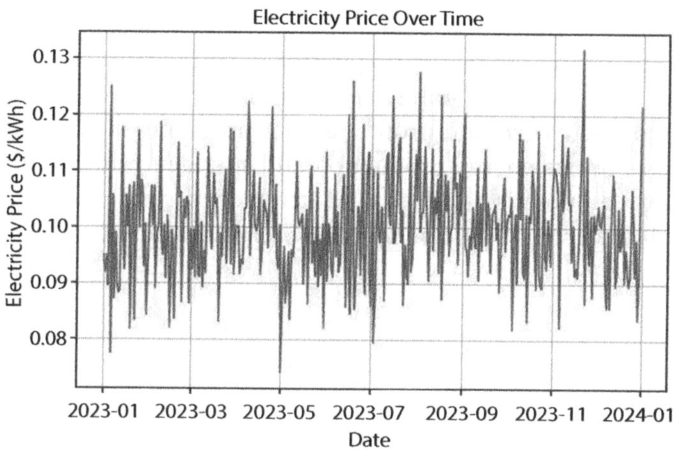

Figure 11.7 Electricity price over a time period of 1 year.

11.4.2 System Designing and Modeling

System designing and modeling are critical steps in developing an efficient and effective hybrid solar-wind-battery energy system. This process begins with identifying the specific energy requirements and site conditions, such as average solar irradiance, wind speed, and local electricity prices. Based on this data, suitable components are selected, including solar panels, wind turbines, and battery storage units, each optimized for the local conditions. Advanced modeling techniques, often incorporating nanotechnology enhancements, are employed to maximize energy capture and conversion efficiency. AI algorithms are integrated to predict energy production, manage storage, and optimize the overall system performance. Detailed financial models are developed to evaluate the economic viability, considering capital and operational costs, revenue generation, and payback periods. Sensitivity analysis will be performed to account for variable conditions, e.g., solar irradiance changing or wind speed changing, in understanding the performance and financial metrics of the system. A robust design of the hybrid system is accomplished through iterative simulations and optimizations that result in a design that is "between" technical efficiency, economic feasibility, and environmental sustainability. This integrated solar-wind-battery system is discussed in this section and Figure 11.8. The key components consist of solar panels, wind turbines, and battery storage in an optimal configuration to use renewable energy generation with optimal storage. Solar and wind energy being promisingly complementary, a continuous reliable supply will be ensured. In this way, AI-driven algorithms manage the operation of solar panels, wind turbines, and storage batteries as real-time data inputs combined with weather forecasts and electricity demand profiles are fed in. The system is constantly learning to optimize energy production and storage towards optimal efficiency, reduced operational costs, and heightened resilience for the total system against fluctuations in renewable energy availability.

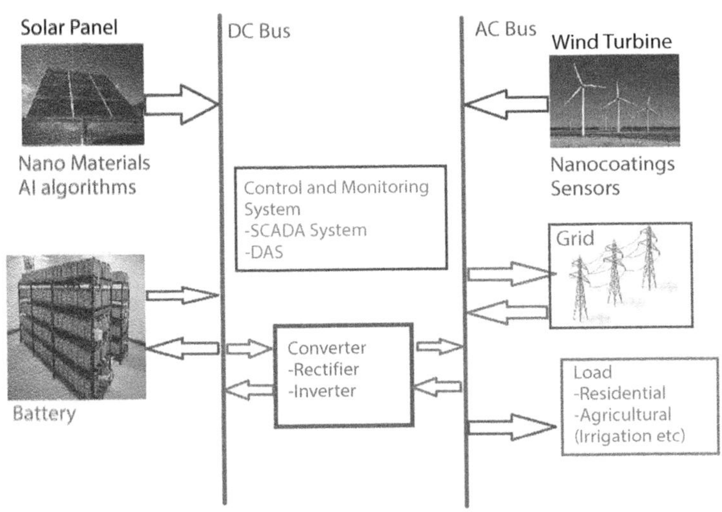

Figure 11.8 Schematic diagram of solar wind battery hybrid system.

(a) **Solar Panels:** Solar panels are equipped with advanced nanomaterials that enhance their efficiency and durability. These nanomaterials are engineered at the molecular level to improve light absorption and electron transport, thereby increasing the conversion efficiency of solar energy into electricity. AI algorithms integrated into the solar panel systems optimize their orientation and tilt angles throughout the day based on real-time weather conditions and solar irradiance data. This optimization ensures maximum energy production and minimizes losses, contributing to the overall performance and reliability of the solar energy subsystem [39–41].

(b) **Wind Turbines:** Wind turbines feature nanocoatings and sensors designed to enhance their aerodynamic performance and operational efficiency. Nanocoatings applied to turbine blades reduce frictional drag and improve airflow, increasing the turbine's energy conversion efficiency. Sensors embedded in the turbine blades and tower monitor wind speed, direction, and turbulence, providing real-time data for AI algorithms to adjust blade pitch angles and rotor speeds. These AI-driven adjustments optimize power output while mitigating structural fatigue, ensuring the safe and efficient operation of the wind energy subsystem [42, 43].

(c) **Battery Storage:** Advanced nanotechnology-enabled battery storage utilizes high energy density, faster charging rates, and longer cycle life batteries. Nanomaterials enhance ion conductivity and stability in the electrodes of the battery to enhance overall battery performance and reliability. The AI algorithm controlling the battery management system monitors storage levels of energy, predicts demand patterns, and allows optimization of cycles of charging and discharging. This intelligent management ensures efficient energy storage and distribution, supporting grid stability and maximizing renewable energy utilization during peak demand periods [44–46].

(d) **Grid:** Solar-wind-battery hybrid systems require grid integration for a reliable and efficient supply of energy. Renewable energy sources will thus be connected to the power grid, ensuring the free flow of electricity from generation sites to consumers [47, 48]. More advanced control systems, like SCADA, monitor this flow and balance generation with energy demand. Nanosensors and AI-driven analytics can enhance integration in grids by providing real-time information about the production, consumption, and reliability of energy. This allows for precise demand forecasting and optimized energy distribution. Grid integration also buffers supply-demand fluctuations, improving resilience and maximizing renewable energy use, reducing fossil fuel reliance, and supporting a sustainable energy future [49–51].

(e) **Control and Monitoring System:** In a hybrid solar-wind-battery energy setup, the SCADA system and data acquisition modules comprise the control and monitoring system. SCADA has centralized control, allowing the operators to review the performance of their systems. They can modify parameters according to the requirements and act upon errors in real-time. Data acquisition modules monitor information coming from sensors around the system such as on solar panels, wind turbines, and batteries. AI algorithms process this data to optimize performance, ensure safety, and support maintenance decisions. Such hybrid cells utilize nanotechnology and artificial intelligence to maximize energy efficiency, reliability, and sustainability in optimizing renewable energy generation, storage, and management toward cleaner and more resilient energy infrastructure [52–54].

11.4.3 Main Outcomes of Hybrid Solar-Wind-Battery System

The hybrid solar-wind-battery system consistently generated reliable energy from both solar and wind sources, with AI predictions demonstrating high accuracy [55]. The financial analysis revealed positive net cash flow and a feasible payback period, demonstrating economic viability. Sensitivity analysis showed the system's robustness to varying solar irradiance levels, maintaining positive financial outcomes. Integrating nanotechnology and AI-enhanced efficiency, contributing to quicker payback periods. Overall, the system proved to be a cost-effective, environmentally friendly solution for sustainable energy generation.

- **Energy Production:** Figure 11.9 illustrates the daily energy production from solar and wind sources, providing insights

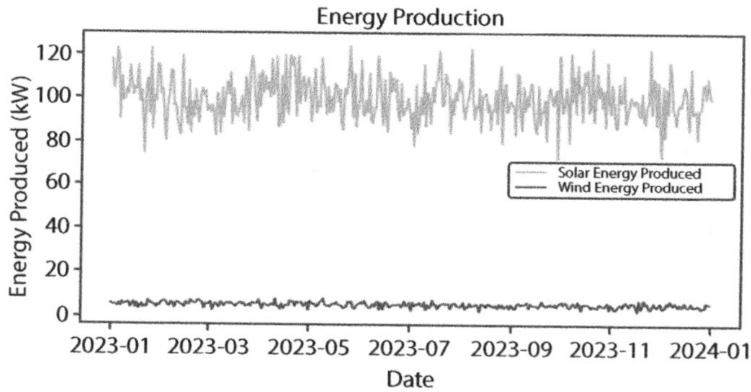

Figure 11.9 Energy production over a period of 1 year.

into the system's renewable energy generation capabilities. The orange line represents the electrical energy generated by solar panels based on solar irradiance, which is a measure of the intensity of sunlight received. The blue line represents the electrical energy generated by wind turbines based on wind speed, indicating the kinetic energy converted into electricity. This graph shows how energy production varies daily due to changes in weather conditions, specifically solar irradiance and wind speed. It also demonstrates how solar and wind energy production can complement each other over time [56, 57]. For instance, solar energy production may peak during daylight hours, while wind energy production might vary based on wind patterns throughout the day and night.

- **Revenue vs. Net Cash Flow:** Comparative analysis of revenue generation versus net cash flow provides valuable financial insights. While revenue trends reflect seasonal variations in energy production and electricity prices, net cash flow accounts for operational costs, maintenance expenditures, and initial investments. Results indicate that despite fluctuating revenue streams, efficient management strategies and cost-effective measures contribute to sustainable financial performance over time. Payback period calculations reveal a positive outlook, suggesting a feasible return on investment within the projected timeframe, supported by steady revenue streams and effective cost management practices. Figure 11.10 compares the revenue generated from selling

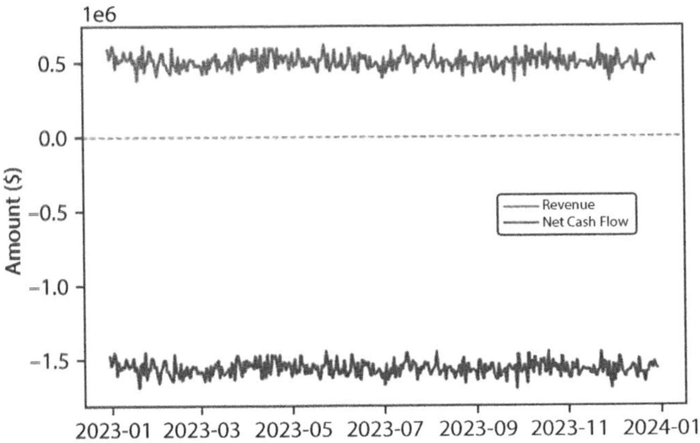

Figure 11.10 Revenue vs net cash flow.

electricity with the net cash flow, which considers both revenue and operational costs. Green Line represents the total income generated from selling the electricity produced by the hybrid system to the grid or consumers. Blue Line reflects the financial health of the system by accounting for all costs (investment, maintenance, operational expenses) subtracted from revenue. It also indicates whether the system is generating sufficient revenue to cover its operating costs and achieve profitability. Positive net cash flow indicates profitability, while negative net cash flow suggests losses. This figure also helps identify when the system starts generating positive cash flow, which is crucial for assessing the financial viability and Return on Investment (ROI).

- **Cumulative Cash Flow:** Figure 11.11 displays the cumulative cash flow over time, illustrating the overall financial performance of the hybrid system from inception. The purple line represents the sum of net cash flow values accumulated over the analyzed period. It provides a long-term perspective on the system's financial performance, showing whether the cumulative cash flow is increasing positively (indicating profitability) or fluctuating and also indicates how quickly the initial investment and ongoing costs are being recouped. A positive cumulative cash flow suggests that the system is on track to recover its investments and start generating profits.

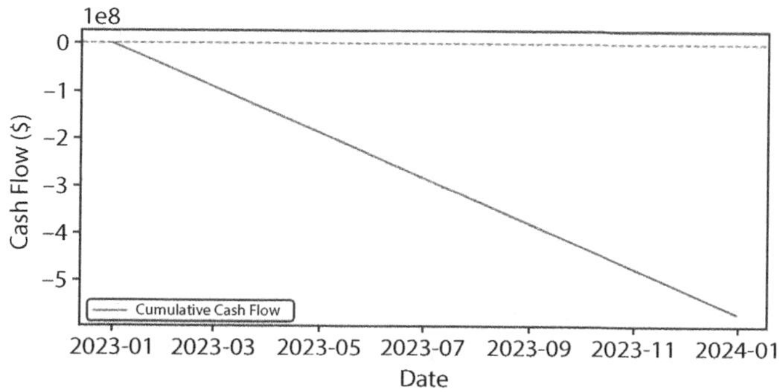

Figure 11.11 Cumulative cash flow.

- **Sensitivity Analysis - Cumulative Cash Flow:** Sensitivity analysis explores the impact of key variables, such as solar irradiance and wind speed, on cumulative cash flow. Variations in these factors demonstrate their influence on energy production and subsequent financial outcomes. Higher solar irradiance levels, for instance, lead to increased energy generation and revenue, thereby positively impacting cumulative cash flow. Conversely, fluctuations in wind speeds may necessitate adjustments in operational strategies to mitigate potential revenue losses. Insights from sensitivity analysis underscore the importance of adaptive management approaches and robust contingency plans to optimize performance under varying environmental conditions. Figure 11.12 demonstrates the impact of sensitivity analysis on the cumulative cash flow, specifically how changes in input parameters affect financial outcomes. Orange Line shows the cumulative cash flow when a specific parameter (e.g., solar irradiance or wind speed) is adjusted or varied. It provides insights into how sensitive the system's financial performance is to changes in key parameters. For example, increasing solar irradiance might lead to higher energy production and revenue, thereby positively influencing cumulative cash flow. It also helps assess potential risks and uncertainties associated with environmental variability (e.g., weather patterns affecting solar irradiance and wind speed) and their impact on financial outcomes.

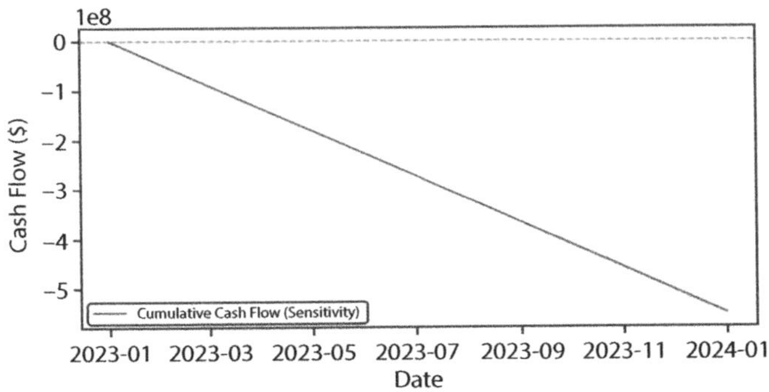

Figure 11.12 Sensitivity analysis-cumulative cash flow.

- **Electricity Price:** Figure 11.13 shows the graphical representation of electricity price variations over time highlighting their impact on revenue generation and financial viability. Analysis reveals periods of price volatility and seasonal fluctuations, influencing overall revenue streams and profitability. Strategies to mitigate risks associated with price variability include diversifying energy sources, leveraging energy storage capabilities, and adopting flexible pricing mechanisms. The findings emphasize the role of market dynamics in shaping operational strategies and underscore

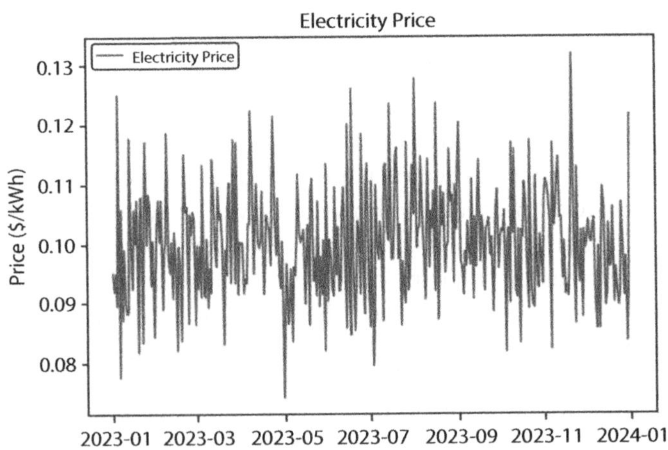

Figure 11.13 Electricity price of electricity generated.

the importance of resilience in navigating price uncertainties to maintain sustainable financial performance. It displays the variation in electricity prices over time, influencing the revenue generated from selling electricity produced by the hybrid system. Red Line represents the cost per kilowatt-hour (kWh) of electricity over the analyzed period. It illustrates how changes in electricity prices directly affect the revenue generated from energy sales. Higher electricity prices generally lead to increased revenue, while lower prices may reduce revenue potential. It reflects market conditions and influences decisions related to energy sales strategies [58–60]. It informs pricing strategies and financial forecasting based on prevailing market rates.

- **Energy Production Comparison:** Figure 11.14 shows daily energy production from solar and wind sources revealing significant insights into system performance. Analysis shows that solar energy production correlates closely with solar irradiance levels throughout the year, demonstrating peak production during summer months and reduced output in winter. Wind energy production, influenced by wind speed variations, exhibits higher output during periods of increased wind speeds, such as in autumn and winter months. These

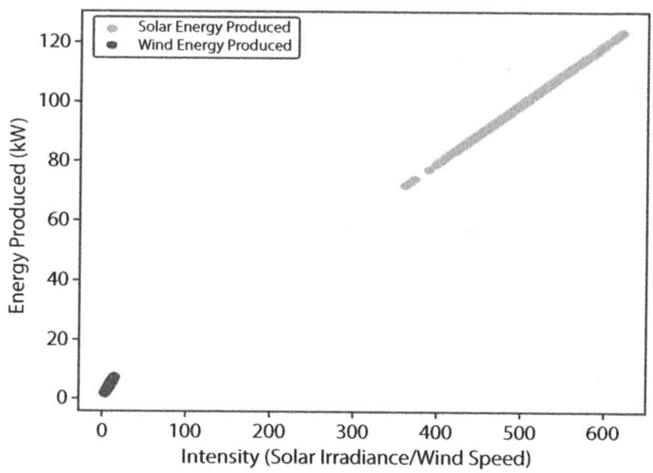

Figure 11.14 Energy production comparison.

findings underscore the importance of weather conditions in determining renewable energy yield and highlight the complementary nature of solar and wind sources in hybrid systems. It compares energy production based on solar irradiance and wind speed intensities, providing insights into their relative contributions to overall energy generation. The X-axis (Intensity) represents solar irradiance (W/m^2) or wind speed (m/s), which are measures of the respective resource intensities. The Y-axis (Energy Produced) represents the corresponding energy produced (kW) by solar panels or wind turbines. It helps optimize resource utilization by understanding the relationship between resource intensity (solar irradiance or wind speed) and energy production. It informs decisions on system design and operational strategies to maximize energy output. It guides strategic decisions based on resource availability and predictability. For instance, if solar irradiance is more consistent in a particular region, it may influence the decision to prioritize solar energy investments over wind energy.

These graphs collectively provide a comprehensive analysis of the hybrid solar-wind-battery energy system, encompassing both technical and financial aspects. They offer insights into daily energy production patterns, financial viability, sensitivity to external factors (such as solar irradiance and electricity prices), and market dynamics. This information is essential for optimizing system performance, making informed investment decisions, and achieving sustainable and efficient energy generation. Adjustments and further analysis can be tailored based on specific project requirements and objectives.

11.5 Results and Discussion

The results focus on several different aspects of integration dynamics, thus showing the potential to increase energy production, achieve maximum financial performance, and reduce the negative environmental impacts of nanotechnology in hybrid systems. Overall insights regarding the ongoing trends in energy production are characterized by the complementary nature of solar and wind sources, and therefore, the importance of hybrid systems in maximizing renewable energy yield. Financial analysis highlights the importance of cost-effective management and strategic investments in ensuring long-term financial sustainability. Sensitivity analysis is useful for insights into how hybrid systems perform under varying environmental conditions concerning firm designs and adaptive strategy towards possible improvement in efficiency and revenue generation.

In summary, the work points up the transformative power of nanotechnology toward a sustainable energy future, offering pivotal insights for policymakers, industry stakeholders, and researchers alike. Future research directions may further optimize nanomaterial applications to strive for even better predictive modeling capabilities and leverage synergies with emerging technologies to accelerate the transition toward a low-carbon energy future. This structured approach will allow for a comprehensive discussion of results, focusing on key findings, their implications, and avenues for further research and development in nanotechnology-driven energy systems.

11.5.1 Energy Efficiency and Sustainability

Energy efficiency and sustainability benefit significantly from the integration of nanotechnology in building construction and transportation sectors. In energy-efficient buildings, nanotechnology introduces lightweight materials like aerogels and nanoporous insulation. Aerogels, known for their high surface area and low thermal conductivity, provide exceptional insulation properties, reducing heat transfer and thus lowering heating and cooling energy requirements. Nanoporous insulation materials enhance thermal resistance without adding bulk, optimizing building envelope efficiency [61, 62].

AI plays a critical role in building energy management systems by leveraging data from sensors and IoT devices to optimize heating, ventilation, and air conditioning (HVAC) operations. AI algorithms analyze occupancy patterns, weather forecasts, and real-time energy consumption data to adjust heating, ventilation, and air conditioning settings dynamically. This approach ensures optimal energy usage while maintaining occupant comfort levels, contributing to significant energy savings and reduced carbon emissions. In transportation, nanotechnology enhances energy efficiency through the development of advanced nanocomposites. Carbon nanotubes and graphene-based nanocomposites are used to manufacture lightweight vehicle components that offer high strength-to-weight ratios. These materials reduce vehicle weight, improving fuel efficiency and extending battery life in electric vehicles (EVs). Additionally, nanomaterials enhance the durability and performance of vehicle batteries, enabling faster charging times and longer operational lifespans [63–65].

AI-driven optimizations further enhance energy efficiency in transportation systems. For instance, AI algorithms in EVs manage battery usage by predicting driving patterns, optimizing regenerative braking, and adjusting power distribution to maximize range and efficiency. AI also plays a role in route optimization, considering factors like traffic conditions, road gradients, and charging station availability to minimize energy consumption during travel. Overall, the combination of nanotechnology and AI is revolutionizing energy efficiency across buildings and transportation. These innovations not only reduce energy consumption and operational costs but also contribute to environmental sustainability by lowering greenhouse gas emissions and promoting the adoption of renewable energy sources in urban infrastructure and mobility solutions [66–70].

11.5.2 Challenges and Future Directions

One of the primary challenges facing nanotechnology in energy applications is scalability and commercialization. While nanomaterials show promising enhancements in energy efficiency and performance, scaling up production to meet industrial demands remains a significant hurdle. Manufacturing processes for nanomaterials often require precise control over particle size, shape, and composition, which can be costly and complex. Additionally, integrating nanotechnology into existing energy infrastructure requires robust testing and validation to ensure reliability and compatibility with operational conditions. Overcoming these scalability challenges requires continued investment in research and development, as well as collaboration between academia, industry, and government agencies to streamline production processes and facilitate widespread adoption.

Another critical consideration is the safety and environmental impact of nanomaterials used in energy applications. Nanoparticles can exhibit unique properties due to their size and surface characteristics, which may pose potential risks to human health and the environment. Proper handling, disposal, and monitoring of nanomaterials throughout their lifecycle are essential to mitigate these risks. Regulatory frameworks and guidelines governing the use of nanotechnology in energy sectors are still evolving, necessitating comprehensive risk assessments and adherence to safety protocols. Research efforts are ongoing to understand the long-term effects of nanomaterial exposure and develop sustainable practices for their safe deployment in energy applications.

11.5.3 Regulatory and Ethical Considerations

The regulatory landscape surrounding nanotechnology in energy applications is complex and varies globally. Regulatory frameworks must balance innovation with safety, environmental protection, and ethical considerations. Issues such as intellectual property rights, data privacy, and equitable access to nanotechnology-enabled energy solutions are also important considerations. Ethical concerns regarding the equitable distribution of benefits and potential socioeconomic impacts of nanotechnology adoption in energy sectors require careful deliberation and stakeholder engagement. Collaborative efforts among policymakers, industry leaders, researchers, and advocacy groups are essential to establish clear guidelines, promote responsible innovation, and address societal concerns related to nanotechnology in energy applications.

Looking ahead, addressing these challenges will pave the way for exciting future directions in nanotechnology for energy applications. Continued advancements in nanomaterial synthesis and characterization will enhance efficiency and performance across renewable energy technologies. Innovations in nanotechnology-enabled energy storage solutions, smart grids, and energy-efficient materials will drive the transition toward sustainable energy systems. Moreover, interdisciplinary research initiatives focusing on nanotechnology, AI, and materials science will unlock new possibilities for enhancing energy conversion, storage, and management capabilities.

Thus, academia-industry-government agency collaboration will hasten the transition of nanotechnology innovations into worldwide energy infrastructures. Education and outreach activities will thus become a critical avenue to raise people's awareness regarding the potential positive, and adverse effects and ethical issues related to nanotechnology for energy applications. It will meet the challenges and fulfill future opportunities leading to the horizon of a revolutionized energy landscape and support environmental sustainability as it guides the world into a low-carbon economy.

11.6 Conclusion

The most important element in changing the energy sector towards efficiency, sustainability, and resilience in energy generation, storage, and management is nanotechnology. Among the explorations of materials made in this study, it was clear that nanomaterials, specifically quantum dots, graphene, or nanowires, are changing solar panels, wind turbines, batteries, and energy-efficient buildings. This allows for the possibility of higher energy conversion efficiency, much-extended battery life span, and far less impact on the environment compared to the ones with traditional technologies.

However, the development of nanotechnology in mainstream energy applications will not be easy work. Scaling up, safety and the complexities of the regulatory environment pose difficult challenges to the responsible deployment of nanotechnology. For now, scaling up is an issue: R&D investment will need to go into processes and cost cuts. Safety considerations demand rigorous testing and monitoring to ensure limited risk related to nanomaterials. Regulatory frameworks had to become accommodating towards innovation but uncompromising on the integrity of public health, environmental integrity, and ethical standards.

Future success will rely on the collaborative action of researchers, policymakers, industry leaders, and stakeholders to tackle the challenges identified above. Therefore, crucial instrumental action toward the actualization of the full potential of nanotechnology in energy applications lies in overcoming these challenges. Future directions in nanotechnology for energy applications include advancing AI-driven optimizations, energy storage technologies, and extending the application of nanomaterials toward smart grids and transportation. Those innovations will hasten the global shift toward a sustainable energy future, reducing greenhouse gas emissions while making energy more secure and fostering economic development. Nanotechnology is this transformative force that can change the future of energy by using nanomaterials and AI-driven technologies to address efficiency, environmental sustainability, and access. With coordinated efforts and responsible stewardship, nanotechnology will unlock cleaner, more efficient, and resilient energy systems that provide and safeguard the benefit to society for future generations.

References

1. Simpa, P., Solomon, N.O., Adenekan, O.A., Obasi, S.C., Nanotechnology's potential in advancing renewable energy solutions. *Eng. Sci. Technol. J.*, 5, 5, 1695–1710, 2024.
2. Malik, S., Muhammad, K., Waheed, Y., Nanotechnology: A revolution in modern industry. *Molecules*, 28, 2, 661, 2023.

3. Adir, O., Poley, M., Chen, G., Froim, S., Krinsky, N., Shklover, J., Schroeder, A., Integrating artificial intelligence and nanotechnology for precision cancer medicine. *Adv. Mater.*, 32, 13, 1901989, 2020.
4. Ghai, D., Tripathi, S.L., Saxena, S., Chanda, M., Alazab, M. (Eds.), *Machine Learning Algorithms for Signal and Image Processing*, John Wiley & Sons, 2022.
5. Zhang, P., Guo, Z., Ullah, S., Melagraki, G., Afantitis, A., Lynch, I., Nanotechnology and artificial intelligence to enable sustainable and precision agriculture. *Nat. Plants*, 7, 7, 864–876, 2021.
6. Kulkarni, S., Bhat, S., Moritz, C.A., Architecting for artificial intelligence with emerging nanotechnology. *ACM J. Emerging Technol. Comput. Syst.*, 17, 3, 1–33, 2021.
7. Aleksenskiy, A.E., Eydelman, E.D., Vul, A.Y., Deagglomeration of detonation nanodiamonds. *Nanosci. Nanotechnol. Lett.*, 3, 1, 68–74, 2011.
8. Pedroso-Santana, S., Sarabia-Saínz, A., Fleitas-Salazar, N., Santacruz-Gómez, K., Acosta-Elías, M., Pedroza-Montero, M., Riera, R., Deagglomeration and characterization of detonation nanodiamonds for biomedical applications. *J. Appl. Biomed.*, 15, 1, 15–21, 2017.
9. Bayda, S., Adeel, M., Tuccinardi, T., Cordani, M., Rizzolio, F., The history of nanoscience and nanotechnology: from chemical–physical applications to nanomedicine. *Molecules*, 25, 1, 112, 2019.
10. Baig, N., Kammakakam, I., Falath, W., Nanomaterials: A review of synthesis methods, properties, recent progress, and challenges. *Mater. Adv.*, 2, 6, 1821–1871, 2021.
11. Shukla, P.K., Sandhu, J.K., Ahirwar, A., Ghai, D., Maheshwary, P., Shukla, P.K., Multiobjective Genetic Algorithm and Convolutional Neural Network Based COVID-19 Identification in Chest X-Ray Images. *Math. Probl. Eng.*, 2021, 1, 7804540, 2021.
12. Mekuye, B. and Abera, B., Nanomaterials: An overview of synthesis, classification, characterization, and applications. *Nano Select*, 4, 8, 486–501, 2023.
13. Altammar, K.A., A review on nanoparticles: characteristics, synthesis, applications, and challenges. *Front. Microbiol.*, 14, 1155622, 2023.
14. Haleem, A., Javaid, M., Singh, R.P., Rab, S., Suman, R., Applications of nanotechnology in medical field: a brief review. *Global Health J.*, 7, 2, 70–77, 2023.
15. Omietimi, H.B., Afolalu, S.A., Kayode, J.F., Monye, S.I., Lawal, S.L., Emetere, M.E., An overview of nanotechnology and its application, in: *E3S Web of Conferences*, vol. 391, EDP Sciences, p. 01079, 2023.
16. Karduri, R.K., The role of artificial intelligence in optimizing energy systems. *Int. J. Adv. Res. Manage. Archit. Technol. Eng.*, 2019.
17. Ahmad, T., Zhu, H., Zhang, D., Tariq, R., Bassam, A., Ullah, F., Alshamrani, S.S., Energetics Systems and artificial intelligence: Applications of industry 4.0. *Energy Rep.*, 8, 334–361, 2022.

18. Ahmad, T., Zhang, D., Huang, C., Zhang, H., Dai, N., Song, Y., Chen, H., Artificial intelligence in sustainable energy industry: Status Quo, challenges and opportunities. *J. Clean. Prod.*, 289, 125834, 2021.
19. Ashfaq, A., Kamran, M., Rehman, F., Sarfaraz, N., Ilyas, H.U., Riaz, H.H., Role of artificial intelligence in renewable energy and its scope in future, in: *2022 5th International Conference on Energy Conservation and Efficiency*, pp. 1–6, IEEE, 2022, March.
20. Javaid, A., Sajid, M., Uddin, E., Waqas, A., Ayaz, Y., Sustainable urban energy solutions: Forecasting energy production for hybrid solar-wind systems. *Energy Convers. Manage.*, 302, 118120, 2024.
21. Razmjoo, A. and Davarpanah, A., Developing various hybrid energy systems for residential application as an appropriate and reliable way to achieve energy sustainability. *Energy Sources Part A*, 41, 10, 1180–1193, 2019.
22. Dincer, I. and Acar, C., Smart energy solutions with hydrogen options. *Int. J. Hydrogen Energy*, 43, 18, 8579–8599, 2018.
23. Natividad, L.E. and Benalcazar, P., Hybrid renewable energy systems for sustainable rural development: Perspectives and challenges in energy systems modeling. *Energies*, 16, 3, 1328, 2023.
24. Farooq, T., Agili, B., He, M., Hybrid energy system using wind, solar & battery storage system. *J. Multidiscip. Eng. Sci. Tech.*, 11, 3, 16648–51, 2024.
25. Chen, L., Hu, Y., Wang, R., Li, X., Chen, Z., Hua, J., Yap, P.S., Green building practices to integrate renewable energy in the construction sector: a review. *Environ. Chem. Lett.*, 22, 2, 751–784, 2024.
26. Serrano, E., Rus, G., Garcia-Martinez, J., Nanotechnology for sustainable energy. *Renew. Sustain. Energy Rev.*, 13, 9, 2373–2384, 2009.
27. Abdin, Z., Alim, M.A., Saidur, R., Islam, M.R., Rashmi, W., Mekhilef, S., Wadi, A., Solar energy harvesting with the application of nanotechnology. *Renew. Sustain. Energy Rev.*, 26, 837–852, 2013.
28. Ganguly, S., Banerjee, D., Kargupta, K., Nanotechnology and nanomaterials for new and sustainable energy engineering, in: *Proceedings of the International Conference Nanomaterials: Applications and Properties*, vol. 1, no. 4, Sumy State University Publishing, pp. 04NEA09–04NEA09, 2012.
29. Sarno, M., Nanotechnology in energy storage: The supercapacitors, in: *Studies in Surface Science and Catalysis*, vol. 179, pp. 431–458, Elsevier, 2020.
30. Deng, J., Lu, X., Liu, L., Zhang, L., Schmidt, O.G., Introducing rolled-up nanotechnology for advanced energy storage devices. *Adv. Energy Mater.*, 6, 23, 1600797, 2016.
31. Awad, A., Ahmed, W., Waleed, M., Nanotechnology for energy storage, in: *Emerging Nanotechnologies for Renewable Energy*, pp. 495–516, Elsevier, 2021.
32. Julien, C., Mauger, A., Vijh, A., Zaghib, K., Julien, C., Mauger, A., Zaghib, K., Nanotechnology for Energy Storage, in: *Lithium Batteries: Science and Technology*, pp. 461–497, 2016.

33. Thirunavukkarasu, M., Sawle, Y., Lala, H., A comprehensive review on optimization of hybrid renewable energy systems using various optimization techniques. *Renew. Sustain. Energy Rev.*, 176, 113192, 2023.
34. Zahraee, S.M., Assadi, M.K., Saidur, R., Application of artificial intelligence methods for hybrid energy system optimization. *Renew. Sustain. Energy Rev.*, 66, 617–630, 2016.
35. Maghami, M.R. and Mutambara, A.G.O., Challenges associated with Hybrid Energy Systems: An artificial intelligence solution. *Energy Rep.*, 9, 924–940, 2023.
36. Asrari, A., Ghasemi, A., Javidi, M.H., Economic evaluation of hybrid renewable energy systems for rural electrification in Iran—A case study. *Renew. Sustain. Energy Rev.*, 16, 5, 3123–3130, 2012.
37. Babatunde, O.M., Adedoja, O.S., Babatunde, D.E., Denwigwe, I.H., Off-grid hybrid renewable energy system for rural healthcare centers: A case study in Nigeria. *Energy Sci. Eng.*, 7, 3, 676–693, 2019.
38. Neves, D., Silva, C.A., Connors, S., Design and implementation of hybrid renewable energy systems on micro-communities: A review on case studies. *Renew. Sustain. Energy Rev.*, 31, 935–946, 2014.
39. Mozumder, M.S., Mourad, A.H.I., Pervez, H., Surkatti, R., Recent developments in multifunctional coatings for solar panel applications: A review. *Sol. Energy Mater. Sol. Cells*, 189, 75–102, 2019.
40. Soomar, A.M., Hakeem, A., Messaoudi, M., Musznicki, P., Iqbal, A., Czapp, S., Solar photovoltaic energy optimization and challenges. *Front. Energy Res.*, 10, 879985, 2022.
41. Al-Shahri, O.A., Ismail, F.B., Hannan, M.A., Lipu, M.H., Al-Shetwi, A.Q., Begum, R.A., Soujeri, E., Solar photovoltaic energy optimization methods, challenges and issues: A comprehensive review. *J. Clean. Prod.*, 284, 125465, 2021.
42. Farrar, N.O., Ali, M.H., Dasgupta, D., Artificial intelligence and machine learning in grid connected wind turbine control systems: A comprehensive review. *Energies*, 16, 3, 1530, 2023.
43. Pandit, R., Astolfi, D., Hong, J., Infield, D., Santos, M., SCADA data for wind turbine data-driven condition/performance monitoring: A review on state-of-art, challenges and future trends. *Wind Eng.*, 47, 2, 422–441, 2023.
44. Wong, K. and Dia, S., Nanotechnology in batteries. *J. Energy Res. Technol.*, 139, 1, 014001, 2017.
45. Deshpande, V.S. and Talele, M.N., Nanotechnology enabled hybrid power system suitable for batteries in hybrid electric vehicle, in: *3rd International Conference on Advances in Electrical, Electronics, Information, Communication and Bio-Informatics*, pp. 33–35, IEEE, 2017, February.
46. Abdalla, A.M., Elnaghi, B.E., Hossain, S., Dawood, M., Abdelrehim, O., Azad, A.K., Nanotechnology utilization in energy conversion, storage and efficiency: a perspective review. *Adv. Energy Convers. Mater.*, 1, 1, 30–54, 2020.

47. Maaruf, M., Khan, K., Khalid, M., Robust control for optimized islanded and grid-connected operation of solar/wind/battery hybrid energy. *Sustainability*, 14, 9, 5673, 2022.
48. Channi, H.K., Sandhu, R., Giri, N.C., Singh, P., Syam, F.A., Comparison of power system flow analysis methods of IEEE 5-bus system. *Indones. J. Electr. Eng. Comput. Sci.*, 34, 1, 11–18, 2024.
49. Channi, H.K., Handa, K., Bhullar, M., Sandhu, R., Kaur, H., Singh, P., Renewable Distributed Generation for Smart Highway Lighting, in: *2024 11th International Conference on Reliability, Infocom Technologies and Optimization (Trends and Future Directions)*, pp. 1–6, IEEE, 2024, March.
50. Channi, H.K., Sandhu, R., Kaur, H., Singh, P., Gaba, S., Kaur, J., An Approach for Load Flow and Load Dispatch Analysis of IEEE-9 Bus System, in: *2023 2nd International Conference on Ambient Intelligence in Health Care*, pp. 01–06, IEEE, 2023, November.
51. Madhavi, K.B., Tripathi, S.L., Saxena, S., Ghai, D., Raj, B., Design and Analysis of 18 nm Multichannel FinFET as Biosensor for Detection of Biological Species. *Silicon*, 15, 14, 6313–6322, 2023.
52. Hentea, M., Improving security for SCADA control systems. *Interdiscip. J. Inf. Knowl. Manage.*, 3, 73, 2008.
53. Bailey, D. and Wright, E., *Practical SCADA for industry*, Elsevier, 2003.
54. Merchán, D.F., Peralta, J.A., Vazquez-Rodas, A., Minchala, L.I., Astudillo-Salinas, D., Open source SCADA system for advanced monitoring of industrial processes, in: *2017 International Conference on Information Systems and Computer Science*, pp. 160–165, IEEE, 2017, November.
55. Ma, T., Yang, H., Lu, L., A feasibility study of a stand-alone hybrid solar–wind–battery system for a remote island. *Appl. Energy*, 121, 149–158, 2014.
56. Khare, V., Nema, S., Baredar, P., Solar–wind hybrid renewable energy system: A review. *Renew. Sustain. Energy Rev.*, 58, 23–33, 2016.
57. Dincer, F., The analysis on photovoltaic electricity generation status, potential and policies of the leading countries in solar energy. *Renew. Sustain. Energy Rev.*, 15, 1, 713–720, 2011.
58. Gabrielli, P., Wüthrich, M., Blume, S., Sansavini, G., Data-driven modeling for long-term electricity price forecasting. *Energy*, 244, 123107, 2022.
59. Wang, D., Gryshova, I., Kyzym, M., Salashenko, T., Khaustova, V., Shcherbata, M., Electricity price instability over time: Time series analysis and forecasting. *Sustainability*, 14, 15, 9081, 2022.
60. Karakatsani, N.V. and Bunn, D.W., Forecasting electricity prices: The impact of fundamentals and time-varying coefficients. *Int. J. Forecast.*, 24, 4, 764–785, 2008.
61. Gharib, N. and Ahmed, B., Paths Towards Energy Efficient Buildings Using Nano Architecture, in: *International Conference on Civil and Architecture Engineering*, 2020, July, vol. 13, no. 13, Military Technical College, pp. 1–12.

62. Chausali, N., Saxena, J., Prasad, R., Nanotechnology as a sustainable approach for combating the environmental effects of climate change. *J. Agric. Food Res.*, 12, 100541, 2023.
63. Delogu, M., Zanchi, L., Dattilo, C.A., Pierini, M., Innovative composites and hybrid materials for electric vehicles lightweight design in a sustainability perspective. *Mater. Today Commun.*, 13, 192–209, 2017.
64. Deng, J., Bae, C., Denlinger, A., Miller, T., Electric vehicles batteries: requirements and challenges. *Joule*, 4, 3, 511–515, 2020.
65. Hofer, J., Wilhelm, E., Schenler, W., Optimal lightweighting in battery electric vehicles. *World Electr. Veh. J.*, 5, 3, 751–762, 2012.
66. Elvin, G., Building green with nanotechnology, in: *Technical Proceedings of the 2007 Cleantech Conference and Trade Show*, pp. 167–170, CRC Press, 2019, August.
67. Pokrajac, L., Abbas, A., Chrzanowski, W., Dias, G.M., Eggleton, B.J., Maguire, S., Mitra, S., Nanotechnology for a sustainable future: Addressing global challenges with the international network4sustainable nanotechnology. *ACS Nano*, 15, 12, 18608–18623, 2021.
68. Tripathi, S.L., Kumar, A., Rao, K.S., Mudimela, P.R. (Eds.), *Nanodevices for Integrated Circuit Design*, John Wiley & Sons, 2023.
69. Kumar, A., Srivastava, S., Saxena, S., Tripathi, S.L., (Ba/Pb) x Sr1− x TiO3 based capacitive sensor with LaNiO3 electrode for higher tunability. *J. Mater. Sci.: Mater. Electron.*, 31, 22, 20387–20399, 2020.
70. Rani, S., Ghai, D., Kumar, S., Reconstruction of simple and complex three dimensional images using pattern recognition algorithm. *J. Inf. Technol. Manage.*, 14, 235–247, 2022.

12

Implementation and Analysis of Various Full Adder Configuration Using Cadence Virtuoso

Spoorthi S.P.[1*], Bharathi S.H.[2], Shipra Upadhyay[3] and Chaithanya D.J.[4]

[1]*Electronics and Communication Engineering, Atria Institute of Technology, Bangalore, India*
[2]*School of Electronics and Communication Engineering, REVA University, Bangalore, India*
[3]*Electronics and Communication Engineering Department, M S Ramaiah Institute of Technology, Bangalore, India*
[4]*Electronics and Communication Engineering, Vidyavardhaka College of Engineering, Mysore, India*

Abstract

The adder is a widely used circuit in digital circuits and a fundamental part of all Arithmetic and Logic Units (ALUs) in microprocessors and microcontrollers. Arithmetic and logical processes, including addition and number multiplication, are carried out by adders among many other tasks. The efficiency of the adders will impact the efficiency of the ALU, which will impact the microprocessor and microcontroller performance. Each technology node's trade-offs are examined in the study; smaller nodes often provide faster speeds and lower power consumption, but they may also have higher leakage currents and more complexity. The goal of this research is to determine the most effective adder designs for particular applications by examining the behavior of adders in these technologies. The development of high-performance, low-power digital system. For this reason, adder design is crucial to digital design. In this research, different types of full adders in 180nm, 90nm, and 45nm technology are analyzed.

Keywords: CMOS, GPDK, full adder, transistor

**Corresponding author*: spoorthisp28@gmail.com

Suman Lata Tripathi, Vrinda Gupta, Sobhit Saxena, Shipra Upadhyay and Sudip Ghosh (eds.) Micro and Nano Semiconductor Devices for Digital, Analog and Sensor Design, (229–242) © 2026 Scrivener Publishing LLC

12.1 Introduction

Adders are vital parts in a lot of computer arithmetic techniques. Apart from computers, they are also utilized by processors for multiple purposes, including the program counter incrementation. It is viewed as adaptable and is now essentially necessary for the bulk of cognitive programs. Adder-based circuits lower the number of excess transistors.

The selection of adders varies depending on the program, even though they are essential in terms of speed, power consumption, and area usage [1]. A digital circuit's primary requirements are:

- low power dissipation
- low power consummation
- low area
- high speed.

12.2 Adders

Adders are sophisticated circuits found in gadgets that are used to add two or more binary digital numbers. Adders are commonly used in large-scale juggling and logically calculating rationale units across various platforms and processing types. They are also employed in various processor parts for tasks such as determining addresses, table lists, and comparable tasks. A distinct kind of complement was utilized for the binary number system to add negative integers [2].

12.2.1 Half Adder

Here, a basic, practical computer foundation is integrated with two logic components. Figure 12.1 shows logic diagram of half adder.

The most basic adder circuit is the half adder. The half adder is a type of combinational arithmetic circuit that adds two numbers and outputs a carry bit (c) and a sum bit (s). A half adder is a type of combination circuit used to add two bits. Sum and carry bits are the output variables, whereas augend and addend bits are the input variables. The two input bits are A and B.

The need for low power, fast speed, low cost, and small footprint in digital circuit design is growing these days [3]. As we continue to reduce the size, leakage current during standby plays a significant role in small scale

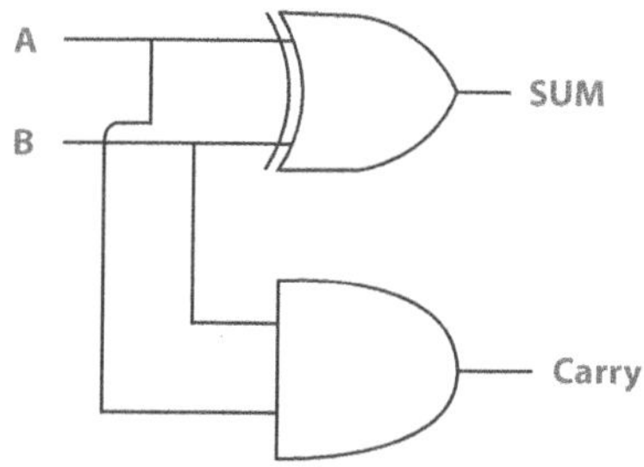

Figure 12.1 Logic diagram of half adder [4].

digital design [2]. The quantity, size, and switching activity of the transistors utilized determine how much power is consumed. In a single publication, several complete adder circuits are constructed and examined using the cadence tool.

There are three binary numbers added in the full adder circuit. It can be utilized to carry out arithmetic and logical operations in the processor chip's ALU.

12.2.2 Full-Adder

Three binary bits, usually denoted as A, B, and C in (carry-in), can be added using a complete adder, a basic digital circuit. Two outputs are generated by the complete adder: the carry-out (Cout) and the sum (S). When it comes to multi-bit addition, the carry out represents the sum output and indicates the most important carry out to the next higher order bit position. When building a complete adder in CMOS,

- Power consummation
- Speed
- Layout
- Noise margin

Power Consumption: The issue if conservation of power is quite sensitive especially where the device is battery operated. Conservative power utilization enhances energy competency as well as reduces heat radiation.
Speed: The speed of the adder entails how fast it can perform some operations. In high-performance applications, there is need for faster processing that is enabled by faster adders.

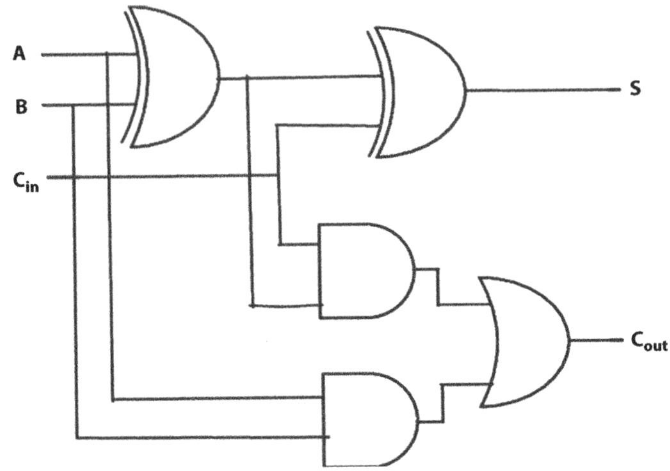

Figure 12.2 Logic diagram of full-adder [4].

Layout: Therefore, the layout design must be very compact to reduce the total surface area of the chip. Also, a compact configuration plays a role in enhancing circuit performance since parasitic problems are minimized. Noise Margin: In case of reliability in operation, there has to be a suitable noise margin. To be able to synthesize the correct result, a stable noise margin filters out noise and voltages in the different circuits. Figure 12.2 presents a full adder logic block and Table 12.1 shows truth table of full adder.

12.2.3 Ripple Carry Adder

A ripple carry adder is a series of full adders used to add multi-bit binary values. Each full adder's carry-in and carry-out are linked to the subsequent higher-order full adder's carry-in. Figure 12.3 depicts ripple carry adder.

12.2.4 Carry Look-Ahead Adder

By lowering the carry propagation latency, a carry look-ahead adder increases adding speed. To calculate the carry signals beforehand based on the inputs, it employs extra logic.

Table 12.1 Truth table of full adder.

Input			output	
A	B	Cin	Sum	Cout
0	0	0	0	0
0	0	1	1	0
0	1	0	1	0
0	1	1	0	1
1	0	0	1	0
1	0	1	0	1
1	1	0	0	1
1	1	1	1	1

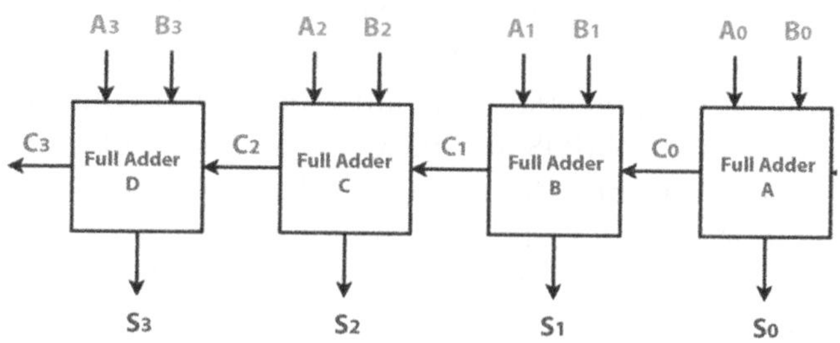

Figure 12.3 Logic diagram ripple carry adder [2].

12.2.5 Carry-Save Adder

When adding many operands, like in multiplication operations, a carry-save adder is utilized. By preserving carries rather than propagating them instantly, it adds three or more numbers at once.

12.2.6 Parallel Prefix Adders

Tree topologies are used by parallel prefix adders, like the Brent-Kung adder and the Kogge-Stone adder, to carry out addition operations. By utilizing parallel processing, these adders are made to reduce the carry propagation time as much as possible.

12.2.7 Serial Adder

A flip-flop is used to store the carry in a serial adder, which uses a single complete adder to add binary integers one bit at a time. In comparison to parallel adders, this kind of adder is slower and simpler.

12.3 CMOS Implementation of Adders

To construct the required logic gates (AND, OR, XOR, NOT), adders in CMOS technology are implemented utilizing complementary NMOS and PMOS transistors. When included into VLSI (Very Large Scale Integration) circuits, CMOS adders are a good choice due to their low power consumption and great noise immunity. Table 12.2 gives information about the number of transistors in different full adder design.

12.3.1 28T Full Adder

A 28-transistor full adder is an example of an ideal design that combines moderate complexity and a respectably high speed, even if CMOS technology highlights the potential capabilities of such circuits. This design generates the required NMOS and PMOS logic gates to finish the adder operation. In actuality, this is a thorough explanation of the machine's circuit and operation. Figure 12.4 shows 28T full-adder.

The optimized design of the 28-transistor full adder effectively integrates CMOS transistors to carry out the addition operation with the least amount of complexity and power consumption. This design is very helpful in low-power digital systems and VLSI circuits, two applications where area and power are crucial.

Table 12.2 Comparison table of full adder.

Full-adder type	**No. of transistor**
28T Design	28
20T	20
14T	14
10T	10

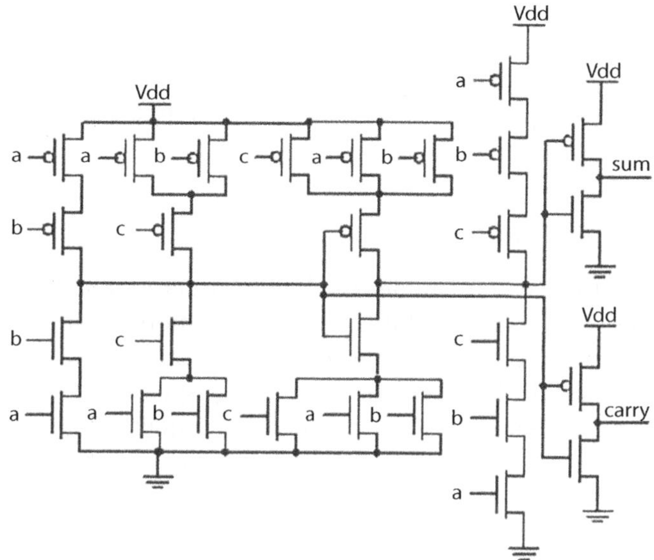

Figure 12.4 28T Full-adder [1].

12.3.2 14T Full Adder

Another type of full adder circuit is called the 14-transistor (14T) full adder which has the capability to add two single-bit binary numbers and the carry input using fourteen transistors. In terms of power consumption and transistor count when compared with other complete adder designs, this design is highly efficient. Figure 12.5 shows 14T full-adder circuit.

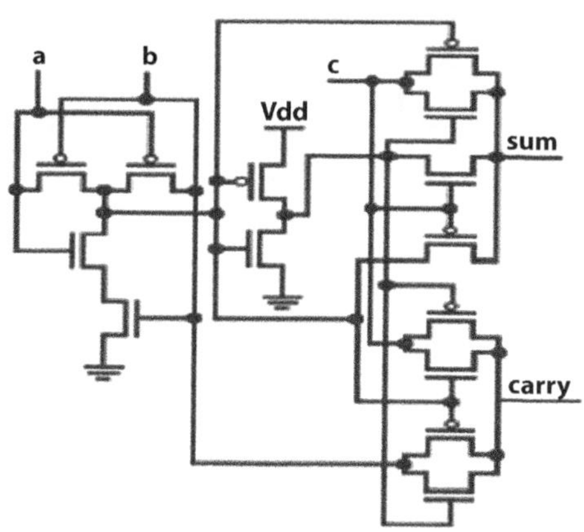

Figure 12.5 14T Full-adder [4].

In the CMOS technology, the most efficient design is the 14T complete adder where few numbers of transistor are used to perform addition of binary numbers. This design is critical for high-speed, low power applications; the speed up and power both directly increases or decreases with number of transistors.

12.3.3 20T Full Adder

The 20-transistor (20T) full adder is a higher-order CMOS full adder circuit than the 14T version, and it is significantly more complex. Typically, this design aims at optimizing the number of transistors while aiming at enhanced modifiable parameters including; speed, power usage, and noise margin. To attain these advancements, a sequential arrangement of Boolean logic gate composed of CMOS is employed. Figure 12.6 shows 20T full-adder circuit.

The application of the 20-transitor complete adder architecture, is therefore, efficient in view of the factors of speed, power consumption, and area. This basic form of the gate uses CMOS logic to perform the XOR, AND and OR functions gated with 20 transistors to obtain the required functions. Resource-intensive applications which compete between resource usage and velocity might undergo an advantage from such architectures.

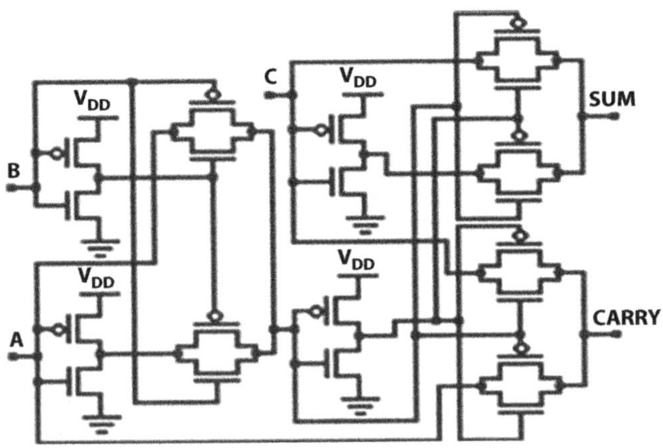

Figure 12.6 20T Full-adder [4].

12.3.4 10T Full-Adder

The challenge in a 10T complete adder is to use only 10 transistors to accomplish the required logic functions (Sum and Carry-out). This frequently entails using transmission gates or passes transistor logic in conjunction with inventive circuit design. An example of a standard 10T complete adder design is shown below. Figure 12.7 shows a 10 transistor full-adder.

For binary addition, the 10-transistor complete adder is a highly optimized design that works well in situations where space and power efficiency are crucial. It makes use of effective CMOS design strategies to obtain the required logic functionalities at the lowest possible transistor cost [7].

12.3.5 8T Full-Adder

Optimizing the circuit to carry out the required logic functions with the fewest possible transistors is the first step towards building an 8T complete adder [8].

The highly optimized 8-transistor full adder uses the fewest possible transistors to accomplish binary addition. The 8T complete adder accomplishes the required functionality with fewer transistors by utilizing effective CMOS design approaches, such as pass transistor logic. Because of this, it's perfect for situations where power consumption and size are important limitations. Figure 12.8 shows 8T full-adder circuit.

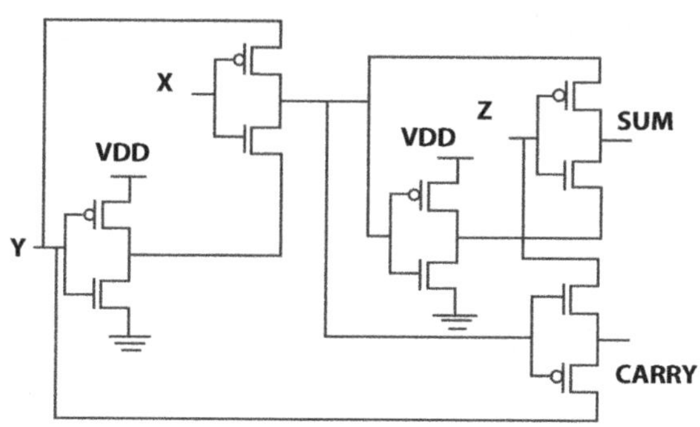

Figure 12.7 10T Full-adder [4].

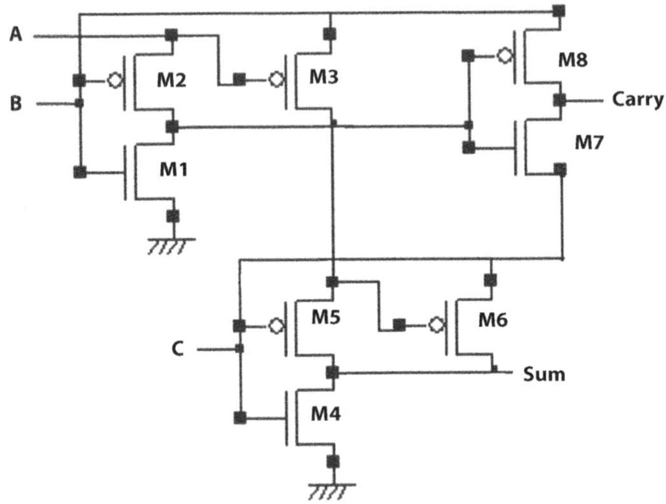

Figure 12.8 8T Full-adder [4].

12.4 Implementation of Full-Adder 28T and 14T

Different typical full adder implementation approaches, including 28T, 14T, and so on, are available. T stands for the total number of transistors in the full adder architecture. Power dissipation becomes increasingly important as scaling towards miniaturization [10].

Battery-driven integrated circuits and applications have shorter battery lives because of leakage power, which also causes significant power dissipation during standby mode. Transistors with various and variable threshold voltages are employed in the design to overcome the aforementioned problem. High threshold transistors tend to be a little bit slower than low threshold transistors, which are used in designs where operating speed is crucial [11–13].

12.4.1 Simulation Results

Using Cadence Virtuoso software, existing full adder circuits are created utilizing 45nm, 90nm, and 180nm technology GPDK tool. The voltage supply is set at 1.2 volts, and the threshold voltage varies depending on the technology. The power consumption and delay of each existing full adder are compared.

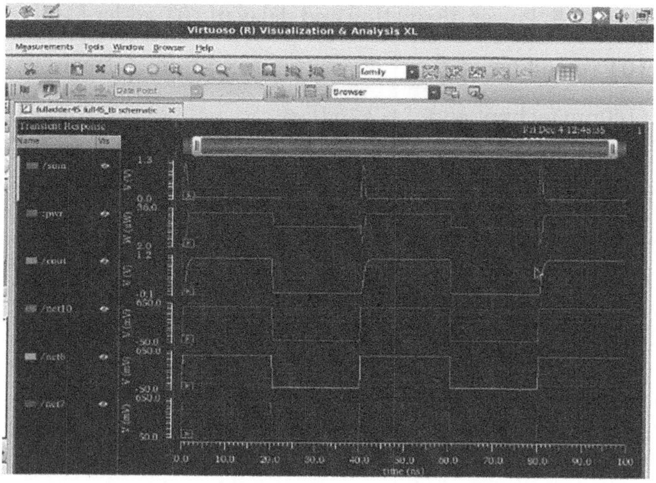

Figure 12.9 Input output waveforms 28T adder in 45nm implementation [4].

Simulation results for a traditional 28-T complete adder employing 180-, 90-, and 45-nm wavelengths. Figure 12.9 shows 28T full-adder waveform on Cadence. Tables 12.3, 12.4, and 12.5 show power dissipation and delay in 28T, 14T and 10T respectively.

Figure 12.10 and Figure 12.11 show 28T adder transient and DC response respectively.

Table 12.3 Power and delay in 28T adder.

28T full adder	Power consumption	Delay
(180nm technology)	24.51E-6	39.04E-9
(90nm technology)	4.432E-6	39.02E-9
(45nm technology)	132E-9	30.01E-12

Table 12.4 Power dissipation and delay 14T adder [4].

14T full adder	Power consumption	Delay
(180nm technology)	4.0E-3	40.02E-9
(90nm technology)	261.1E-6	70.01E-12
(45nm technology)	23.02E-6	10E-12

Table 12.5 Power dissipation and delay in 10T adder [4].

10T full adder	Power consumption	Delay
(180nm technology)	20.62E-6	6.20E-12
(90nm technology)	21.1E-6	4E-18
(45nm technology)	1.4.02E-6	10E-15

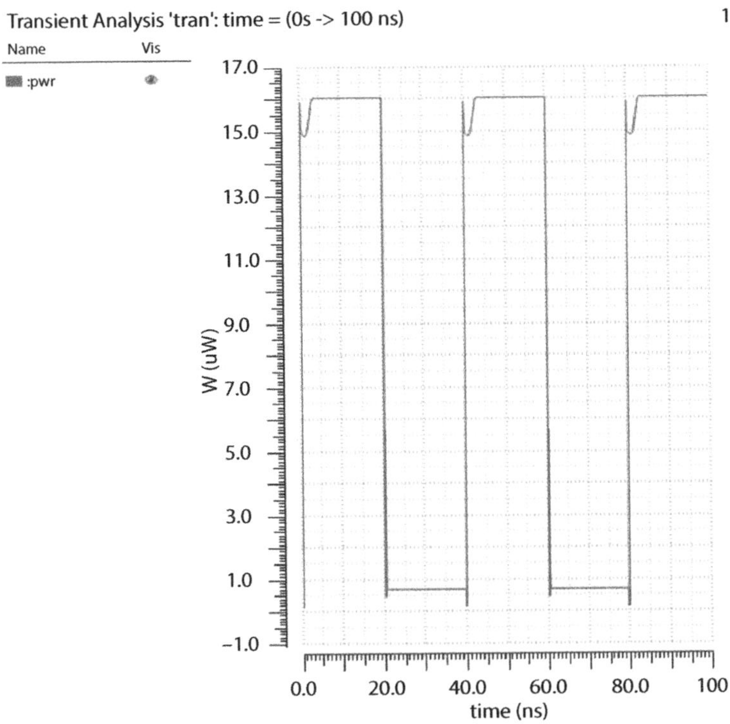

Figure 12.10 Power of 28T full-adder in 180nm technology [4].

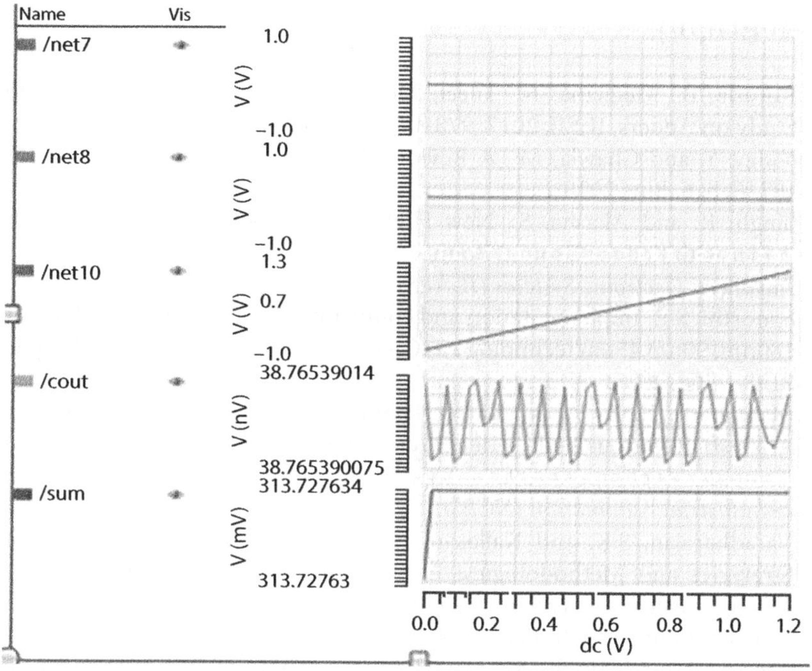

Figure 12.11 DC analysis of 28T [4].

12.5 Conclusion

The 180nm, 90nm, and 45nm technologies were used in the Cadence Virtuoso design of the current full adder. Threshold voltage is 1.8 volts and voltage supply is 2.5 volts for 180nm technology. The 45nm technology uses less power because of reduced supply voltage, lower capacitance, and innovations that mitigate leakage, making it a better choice for power-sensitive applications compared to the older 180nm and 90nm technology. In comparison to the other two technologies, 45nm technology uses a relatively little amount of power.

As technology advances in descending order, the delay decreases. Additional analysis can be done by taking the area into account, comparing the entire adder with various technologies, such as 65 nm, 32 nm, and 22 nm, and evaluating the performance of various digital circuits.

Bibliography

1. Koyada, B., Meghana, N., Jaleel, Md.O., Jeripotula, P.R., A Comparative Study on Adders, IEEE, 2017, 978-1-5090-4442-9/17/$31.00 c.
2. Yadav, Y. and Chawla, M., A Review On Adders In Digital Circuits. *2018 JETIR*, 5, 6, June 2018.
3. Singh, R. and Akasha, S., New High-Performance. Low Power 4bit Full Adder with Reduce Ground Bounce Noise. *2013 International Conference on Advanced Electronic Systems (ICAES)*.
4. Spoorthi, S.P., Implementation and comparative analysis of full adders using 180nm, 90nm 45nm technology in CADENCE. *Mater. Today Proc.*, 80, part3, 2198–2206, 2023.
5. Manjunath, K.M., Abdul Lateef Haroon, P.S., Pagi, A., Ulaganathan, J., Analysis of Various Full-Adder Circuits in Cadence. *International Journal of Computer Applications (0975 – 8887) National Conference on Power Systems & Industrial Automation (NCPSIA 2015)*.
6. Varadharajan, S.K. and Nallasamy, V., Low Power VLSI Circuits Design Strategies and Methodologies: A Literature Review. *IEEE Conference on Emerging Devices and Smart Systems (ICEDSS 2017)*, Mahendra Engineering College, Tamilnadu, India, 3-4 March 2017.
7. Nigam, A. and Singh, R., Comparative analysis of 28T full adder with 14T full adder using 180nm. *Int. J. Eng. Sci. Adv. Res.*, 2016 March, ISSN NO:2395-0730.
8. Irfan Khan, P.A., Mishra, R.S., Dhariwal, S., Comparative Analysis of Different Types of Full Adders using 180nm ad 90nm Technology. *International Journal of Computer Applications (0975 – 8887) International Conference on Advances in Emerging Technology (ICAET 2016)*.
9. Gupta, R., Pandey, S.P., Akashe, S., Vidyarthi, A., Analysis and Optimisation of Active Power and Delay of 10T Full Adder using Power Gating Technique at 45nm Technology. *IOSR J. VLSI Signal Process.*, Mar-Apr2013.
10. Agarawal, V., Shiravastav, R., Khandelwal, S., Akashe, S., Performance Analysis of 28T Full Adder using SVL Technique for Reducing the Leakage Current at 45nm Technology. *IOSR J. VLSI Signal Process.*, Jul-Aug. 2015.
11. Harsitha, S. and Krishnarao, G., Design and Comparison of 1bit Full Adder in GPDK 180nm and 45nm Technology. *Int. J. Adv. Res. Electr. Electron. Instrum. Eng.*, 7, 4, April – 2018.
12. Arora, S. and Tripathi, S.L., A comprehensive analysis of ultra-low power GNRFET based 20T hybrid full adder for computing applications. *Phys. Scr.*, 99, 085008, 2024, DOI 10.1088/1402-4896/ad5c0b.
13. Arora, S., Tripathi, S.L., Raj, B., High speed and low power hybrid adder using 22nm SG-TG FinFET. *Phys. Scr.*, DOI 10.1088/1402-4896/ad92bc.

13

Process Corner Analysis of 4-Bit Look Up Table (LUT) Using 90nm CMOS Technology

Talla. Narayana Swami[1], Shiridi Sravanthi[1], Suman Lata Tripathi[2]* and Yuli Sun Hariyani[3]

[1]*VLSI Design, School of Electronics and Electrical Engineering, Lovely Professional University, Phagwara, India*
[2]*Department of E & TC, Symbiosis Institute of Technology (SIT), Symbiosis International (Deemed University), Pune, India*
[3]*School of Electrical Engineering, Telkom University, Bandung, Indonesia*

Abstract

This chapter presents the design and analysis of a 4-bit Look-Up Table (LUT) using the Cadence Virtuoso tool for digital circuit applications. The LUT is a fundamental component in Field-Programmable Gate Arrays (FPGAs), enabling flexible implementation of logic functions. In this design, the 4-bit LUT is implemented at the transistor level using CMOS technology. The schematic design process involves the creation of memory cells (D flipflop) and multiplexers to realize the LUT's functionality. The primary focus of this work includes power and delay optimization for the 4-bit LUT. Power consumption and delay are critical metrics for evaluating the performance of digital circuits, particularly in low-power applications. Post-layout simulations are conducted to analyze the power dissipation and propagation delay. The power is measured in terms of dynamic and static components, and the delay is assessed from input to output transitions under different load conditions.

Keywords: LUT, multiplexer, CMOS, transient and average power, delay, cadence virtuoso

**Corresponding author*: tri.suman78@gmail.com

Suman Lata Tripathi, Vrinda Gupta, Sobhit Saxena, Shipra Upadhyay and Sudip Ghosh (eds.) Micro and Nano Semiconductor Devices for Digital, Analog and Sensor Design, (243–260) © 2026 Scrivener Publishing LLC

13.1 Introduction

Semiconductor IC design and fabrication is a complex process and involves high fabrication cost. Prefabrication design and analysis help in to avoid any post fabrication IC failures. Reconfigurable architectures help in design and performance evaluation of ICs to avoid such loss after fabrication. Field programmable gate arrays (FPGAs) are useful in implementation of several digital and analog ICs [1]. Reconfigurable blocks are important part of FPGA's that are made of Look up tables (LUTs). LUTs enable the efficient implementation of logic functions by storing precomputed outputs for all possible input combinations [2, 3]. A basic 2-input LUT has two levels of 2x1 multiplexer with 4 input register. An optimum design of multiplexer tree at block or transistor level reduces the overall power consumption of these processors. For commercial purpose, 4 or more input LUTs are preferred. A 4-bit LUT, for instance, can realize any function with four inputs, making it a versatile component in digital designs. The 4-bit LUT design is implemented using CMOS technology [4, 5], where the LUT stores 16 registers corresponding to the 16 combinations of 4 input LUT. A multiplexer-based structure is used to design a given function for specific values stored in registers at input stage. The power and delay characteristics of the designed LUT are critical for evaluating its performance, especially in low-power and high-speed applications. Delay refers to the time taken for a signal to propagate through the circuit, which directly impacts the operational speed of the system.

To ensure the stability of the LUT design for varying environmental and process conditions, corner analysis is performed. Corner analysis simulates the circuit under different fabrication process corners (e.g., fast-fast (FF), slow-slow (SS), typical-typical (TT)) and varying voltage and temperature conditions. This analysis helps in understanding the worst-case performance of the LUT in terms of power and delay.

13.2 Look-Up Table (LUT)

The LUT is a fundamental building block in digital systems, particularly in programmable logic devices such as Field-Programmable Gate Arrays (FPGAs). LUTs are used to implement combinational logic functions efficiently by storing precomputed output values for all possible input combinations.

In a general sense, an n-bit LUT can implement any Boolean function with n inputs by storing 2^n output values. For example, a 4-bit LUT, which is the focus of this design, can map 16 different input combinations (from 0000 to 1111) to their respective output values. The LUT behaves as a truth table where each row corresponds to a unique input combination, and the stored value at that location is the desired output.

LUTs are crucial in FPGAs because they provide flexibility in implementing various logic circuits without requiring the circuit to be hardwired. By changing the contents of the LUT, different logic functions can be realized without altering the hardware itself. As the number of input bits increases, the size of the LUT increases exponentially, which can impact the area, power consumption, and delay of the circuit.

A 4-bit LUT can realize any combinational logic function of 4 inputs. With 4 inputs, there are $2^4 = 16$ possible input combinations. The LUT stores a precomputed output for each combination of the inputs. Consider the inputs variables A, B, C, D (4 input bits) and Output F (Function output). The truth table shown in Table 13.1 for the 4-bit LUT has 16 rows, one for each input combination. The specific values of the output (F) will depend on the particular function the LUT is designed to implement. Below is a generic truth table where the output values can be defined based on the desired logic function.

General block diagram of LUT:

The LUT consists of memory elements and a multiplexer (MUX) that selects the appropriate stored value based on the input combination. The block diagram for an n-bit LUT generally follows the same structure, regardless of the number of inputs.

For a 4-bit LUT, the block diagram will have: 4 input bits (A, B, C, D), 16 stored values (since corresponding to the possible input combinations. A 16-to-1 multiplexer to select the appropriate output value based on the inputs as shown in Figure 13.1. Any 4 variable Boolean expression can be implemented with one block of 4-bit LUT by giving a suitable value to input registers. Multiple blocks of 4bit LUT are required for implementation of Boolean functions for higher input variables (5 or more input). In this chapter two different Boolean functions are designed and performance are evaluated at different process corners.

Table 13.1 4-bit LUT.

A	B	C	D	F
0	0	0	0	F(0)
0	0	0	1	F(1)
0	0	1	0	F(2)
0	0	1	1	F(3)
0	1	0	0	F(4)
0	1	0	1	F(5)
0	1	1	0	F(6)
0	1	1	1	F(7)
1	0	0	0	F(8)
1	0	0	1	F(9)
1	0	1	0	F(10)
1	0	1	1	F(11)
1	1	0	0	F(12)
1	1	0	1	F(13)
1	1	1	0	F(14)
1	1	1	1	F(15)

13.3 Basic Blocks Used in Design of LUT at 90nm CMOS

2-bit LUT are designed by designing 2x1 multiplexer using transistor level design of NOT Gate ($Y = \overline{A}$), AND Gate ($Y = AB$) and NOR Gate ($Y = \overline{(A+B)}$) and joining them as blocks. Also 2x1 multiplexer can be designed at transistor level using different logic style such CMOS logic, pass transistor logic (PTL), CMOS transmission-gate (CMOS TG), Pseuodo NMOS logic and dynamic logic etc. PTL logic is that which uses lowest number of transistors resulting in lesser on-chip area. Figures 13.2 (a) & (b) show the schematic of 2x1 multiplexer using logic gate and NMOS transtors respectively.

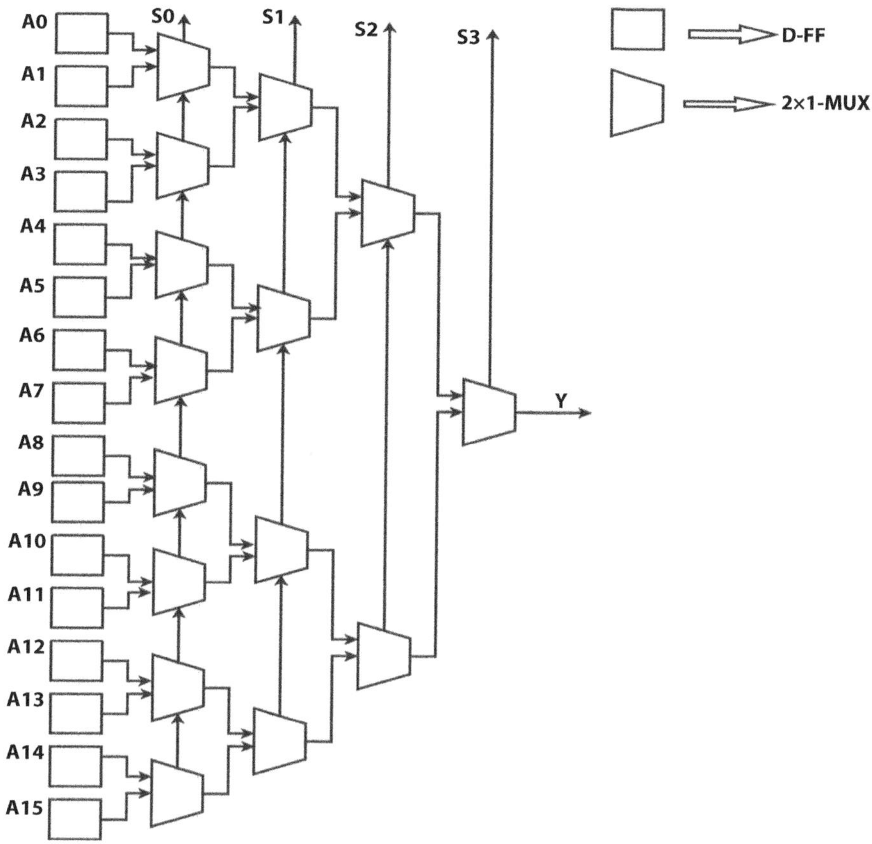

Figure 13.1 Block diagram of 4-bit LUT.

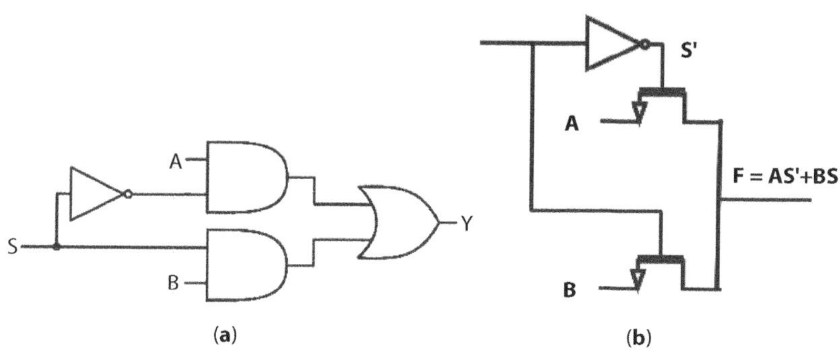

Figure 13.2 (a) 2x1 Mux (F=AS'+BS) at block level; (b) 2x1 Mux using PTL.

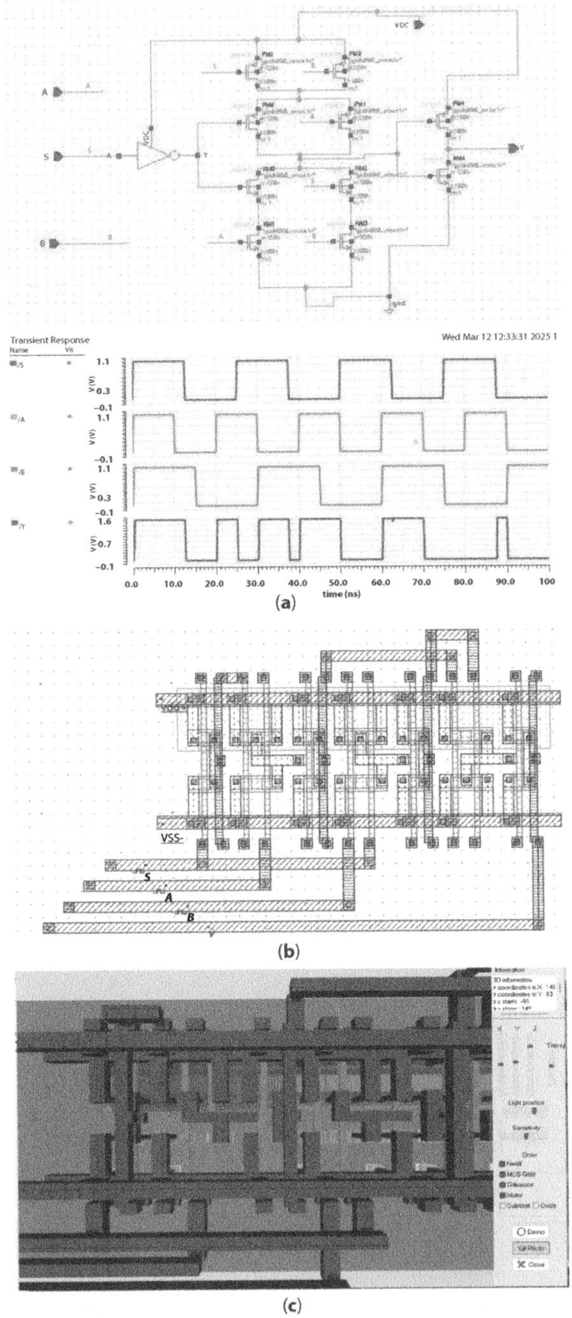

Figure 13.3 (a) 2x1-multiplexer circuit, and wave form; layout of 2x1 multiplexer at 90nm technology; (b) 2D view; (c) 3D view.

13.3.1 2X1 Multiplexer

A multiplexer is an example combinational digital circuit made with logic gates. It is based on multiple input lines and single output line including control or select inputs depending on number of input. Logic gate level 2x1 multiplexer block is shown in Figure 13.3 (a). For N input lines, N=2^m, m selection lines are required. It is an important block various in digital ICs such as in digital processors performaning arithmatic and logoic operations.

Figures 13.3 (b) & (c) present a 2D & 3D view 2x1 multiplexer layout at 90nm technology respectively. Layout simulation of 2x1 multiplexer is showing error free results with low leakage current.

13.3.2 D-Flip Flop (DFF)

It acts as a 1-bit memory element in design of LUT. Flip Flop is an example of sequential circuit that can store one bit data in memory components as shown in Figure 13.4. A flip flop has two stat represented by "SET" and

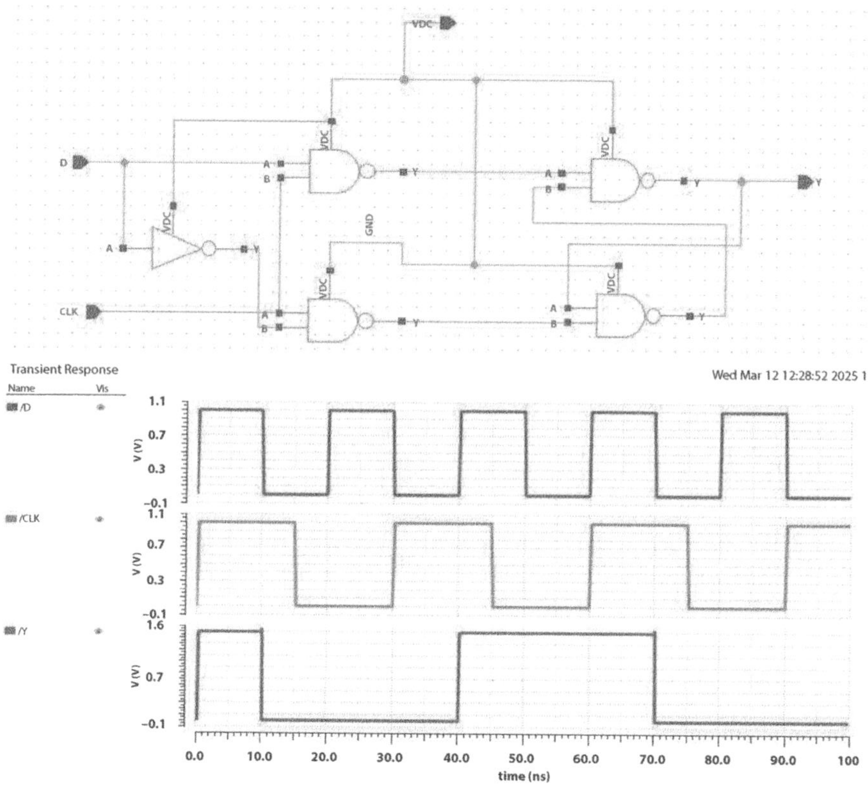

Figure 13.4 D-flip flop circuit, and waveform.

"RESET" showing binary values 0 and 1. The flip flop remains in its current state until it receives a signal that switches it to opposite state. The role of a clock signal is to trigger the flip flop to change state from 0-1 or 1-0. D flip flop is also known as delay flip flop that is used to store single bit of data.

13.3.3 Schematic of 4 Bit-LUT

Figure 13.5 present a block level design of 4-input LUT using 90nm CMOS logic. 4-input LUT is designed using 21 numbers of 2x1 multiplexer tree designed at transistor level using CMOS technology.

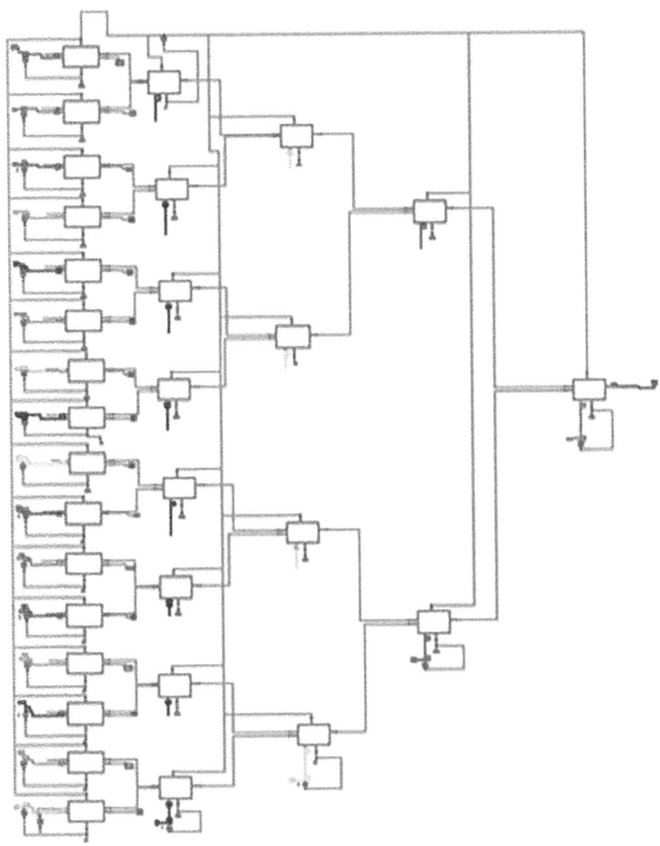

Figure 13.5 Schematic of 4-bit LUT in cadence.

13.3.3.1 Functions Implementation Using 4-Bit LUT

Figure 13.6 shows a 4-input AND logic block. The boolean function for four input and gate is:

$$F = ABCD$$

Figure 13.7 and Figure 13.8 show a transient response for Boolean function 1 for given inputs and select lines with output respectively. Here, input variable are A, B, C, D and output variable is F.

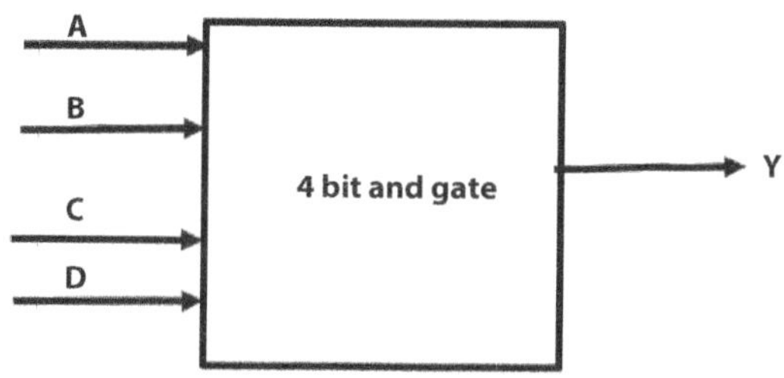

Figure 13.6 Block diagram of 4-bit AND gate.

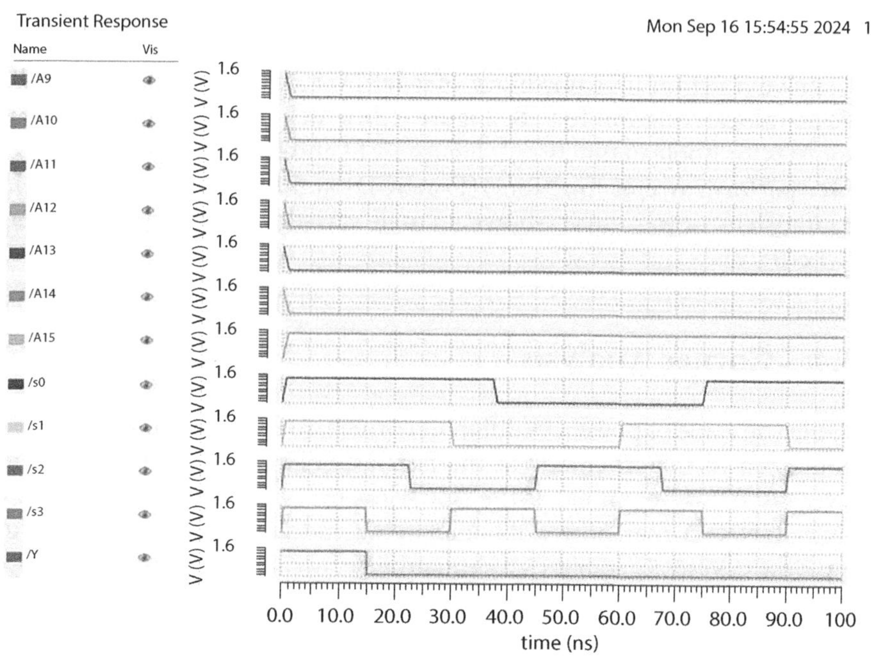

Figure 13.7 Waveform for function 1 (ABCD) with inputs.

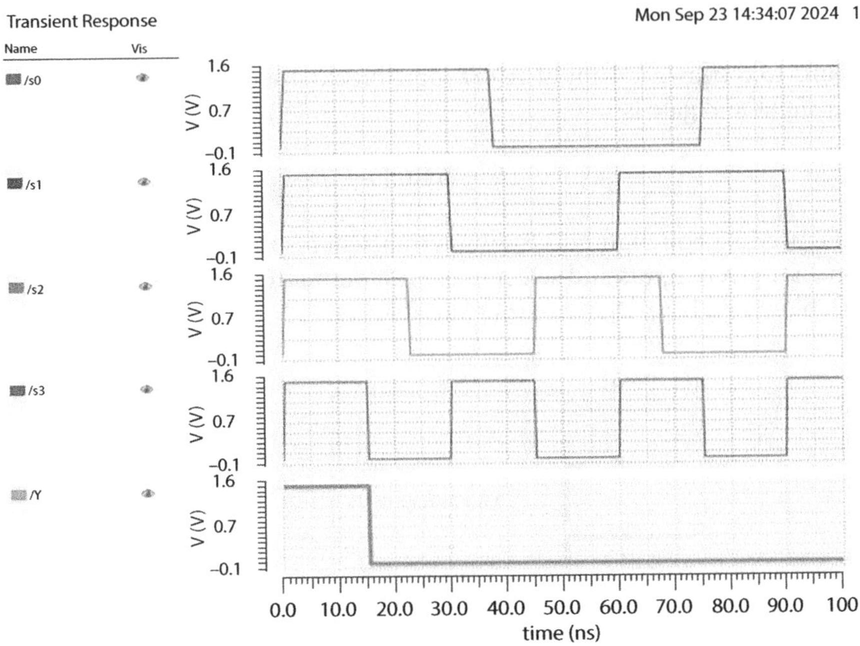

Figure 13.8 Waveform for function 1 (ABCD) with select inputs.

(i) F = A'BCD

The above function uses one block of 4-bit LUT designed using 2x1 multiplexer blocks. Figure 13.9 & Figure 13.10 demonstrate the transient responses for Boolean function 2. Figure 13.11 and Figure 13.12 present the power performance of function 1(F=ABCD) and function 2 (F=A'BCD). The transient power curve shows a small variation with average power 111.1×10^{-6} W.

13.4 Corner Analysis

Among the value of powers in different corners, maximum power (154.7E-6) obtained in the schematic at the fast-fast (FF) corner. The minimum power (70.07E-6) obtained at the slow-slow (SS) corner. At the nominal-nominal (NN) corner, power obtained in the middle range between FF and slow-slow (SS) [6, 7]. Table 13.2 shows the delay calculation of function 1.

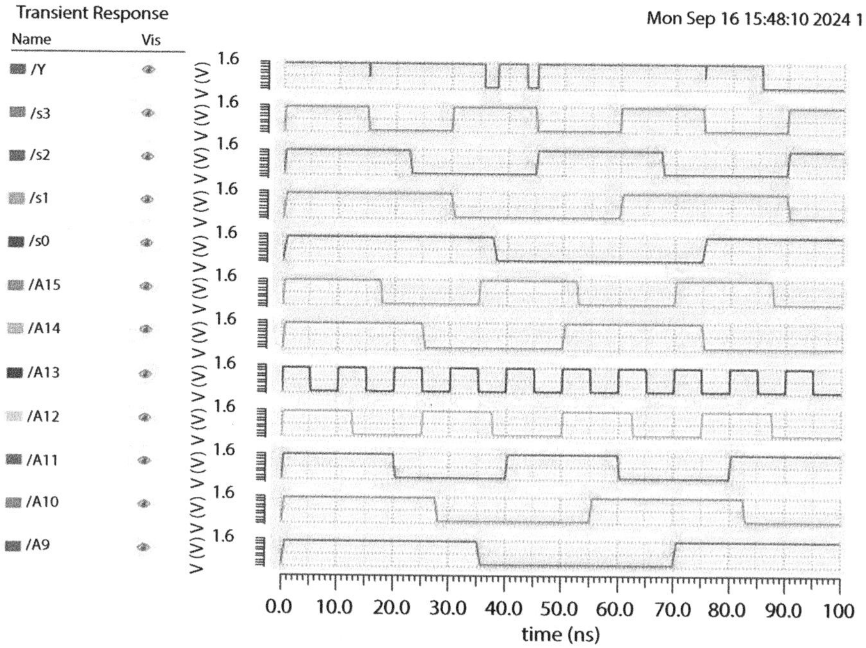

Figure 13.9 Waveform for function 2 (A'BCD) with inputs.

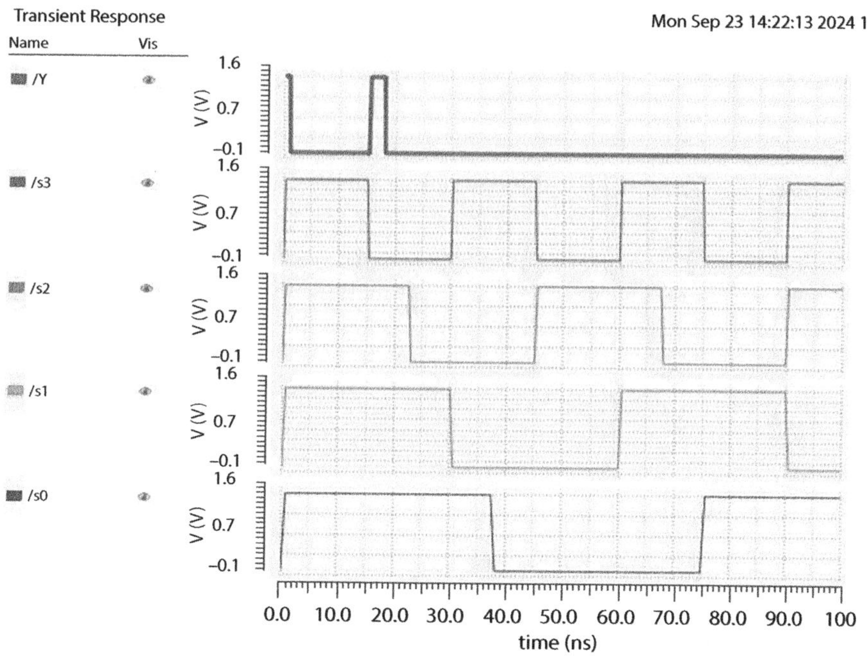

Figure 13.10 Waveform for function 2 (A'BCD) with select lines.

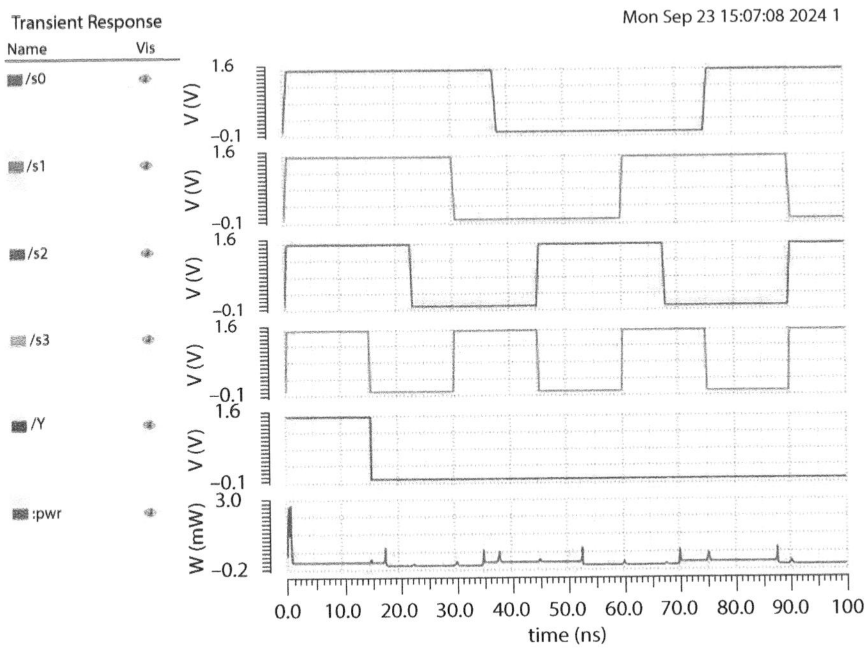

Figure 13.11 Waveform of power for function 1 (F=ABCD).

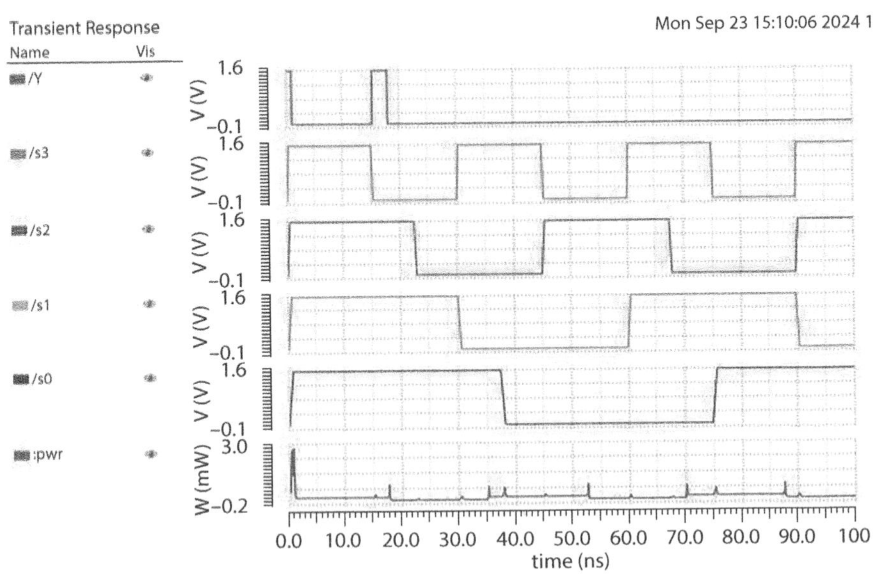

Figure 13.12 Waveform of power for function 2 (F=A'BCD).

Table 13.2 Average power of Boolean function F=ABCD.

Function	Power at NN (μW)	Power at SS (μW)	Power at FF (μW)	Power at FS (μW)
F=ABCD	111.1	70.07	154.7	143.1
F=A'BCD	111.1	75.98	154.6	143.0

Table 13.3 Delay calculations for function 2.

S. no.	Input	Output	Rising – falling (ns)
1.	S0	Y	14.88
2.	S1	Y	14.95
3.	S2	Y	15.02
4.	S3	Y	15.10

These delay values had calculated at the NN corner. The delay values calculated between one input signal transition to output signal transition. The variations between the different delay values were very small. For industrial purpose, 4 or more than 4-bit LUT are used to design FPGAs to accommodate large data processing used in several signal processing applications [8–12]. Table 13.3 shows delay calculations for function F=A'BCD.

13.5 Applications

The LUTs are part of several processors and digital circuit that are listed here:

13.5.1 Implementation of Digital Logic Functions

n-input Boolean function can be implemented through multiple blocks of LUT's. These LUTs can be reconfigured for any digital function [13], by providing suitable input values to the input registers of LUT.

13.5.2 DSP Processors

High performance adders, subtractors and multipliers are mainly part of DSP processors [3, 14–16]. LUTs are useful in designing these blocks and analyzing them in terms of power, area and delay performances.

13.5.3 Signal and Image Processing

The audio, image or video signals are also analyzed using LUT based reconfigurable architectures and digital filters for performance optimisation [17–21].

The LUT performance mainly depends on the configurable logic block which is made using 2x1 Multiplexer designed with CMOS logic. Use of PTL logic helps in reducing transistor count and optimise the IC area. Use of the FinFET technology [22–25] available up to 3nm technology node, helps in improving the overall delay and power consumption of transistor result in overall performance improvement of LUTs and overall FPGA.

13.6 Conclusion

The design and implementation of a 4-bit Look-Up Table (LUT) demonstrate the versatility and efficiency of LUTs in digital logic systems, particularly in reconfigurable hardware like FPGAs. By precomputing and storing output values for all possible input combinations, the LUT offers a fast and flexible way to realize any combinational logic function. The 4-bit LUT can implement any Boolean function with 4 inputs by storing 16 precomputed output values. This flexibility makes it a crucial building block in programmable logic devices, where functions may need to change dynamically without altering hardware. The transistor-level implementation of the LUT, particularly with D-FF memory cells and efficient multiplexers, allows for optimizations in power consumption and delay. Proper design choices in these areas help meet the requirements of low-power, high-speed digital systems. Through detailed power, delay, and corner analysis, the 4-bit LUT is validated for reliable operation across varying process, voltage, and temperature conditions, ensuring in real-world applications. In conclusion, the 4-bit LUT offers a fundamental and highly adaptable solution for implementing digital logic. Its speed, flexibility, and scalability make it an indispensable component in modern digital design, particularly for applications requiring reconfigurable logic and efficient combinational function realization. The results of the project highlight the importance of balancing performance metrics such as power, delay, and area in practical circuit design.

References

1. Tripathi, S.L., Saxena, S., Sinha, S.K., Patel, G.S., *Digital VLSI Design Problems and Solution with Verilog*, John Wiley & Sons, Ltd., USA, 2021, DOI:10.1002/9781119778097 ISBN: 978-1-119-77804-2.
2. Tripathi, S.L., Kumar, A., Srinivasa Rao, K., Mudimela, P.R., *Nanodevices for Integrated Circuit Design*, Wiley & Scrivener Publishing LLC, USA, October-2023, DOI:10.1002/9781394186396.
3. Tripathi, S.L., Kumar, A., Mahmud, M., FPGA for secured hardware & IP ownership. *2022 IEEE International Conference of Electron Devices Society Kolkata Chapter (EDKCON)*, Kolkata, India, pp. 184–189, 2022, doi: 10.1109/EDKCON56221.2022.10032935.
4. Tripathi, S.L., Design, synthesis and implementation of 8-bit accumulator using FPGA Board, pp. 1–5, ACM New York, NY, USA, 2024, https://doi.org/10.1145/3647444.3652476.
5. Tripathi, S.L. and Mahmud, M., *Explainable Machine Learning Models and Architecture*, Wiley & Scrivener Publishing LLC, USA, 2023, https://doi.org/10.1002/9781394186570.
6. Harris, D.M. and Weste, N., *CMOS VLSI Design: A Circuits and Systems Perspective*, 4th, Edison Pearson, 2010.
7. Yeap, G.K., *Practical Low Power Digital VLSI Design*, 1st, Edison, Springer, 2012, https://doi.org/10.1007/978-1-4615-6065-4.
8. Prasanna, D.L. and Tripathi, S.L., A Review on Tongue Based Assistive Technology, Devices and FPGA Processors Using Machine Learning Module. *Wirel. Pers. Commun.*, 134, 151–170, 2024, https://doi.org/10.1007/s11277-024-10897-8.
9. Tripathi, S.L., Prasanna, D.L., Mahmud, M., A Low-Power Audio Processing Using Machine Learning Module on FPGA and Applications, in: *Explainable Machine Learning Models and Architectures*, CRC Press, Taylor & Francis, Boca Raton, UK, 2023, https://doi.org/10.1002/9781394186570.ch4.
10. Raj, B., Tripathi, S.L., Chaudhary, T., Rao, K.S., Singh, M., *Integrated Devices for Artificial Intelligence and VLSI*, Wiley Scrivener Publishing LLC, USA, 15 August 2024, ISBN:9781394205158, DOI:10.1002/9781394205158.
11. Tripathi, S.L., Saxena, S., Mohapatra, S.K. (Eds.), *Advanced VLSI Design and Testability Issues*, 1st ed., CRC Press, Boca Raton, UK, 2020, https://doi.org/10.1201/9781003083436.
12. Kumar, A., Tripathi, S.L., Rao, K.S., *Machine Learning for VLSI Chip Design*, Wiley Scrivener Publishing LLC, USA, 2023, DOI:10.1002/9781119910497.
13. Tripathi, S.L., Design, synthesis and implementation of 8-bit accumulator using FPGA Board, in: *Proceedings of the 5th International Conference on Information Management & Machine Intelligence*, 2023, November, pp. 1–5.

14. Haghi, P., Kamal, M., Afzali-Kusha, A., Pedram, M., O&xxxx2074;-DNN: A Hybrid DSP-LUT-Based Processing Unit With Operation Packing and Out-of-Order Execution for Efficient Realization of Convolutional Neural Networks on FPGA Devices. *IEEE Trans. Circuits Syst. I Regul. Pap.*, 67, 9, 3056–3069, Sept. 2020, doi: 10.1109/TCSI.2020.2986350.
15. Satyanarayana, B.V.V., Lakshmi, B.K.S., Kumar, G.P., Srinivas, K., LUT-Based Area-Optimized Accurate Multiplier Design for Signal Processing Applications, in: *Cognitive Computing and Cyber Physical Systems. IC4S 2023, Lecture Notes of the Institute for Computer Sciences, Social Informatics and Telecommunications Engineering*, vol. 536, P. Pareek, N. Gupta, M.J.C.S. Reis (Eds.), Springer, Cham, 2024, https://doi.org/10.1007/978-3-031-48888-7_26.
16. Pradhan, D., Meher, P.K., Meher, B.K., FPGA-Specific Efficient Designs of Digit-Serial Multiplier for Galois Field GF(2m). *Circuits Syst. Signal Process.*, 2024, https://doi.org/10.1007/s00034-024-02886-w.
17. Tripathi, S.L., Mahmud, M., Balas, V.E., FPGA implementation for explainable machine learning and deep learning models to real-time problems, in: *Machine Learning Models and Architectures for Biomedical Signal Processing*, pp. 449–471, Academic Press, 2025.
18. Prasanna, D.L., Tripathi, S.L., Mahmud, M., Wijayanto, I., High attenuation electromagnetic interface filter for effective processing of audio signals. *Eng. Res. Express*, 6, 3, 035309, 2024.
19. Ma, Y., Cao, Y., Vrudhula, S., Seo, J.-S., Optimizing the convolution operation to accelerate deep neural networks on FPGA. *IEEE Trans. Very Large Scale Integr. VLSI Syst.*, 26, 7, 1354–1367, Jul. 2018.
20. Syed, F., Ali, W., Kumar, A., Bakhsh, F.I., Implementation of FIR Digital Filters on FPGA Board for Real-Time Audio Processing. *2023 International Conference on Technology and Policy in Energy and Electric Power (ICT-PEP)*, Jakarta, Indonesia, pp. 233–237, 2023, doi: 10.1109/ICT-PEP60152.2023.10351182.
21. Al-Dulaimi, M.A.A., Wahhab, H.A., Amer, A.A., Design and Implementation of Communication Digital FIR Filter for Audio Signals on the FPGA Platform. *J. Commun.*, *18*, 2, 89–96, 2023.
22. Tripathi, S.L., Mishra, R., Mishra, R.A., Characteristic comparison of connected DG FINFET, TG FINFET and Independent Gate FINFET on 32 nm technology. *2012 2nd International Conference on Power, Control and Embedded Systems*, Allahabad, India, pp. 1–7, 2012, doi: 10.1109/ICPCES.2012.6508037.
23. Tripathi, S.L., Mishra, R., Narendra, V., Mishra, R.A., High performance Bulk FinFET with bottom spacer. *2013 IEEE International Conference on Electronics, Computing and Communication Technologies*, Bangalore, India, pp. 1–5, 2013, doi: 10.1109/CONECCT.2013.6469282.

24. Verma, S. and Tripathi, S.L., Impact & Analysis of Inverted-T shaped Fin on the Performance parameters of 14-nm heterojunction FinFET. *Silicon*, 14, 9441–9451, 2022, https://doi.org/10.1007/s12633-022-01708-5.
25. Samal, A., Tripathi, S.L., Mohapatra, S.K., A Journey from Bulk MOSFET to 3 nm and Beyond. *Trans. Electr. Electron. Mater.*, 21, 443–455, 2020, https://doi.org/10.1007/s42341-020-00222-y.

14

Designing and Small Signal Analysis of Common Source Amplifier Using GaN Based HEMT

Yogesh Kumar Verma

Department of Electronics and Communication Engineering, Lovely Professional University, Phagwara, India

Abstract

A common source amplifier is designed and the frequency response of this amplifier is also analyzed. The amplified and inverted output is obtained across the drained terminal. The impact of parasitic capacitances is included in the analysis. The gain of the amplifier is noticed to be decreasing at higher magnitude of frequency. The AlGaN/GaN HEMT provides better RF performance as compared to MOSFET. The AlGaN/GaN HEMT is biased in active mode and common-source configuration to analyze its performance as an amplifier. In this brief, the designing of amplifier using AlGaN/GaN HEMT is performed. The frequency is varied from 10 Hz to 1 MHz to analyze the behavior of HEMT as an amplifier. The HEMT is considered in active mode for its application in designing of amplifier regarding RF applications.

Keywords: Common-source configuration, AlGaN, GaN HEMT, amplifier, frequency-response

14.1 Introduction

The mobility of electrons at the hetero-interface of HEMT depends on several factors such as: quality of material, electric field, temperature, charge density, and interface quality. The higher mobility of electrons at the

Email: yogeshkvermaec@gmail.com

Suman Lata Tripathi, Vrinda Gupta, Sobhit Saxena, Shipra Upadhyay and Sudip Ghosh (eds.) Micro and Nano Semiconductor Devices for Digital, Analog and Sensor Design, (261–272) © 2026 Scrivener Publishing LLC

heterointerface is achieved due to reduced scattering [1–20]. Other than higher electron mobility, HEMT offers other merits such as higher breakdown voltage, better thermal performance, higher power density, reduced gate charge, higher switching speed, low-on resistance, higher efficiency, improved linearity, and lower noise figure, as compared to Si based n-channel enhancement type MOSFET. The sub-threshold slope is significant electrical parameter responsible for examining the switching performance of any device [21–27]. The common-source amplifier is used as voltage amplifier or transconductance amplifier. Due to the Miller effect, the bandwidth of common-source amplifier is low. The model includes frequency dependent capacitances. The gate-source capacitance is nearly constant in triode region but it reduces significantly in saturation region. The operating regions include cut-off, triode, and saturation region. In this brief, the designing of amplifier using AlGaN/GaN HEMT is performed. The frequency is varied from 10 Hz to 1 MHz to analyze the behavior of HEMT as an amplifier.

The High Electron Mobility Transistor (HEMT) is a critical device in high-frequency, high-power, and low-noise applications. The combination of high electron mobility, superior power handling, and low noise makes HEMTs ideal for a range of industries, including telecommunications, radar, space systems, and power electronics. The HEMTs are prominent contender for 5G and quantum computing applications. HEMTs have significant properties for cryogenic applications. They can be used for image systems and security applications. They are popularly used for wireless power transmission systems. HEMTs are recently explored for security screening and industrial applications. High bandwidth data transmission is possible due to the ability of HEMTs to operate at higher frequency. The applications of HEMT can be extended to radar, aerospace, and SAR. The HEMTs can be effectively used in low noise amplifier and high-power amplifiers due to their high frequency ability. GaN HEMT MMICs are recently explored extensively. 12.2 watts/mm is the recently reported power density using HEMTs. Higher thermal stability and higher breakdown voltage is possible by the use of HEMT is MMIC technology. The GaN-based HEMT are significantly used in power amplifier and 5G technology. InP MMICs are being used in millimeter-wave and terahertz applications, particularly for next-gen radar and communication systems. One of the most significant advancements is the ability of MMICs to operate at millimeter-wave (30–300 GHz) and terahertz (0.1–10 THz) frequencies. These frequencies are crucial for 5G, 6G, and advanced radar systems. Recent breakthroughs in e-beam lithography and material engineering have allowed the design of MMICs that operate at these extremely high frequencies with high efficiency and low power loss.

The HEMT is considered in active mode for its application in designing of amplifier regarding RF applications.

The voltage-gain is calculated using the following equation:

$$A_v = -g_m \cdot \frac{r_d \cdot R_L}{r_d + R_L}$$

14.2 Device Structure

The electrons in the 2-DEG are not scattered by impurity scattering since they are confined in a channel away from the dopants. The electron mobility is less in MOSFET as compared to HEMT due to scattering generated by silicon substrate and impurities in the channel. The wide bandgap materials allow for higher power density due to the ability to operate at higher voltages and higher currents without breakdown. The use of wide bandgap materials in HEMT provides higher breakdown voltage as compare to MOSFET. Figures 14.1 and 14.2 represent the device structure and circuit of GaN-based HEMT respectively. The AlGaN layer is used as a barrier layer and GaN layer is used as a channel layer [1–3]. The thermal conductivity is also enhanced by the use of buffer layer. The cap layer passivates the surface, stabilizes the 2-DEG, reduces surface traps, and protects

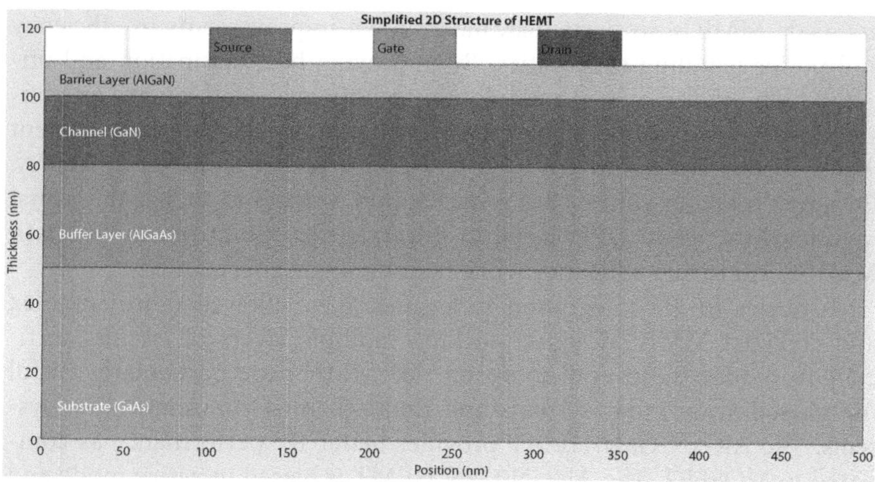

Figure 14.1 Structure of high electron mobility transistor.

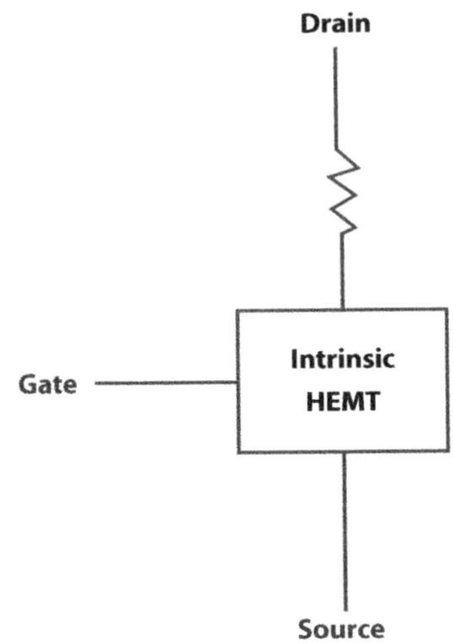

Figure 14.2 Circuit diagram of HEMT.

the device from environmental damage. The cap layer provides surface protection, enhancement of 2-DEG density, passivation of surface states, improvement of gate control, reduction of gate leakage, thermal stability, reduction of parasitic capacitance, and provides uniformity in fabrication [28–37].

MMICs operating in the sub-terahertz and terahertz range are emerging for next-generation communication systems (6G), which will demand enormous bandwidths and very low latency. One of the key trends in MMIC development is the push for higher efficiency in power amplifiers. GaN MMICs, in particular, have shown improvements in efficiency and power handling capabilities. By improving heat dissipation and utilizing high-efficiency load modulation techniques (e.g., Doherty amplifiers), MMICs can handle higher power with reduced thermal management issues. These techniques are used to increase power amplifier efficiency. Envelope tracking adjusts the power supply voltage to match the signal requirements, while DPD helps to linearize the output of amplifiers in MMICs, improving efficiency while maintaining signal quality.

Advances in 3D integration techniques have allowed more complex and compact MMIC designs, stacking multiple layers of circuits vertically to reduce the overall footprint. 3D MMICs are particularly useful for phased-array radar systems and multi-channel communication systems. The AlGaN/GaN HEMT provides better RF performance as compared to MOSFET. The AlGaN/GaN HEMT is biased in active mode and common-source configuration to analyze its performance as an amplifier.

In this brief, the designing of amplifier using AlGaN/GaN HEMT is performed. The frequency is varied from 10 Hz to 1 MHz to analyze the behavior of HEMT as an amplifier.

14.3 Results and Discussions

Transconductance, threshold-voltage, drain current, noise-figure, power dissipation, and output conductance are the important parameters of interest in designing of amplifier based on AlGaN/GaN HEMT. The small signal parameters of amplifier designed using HEMT are voltage-gain, input-impedance, output-impedance, gain-bandwidth product. The different performance parameters of GaN-based amplifier include voltage gain, transconductance, input impedance, output impedance, bandwidth, power gain, noise figure, output-range, drain current, power dissipation, breakdown voltage, and stability. The peak value of small-signal input voltage occurs at 0.25 and 0.75 ms in first positive and negative cycle respectively. Figure 14.3 represents the sinusoidal input voltage applied to determine the frequency response. The output of common source amplifier designed using HEMT is inverted and amplified as evident from Figure 14.4.

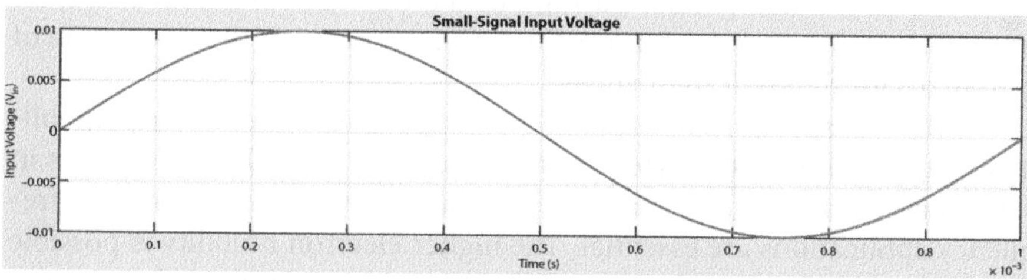

Figure 14.3 Input voltage (volts).

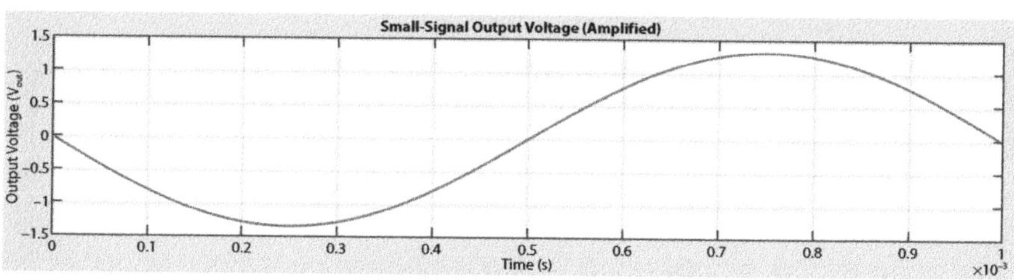

Figure 14.4 Amplified output voltage (volts).

The parasitic capacitance needs to be modeled as these capacitances affect the input and output impedances of the amplifier. HEMT-based MMICs are being explored for their application in quantum computing. The low-noise and high-frequency performance of HEMTs make them suitable for cryogenic amplification and signal processing in quantum processors. HEMT-based MMICs are integral to automotive radar systems used in autonomous vehicles, where high-frequency operation and low-noise performance are critical for detecting objects and ensuring safe navigation. HEMT-based MMICs are crucial in 5G infrastructure, particularly in mm-wave bands for base stations and user equipment. The AlGaN/GaN HEMT provides better RF performance as compared to MOSFET. The AlGaN/GaN HEMT is biased in active mode and common-source configuration to analyze its performance as an amplifier. In this brief, the designing of amplifier using AlGaN/GaN HEMT is performed. The frequency is varied from 10 Hz to 1 MHz to analyze the behavior of HEMT as an amplifier. With the development of 6G technologies, operating in higher frequency ranges (above 100 GHz), HEMTs are expected to play an even larger role due to their high-frequency capabilities and power efficiency. New techniques in epitaxial layer growth have led to reduced defect densities in GaN and InP HEMTs, improving the overall reliability and performance of MMICs in high-power applications. Extreme environment applications are also possible using HEMT due their inherent ability to operate in cryogenic temperature. Figure 14.5 represents the calculation of gain in dB for different frequencies.

HEMTs are successfully used in SoC design. The higher electron mobility and higher breakdown voltage is possible after using HEMT devices in solid-state electronics. HEMTs are popularly used where the higher frequency applications are essential. The higher electron mobility is possible due to the reduction in the scattering process that occurs in the conventional MOSFET. The flip-chip designs are possible after the introduction of HEMT devices in solid-state electronics and their interesting operating

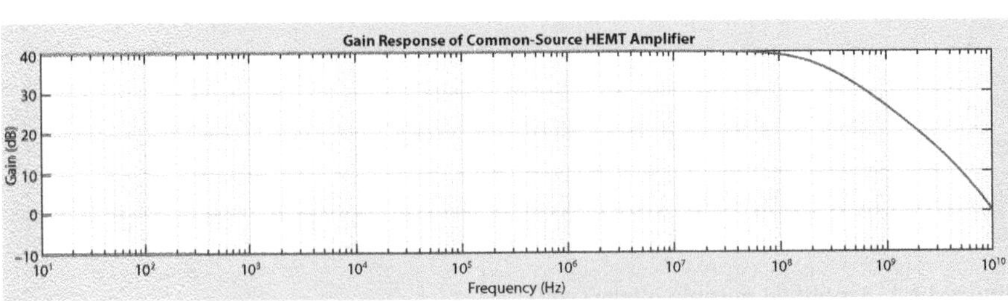

Figure 14.5 Calculation of gain (dB).

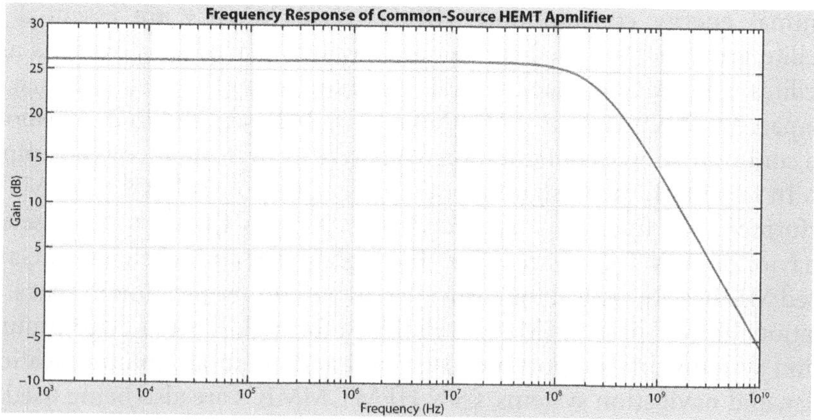

Figure 14.6 Frequency response of HEMT-based amplifier.

feature and ability. There is no significant degradation in the properties of HEMT when they operate at extreme environment conditions. Figure 14.6 represents the frequency response of CS HEMT amplifier.

The improvement in passivation technology assists in improving the performance of HEMT device. The low-noise amplifiers are possible interesting features of HEMT leading to the new evolution in this era. HEMTs are explored in the device level from the last few years; however, in this work we have extended the features of HEMT to operate as a common-source amplifier at the circuit-level design. The gain vs. frequency and phase vs. frequency plots are realized after considering the effects of parasitic resistances and capacitances in the high-frequency model. The AlGaN/GaN HEMT provides better RF performance as compared to MOSFET. The AlGaN/GaN HEMT is biased in active mode and common-source configuration to analyze its performance as an amplifier. In this brief, the designing of amplifier using AlGaN/GaN HEMT is performed. The frequency is varied from 10 Hz to 1 MHz to analyze the behavior of HEMT as an amplifier. The intrinsic HEMT is incorporated with the parasitic resistances so that there is good resemblance between the proposed model and real data obtained after the realizing the design through practical approach.

The phase starts at 0° at low frequencies and then becomes increasingly negative as the frequency rises. HEMT-based MMICs are used in IoT devices that require reliable communication in harsh environments, such as industrial IoT systems deployed in remote or high-temperature areas (e.g., oil rigs, factories). The high efficiency and frequency performance of HEMTs ensure that IoT devices maintain connectivity with minimal energy consumption. HEMT-based MMICs are essential in satellite transponders used for signal relay between ground stations and satellites. The AlGaN/GaN HEMT provides better RF performance as compared to MOSFET. The AlGaN/GaN HEMT is biased in active mode and common-source configuration to analyze its performance as an amplifier. In this brief, the designing of amplifier using AlGaN/GaN HEMT is performed. The frequency is varied from 10 Hz to 1 MHz to analyze the behavior of HEMT as an amplifier. The high power and efficiency of GaN-based MMICs are critical for ensuring reliable communication in space. In aviation, HEMT-based MMICs are utilized for high-frequency communication systems that support air-to-ground and air-to-air communication, radar, and navigation systems. GaN HEMT MMICs are also being used in wireless power transfer systems, where efficient power amplification and high-frequency operation are crucial for transferring energy wirelessly over short distances. HEMT-based MMICs, particularly those using GaN, are being applied in power electronics for high-efficiency power conversion systems. These systems are used in data centers, renewable energy installations, and electric vehicles (EVs), where GaN HEMTs provide superior performance over silicon-based devices in terms of power handling, efficiency, and switching speed. HEMT-based MMICs are also being explored for use in quantum sensing systems, where ultra-sensitive measurements of physical properties (e.g., magnetic fields, temperature) are essential for emerging technologies like quantum imaging and quantum radar. In scientific research and industrial applications, terahertz spectroscopy powered by HEMT-based MMICs is used for chemical analysis, pharmaceutical quality control, and gas detection. The high sensitivity and resolution of THz MMICs make them invaluable in these fields. HEMT-based MMICs operating in the terahertz range (0.1–10 THz) are crucial for future wireless communication systems that require ultra-high data rates. Figure 14.7 represents the calculation of phase for different frequencies.

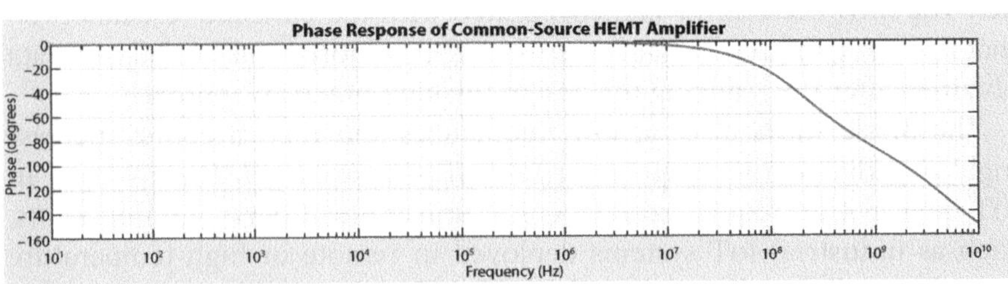

Figure 14.7 Calculation of phase (degree).

14.4 Conclusion

The HEMT-based amplifiers are capable to amplify small signal input levels to higher output levels. The input signal is applied at the gate terminal and the output is analyzed at the drain terminal. The transconductance type of amplification is noticed. The input is applied as voltage at the gate and the output is current at the drain terminal. The stability of the amplifier can be improved using the feedback technologies. This process not only linearizes the transconductance but also stabilizes the gain. The characteristics of the amplifier design are similar to band-pass. In this brief, the designing of amplifier using AlGaN/GaN HEMT is performed. The frequency is varied from 10 Hz to 1 MHz to analyze the behavior of HEMT as an amplifier. The HEMT is considered in active mode for its application in designing of amplifier regarding RF applications. The gain of the amplifier is noticed to be decreasing at higher magnitude of frequency. The AlGaN/GaN HEMT provides better RF performance as compared to MOSFET. The features of HEMTs can be extended further in the field of robotics owing to their applications in sensors due to their ability to perform higher frequency operations. The ability of HEMTs to operate quite successfully in cryogenic temperature as well as their higher breakdown ability makes them a potential contender in MMIC designing and fabrication processes.

References

1. Ambacher, O., Smart, J., Shealy, J.R., Weimann, N.G., Chu, K., Murphy, M., Schaff, W.J., Eastman, L.F., Two-dimensional electron gases induced by spontaneous and piezoelectric polarization charges in N- and Ga-face AlGaN/GaN heterostructures. *J. Appl. Phys.*, 85, 6, 3222–3233, 1999.
2. Ghosh, S., Dasgupta, A., Khandelwal, S., Agnihotri, S., Chauhan, Y.S., Surface-potential-based compact modeling of gate current in AlGaN/GaN HEMTs. *IEEE Trans. Electron Devices*, 62, 2, 443–448, 2015.
3. Ahsan, S.A., Ghosh, S., Sharma, K., Dasgupta, A., Khandelwal, S., Chauhan, Y.S., Capacitance modeling in dual field-plate power GaN HEMT for accurate switching behavior. *IEEE Trans. Electron Devices*, 63, 2, 565–572, 2016.
4. Ganeriwala, M.D., Yadav, C., Mohapatra, N.R., Khandelwal, S., Hu, C., Chauhan, Y.S., Modeling of Charge and Quantum Capacitance in Low Effective Mass III-V FinFETs. *IEEE J. Electron Devices Soc.*, 4, 6, 396–401, 2016.
5. Yadav, C., Duarte, J.P., Khandelwal, S., Agarwal, A., Hu, C., Chauhan, Y.S., Capacitance Modeling in III–V FinFETs. *IEEE Trans. Electron Devices*, 62, 11, 3892–3897, 2015.

6. Yadav, C., Kushwaha, P., Khandelwal, S., Duarte, J.P., Chauhan, Y.S., Hu, C., Modeling of GaN-based normally-off FinFET. *IEEE Electron Device Lett.*, 35, 6, 612–614, 2014.
7. Khandelwal, S. and Fjeldly, T.A., Analysis of drain-current nonlinearity using surface-potential-based model in GaAs pHEMTs. *IEEE Trans. Microwave Theory Tech.*, 61, 9, 3265–3270, 2013.
8. Khandelwal, S., Chauhan, Y.S., Fjeldly, T.A., Analytical modeling of surface-potential and intrinsic charges in AlGaN/GaN HEMT devices. *IEEE Trans. Electron Devices*, 59, 10, 2856–2860, 2012.
9. Ye, J.D., Pannirselvam, S., Lim, S.T., Bi, J.F., Sun, X.W., Lo, G.Q., Teo, K.L., Two-dimensional electron gas in Zn-polar ZnMgO/ZnO heterostructure grown by metal-organic vapor phase epitaxy. *Appl. Phys. Lett.*, 97, 11, 2010–2012, 2010.
10. Khan, M.A., Singh, R., Mukherjee, S., Kranti, A., Buffer Layer Engineering for High ($\geq 10^{13}$ cm^{-2}) 2-DEG Density in ZnO-Based Heterostructures. *IEEE Trans. Electron Devices*, 64, 3, 1015–1019, 2017.
11. Khandelwal, S., Goyal, N., Fjeldly, T.A., A physics-based analytical model for 2DEG charge density in AlGaN/GaN HEMT devices. *IEEE Trans. Electron Devices*, 58, 10, 3622–3625, 2011.
12. Tampo, H., Shibata, H., Maejima, K., Yamada, A., Matsubara, K., Fons, P., Kashiwaya, S., Niki, S., Chiba, Y., Wakamatsu, T., Kanie, H., Polarization-induced two-dimensional electron gases in ZnMgO/ZnO heterostructures. *Appl. Phys. Lett.*, 93, 20, 11–14, 2008.
13. Kuhn, K.J., Giles, M.D., Becher, D., Kolar, P., Kornfeld, A., Kotlyar, R., Ma, S.T., Maheshwari, A., Mudanai, S., Process technology variation. *IEEE Trans. Electron Devices*, 58, 8, 2197–2208, 2011.
14. Shur, M.S., Gaska, R., Khan, A., Simin, G., Wide band gap electronic devices, in: *Proceedings of the Fourth IEEE International Caracas Conference on Devices, Circuits and Systems (Cat. No. 02TH8611)*, Oranjestad Aruba Netherlands, pp. D051–D051, 2002.
15. Anderson, R.L., Germanium-gallium arsenide heterojunctions. *IBM J. Res. Dev.*, 4, 3, 283–287, 1960.
16. Dingle, R., Störmer, H.L., Gossard, A.C., Wiegmann, W., Electron mobilities in modulation-doped semiconductor heterojunction superlattices. *Appl. Phys. Lett.*, 33, 7, 665–667, 1978.
17. Witkowski, L.C., Drummond, T.J., Stanchak, C.M., Morkoc, H., High mobilities in $Al_xGa_{1-x}As$-GaAs heterojuntions. *Appl. Phys. Lett.*, 37, 11, 1033–1035, 1980.
18. Wang, W.I., Wood, C.E.C., Eastman, L.F., Extremely high electron mobilities in modulation-doped GaAs-$Al_xGa_{1-x}As$ heterojunction superlattices. *Electron. Lett.*, 17, 1, 36–37, 1981.
19. Delagebeaudeuf, D. and Linh, N.T., Metal-(n) AlGaAs-GaAs two-dimensional electron gas FET. *IEEE Trans. Electron Devices*, 29, 6, 955–960, 1982.

20. Ambacher, O., Foutz, B., Smart, J., Shealy, J.R., Weimann, N.G., Chu, K., Murphy, M., Sierakowski, A.J., Schaff, W.J., Eastman, L.F., Two dimensional electron gases induced by spontaneous and piezoelectric polarization in undoped and doped AlGaN/GaN heterostructures. *J. Appl. Phys.*, 87, 1, 334-344, 2000.
21. Shur, M.S., GaN based transistors for high power applications. *Solid State Electron.*, 42, 12, 2131–2138, 1998.
22. Look, D.C., Recent advances in ZnO materials and devices. *Mater. Sci. Eng., B*, 80, 1–3, 383–387, 2001.
23. Özgür, U., Alivov, Y.I., Liu, C., Teke, A., Reshchikov, M.A., Dogan, S., Avrutin, V., Cho, S.-J., Morkoc, H., A comprehensive review of ZnO materials and devices. *Appl. Phys.*, 98, 4, 041301, 2005.
24. Meyer, B.K., Alves, H., Hofmann, D.M., Kriegseis, W., Forster, D., Bertram, F., Christen, J., Hoffmann, A., StraBburg, M., Dworzak, M., Haboeck, U., Rodina, A.V., Bound exciton and donor-acceptor pair recombinations in ZnO. *Phys. Status Solidi Basic Res.*, 241, 2, 231–260, 2004.
25. Singh, R., Khan, M.A., Mukherjee, S., Kranti, A., Analytical Model for 2DEG Density in Graded MgZnO/ZnO Heterostructures with Cap Layer. *IEEE Trans. Electron Devices*, 64, 9, 3661–3667, 2017.
26. Singh, R., Khan, M.A., Mukherjee, S., Kranti, A., Role of Surface States and Interface Charges in 2DEG in Sputtered ZnO Heterostructures. *IEEE Trans. Electron Devices*, 65, 7, 2850–2854, 2018.
27. Yadav, C., Ganeriwala, M.D., Mohapatra, N.R., Agarwal, A., Chauhan, Y.S., Compact Modeling of Gate Capacitance in III-V Channel Quadruple-Gate FETs. *IEEE Trans. Nanotechnol.*, 16, 4, 703–710, 2017.
28. Ferain, I., Colinge, C.A., Colinge, J.-P., Multigate transistors as the future of classical metal-oxide-semiconductor field-effect transistors. *Nature*, 479, 7373, 310–316, 2011.
29. He, L., Chiang, T.K., Liou, J.J., Zheng, W., Liu, Z., A new analytical subthreshold potential/current model for quadruple-gate junctionless MOSFETs. *IEEE Trans. Electron Devices*, 61, 6, 1972–1978, 2014.
30. Lee, C.-T., Chen, W.-S., Lee, H.-Y., Quadruple Gate-Embedded T Structured GaN-Based Metal-Oxide-Semiconductor High-Electron Mobility Transistors. *IEEE J. Electron Devices Soc.*, 1, 6, 63–67, 2018.
31. Sharma, D. and Vishvakarma, S.K., Analyses of DC and analog/RF performances for short channel quadruple-gate gate-all-around MOSFET. *Microelectron. J.*, 46, 8, 731–739, 2015.
32. Li, M. and Wang, Y., 2-D Analytical Model for Current–Voltage Characteristics and Transconductance. *IEEE Trans. Electron Devices*, 55, 1, 261–267, 2008.
33. Manual, A.U., 3-D device simulator, Silvaco Inc., St. Clara, CA, USA, 2012.
34. Mimura, T., Hiyamizu, S., Fujii, T., Nanbu, K., A new field-effect transistor with selectively doped GaAs/n-$Al_xGa_{1-x}As$ heterojunctions. *Jpn. J. Appl. Phys.*, 19, 5, L225, 1980.

35. Wu, Y.-F., *et al.*, Very high breakdown voltage and large transconductance realized on GaN heterojunction field effect transistors. *Appl. Phys. Lett.*, 69, 10, 1438–1440, 1996.
36. Khan, M.A., *et al.*, AlGaN/GaN metal oxide semiconductor heterostructure field effect transistor. *IEEE Electron Device Lett.*, 21, 2, 63–65, 2000.
37. Albrecht, J.D., Wang, R.P., Ruden, P.P., Farahmand, M., Brennan, K.F., Electron transport characteristics of GaN for high temperature device modeling. *J. Appl. Phys.*, 83, 9, 4777–4781, 1998.

15

The 5th Generation: Major Implementation, Challenges and Massive MIMO Technology

Rashmi Roges[1], Sandeep Sharma[1] and Praveen Kumar Malik[2*]

[1]*Department of Electronics and Communication, Bhagwan Parshuram Institute of Technology, Rohini, New Delhi, India*
[2]*Department of Electronics and Communication, Lovely Professional University, Phagwara, Jalandhar, Punjab, India*

Abstract
This article is about the 'talk of the decade,' 5G technology. The one of the most awaited technologies, 5G, promises very high bandwidth and data rates with almost zero latency experience during video streaming, and very low latency supporting real-time applications like telemedicine, virtual and augmented reality and smart systems. When the technology till the 4th generation were human centric, 5G is a blend of human and things centric services. In 5G, internet enabled things are promised with intelligence and smartness to make their own suggestions and at times even decisions. While 5G comes with a big bag of promises, there are several challenges that need to be addressed. There is also a technological evolution brimming around 5G mobile technology, which makes the 5G dreams comes true. The chapter points out the major hard milestones which must be crossed before reaping the full benefits of 5G. It also discusses one of the basic enabling technologies of 5G, the 'Massive MIMO'.

Keywords: 5G, Massive MIMO, security, antenna, URLLC

15.1 Introduction

5G communication outstands 4G with support to AI and extended reality. With the modern technologies like the Internet of Things, virtual reality,

**Corresponding author*: malikbareilly@gmail.com

Suman Lata Tripathi, Vrinda Gupta, Sobhit Saxena, Shipra Upadhyay and Sudip Ghosh (eds.)
Micro and Nano Semiconductor Devices for Digital, Analog and Sensor Design, (273–288) © 2026
Scrivener Publishing LLC

autonomous vehicles etc conquering the hearts of consumers, the wireless technology had to nurture correspondingly to maintain them. This is achieved by the promised higher data rates (10 times of 4G) and low latency (one tenth of 4G) made possible through several key technologies such as massive MIMO (use of large number of transmitting and receiving antenna), small cell (assignment of smaller area), mm Wave and THz technology (use of higher frequency spectrum), NOMA (an efficient multiple access technique) etc. The increasing use of mobile devices and radio technologies forced evolution of still better wireless standards. 5G evolved with higher speed and better quality of service compared to its previous generations, to support not just mobile phones but various upcoming technologies which demands high performing wireless standards for their efficient implementation [1, 2].

The past 4-5 decades has seen vigorous development in cellular technology. 1970s and 80s saw the first generation of cellular technology. The bulky cell phones (nicknamed as 'brick') supported only voice communication like a normal telephone. The system provided mobility and was based on the traditional analogue technology. The device was very costly and had poor battery backup. Technologies such as the NMT (Nordic Mobile Telephony), TACS (Total Access Communication System), ETACS (European Total Access Communication System) and AMPS (Advanced Mobile Phone System) comes under 1st Generation systems. 1G systems supported a speed of up to 2Kbps with circuit switching and FDMA (Frequency Division Multiple Access) for multiplexing. Early 1990s saw a leap into the digital technology and this led to the evolution of a new cellular generation, the 2G, which enabled both voice and data communication (text messages and multimedia messages). Voice communication of 2G was supported by GSM technology (originally Groupe Special Module changed to Global System for Mobile communication, after it was accepted globally). 2G employed both circuit and packet switching while employing TDMA (Time Division Multiple Access) and CDMA (Code Division Multiple Access) for multiplexing. 2G could support a mobile data rate of up to 64Kbps. The advanced versions of 2G are the 2.5G, with GPRS (General Packet Radio Service) technique, i.e., a technique which supports packet communication through air interface, and 2.75G, with EDGE which stands for Enhanced data rate for GSM Evolution, with enhanced data rate of 2.5G.

With 3G, Internet browsing came handy in mobile phones. The advancements in digital technologies paved way for the better and faster communication and internet services in handset. It was technologies like HSPA (High Speed Packet Access) and HSPA+ that enabled 3G. MIMO

(Multiple Input Multiple Output) technology was also first introduced in 3G to attain better performing radio links. 3G was based on UMTS (Universal Mobile Terrestrial/Telecommunication System) and supported a data rate of up to 384Kbps. The data rates were 170Kbps with EDGE and 56-64Kbps with GSM. 3G enabled video calling with its enhanced data rates. WCDMA (Wideband Code Division Multiple Access) access technique was employed to achieve efficient radio resource allocation between multiple users. By 2010, 4G technology came into market supported by LTE (Long Term Evolution) and WiMAX. 4G converted the mobile phones into mini computers with IP and Packet Switching and provided a computer in hand. 4G systems employed OFDMA (Orthogonal Frequency Division Multiple Access) for multi-access and supported a speed till 1Gbps for stationer users and more than 100Mbps for mobile users. It's 4G that delivered the cellular technology that we are familiar with today.

But as the number of users for wireless systems augmented exponentially, 4G is no more adequate to deliver the demands of the present-day wireless systems. The explosive growth in the utilization of wireless technology propelled the activation of 5G. 5G was designed to handle about 1000 times more traffic at 10 times better speed than 4G. The modern technologies like the Internet of Things, Telemedicine, VR ("Virtual Reality") and AR ("Augmented Reality") are to be based on 5G. A walk through the generations of cellular technology as to a layman is suggested in Figure 15.1.

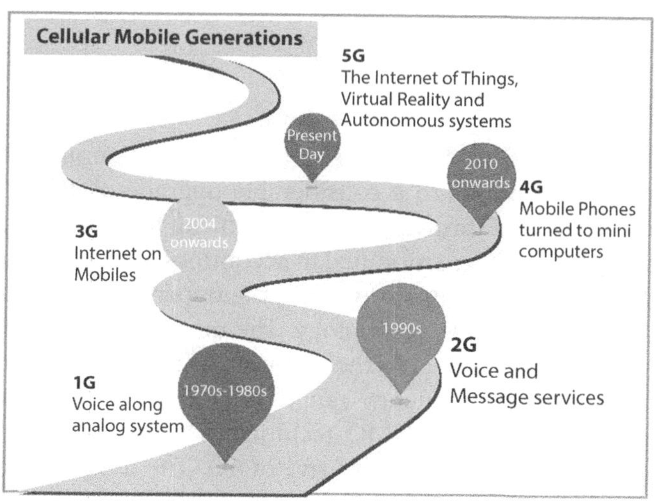

Figure 15.1 Cellular technology generations in layman's language.

15.2 Major Challenges Faced in 5G Implementation

There are some major challenges that need to be addressed while embracing 5G technology. Some of them are discussed below.

15.2.1 Infrastructure

The fifth generation of mobile networks promises high data throughput and very low latency. Though 5G is not a brand-new technology and can be seen as an evolution of 4G, and the already deployed infrastructure of 4G can be upgraded to support 5G, it still needs good amount of extra infrastructure. Massive MIMO and small cell technology needs to be implemented specially to support the mm wave technology introduced to implement and attain full benefits of 5G technology [3].

15.2.2 Cost

Implementation of 5G demands new infrastructure which directs to an additional investment of money and resources to the existing 4G system. Spectrum acquirement and large-scale optical cable deployment along with installation of new towers will consume both time and money. The introduction of mm wave technology in 5G mandates small cell technology to maintain the range and quality of operation. Small cells will require additional cell in every few meters to guarantee the overall performance of the system [4, 5].

15.2.3 Testing of 5G

Testing technology faces a new challenge with every upcoming technology. 5G implementation points to a considerable upgrade of 4G by enabling higher bandwidths and throughput while assuring lower latency. The system architecture is therefore modified to accommodate higher bandwidths which are in turn achieved majorly using the upper frequency range proposed for 5G, the mm wave technology. The mm wave technology, being new, aims at carrier the signal to the end user while compensating for the inherent losses in this frequency range. Beamforming along with gain enhancement with massive MIMO technique is used to compensate for the propagation losses in the high band of 5G. The design engineers are constantly challenged to test the 5G devices at the new frequencies, bandwidths, and performance [6, 7]. Figure 15.2 summarizes the operational bands of 5G with their characteristics.

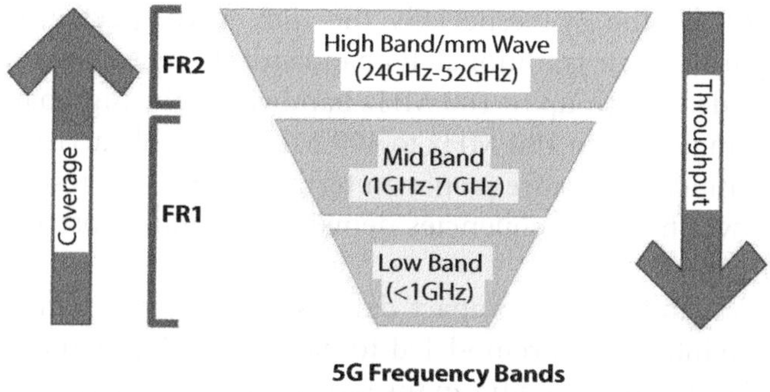

Figure 15.2 5G operational spectrum and its characteristics.

Testing of beam formation in a multiple antenna radiating or receiving system economically is a big challenge in 5G. A radio frequency system under study so far included input power delivered to an amplifying system tested for its output power or gain. With massive MIMO and beamforming in picture, phase of the wave, in addition to the gain, needs to be tested. The RF transmitter side is likely to create steered beams with different phases and gain. The phase and gain of these transmitters and receivers are to be investigated to characterize the operation of the system. Additionally, the performance of this new architecture with other active components of the 5G system also needs to be analyzed. A comparison of testing scenario for a single antenna to a multiple antenna system is shown in Figure 15.3.

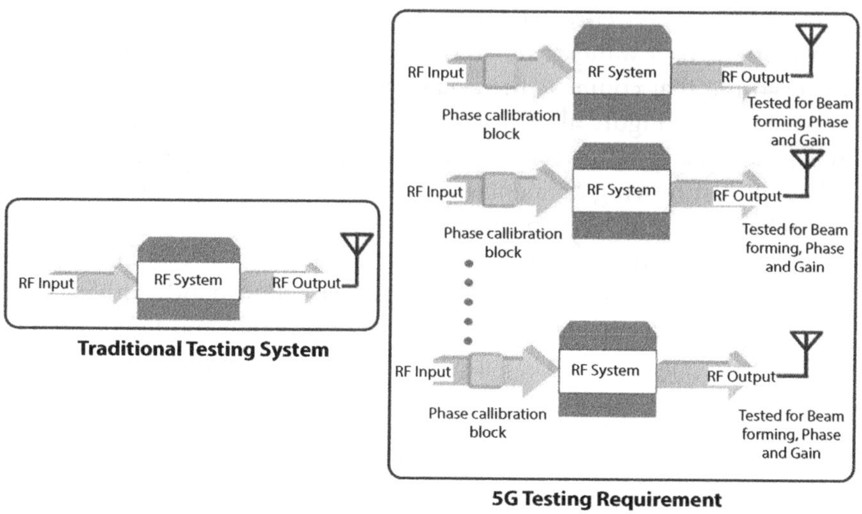

Figure 15.3 Testing challenge of 5G.

Bandwidth for the 5G, especially the mm wave spectrum, is much larger compared to the traditional 4G bandwidths. When the lower band, (sub-7 GHz) deals with up to 100 MHz bandwidth, the high bands have a bandwidth ranging from 100 MHz to 400 MHz. The subcarriers proposed are also at a much higher frequency in terms of GHz. So, testing of these higher bandwidths and frequencies demands advanced instruments and testing equipment.

5G also demands more compact devices, especially handheld devices. The whole architecture is remodeled to have printed antenna on the top package and rest of the circuitry in the layers beneath. The characterization of these devices for real world performance needs additional connectors to test these miniaturized devices. Addition of these connectors to test these miniaturized 5G devices for their wholesome performance with the other role players of the same system is another big challenge faced by testing engineers.

Testing of devices during high mobility promised by 5G is also another challenge. The antenna and its radiation and reception need to be characterized from any angle and direction to ensure performance of the device during 3D motion in space. The production flaws, while producing millions of devices, also needs to be addressed and tested cost effectively.

15.2.4 5G Backhaul

Two important points in a wireless network is the access network, which refers to the base stations to which the mobile phones and other devices are connected and the core network, referring to the point where the actual data which the user wants to download or upload is kept. Backhaul is simply everything that connects the access network part to the core network part as shown in Figure 15.4. The backhaul of the network can either be wired (build using optical fibre network) or wireless. Though backhaul is all about just moving data and information from one point to another, wireless backhaul for 5G is specially challenging.

The 5G requirements can be summarized as 1-10Gbps connection between devices, latency lesser than 1ms, very high bandwidth, almost 1000 times more connected devices, energy efficiency and better battery life. In short, 5G is expected to support 10 times more applications and 100 times more devices at 1000 times more bandwidth compared to the existing 4G services. So, in 5G, each service will have to share the backhaul with maximum efficiency while optimizing the QoE (Quality of Experience) of each application [8, 9].

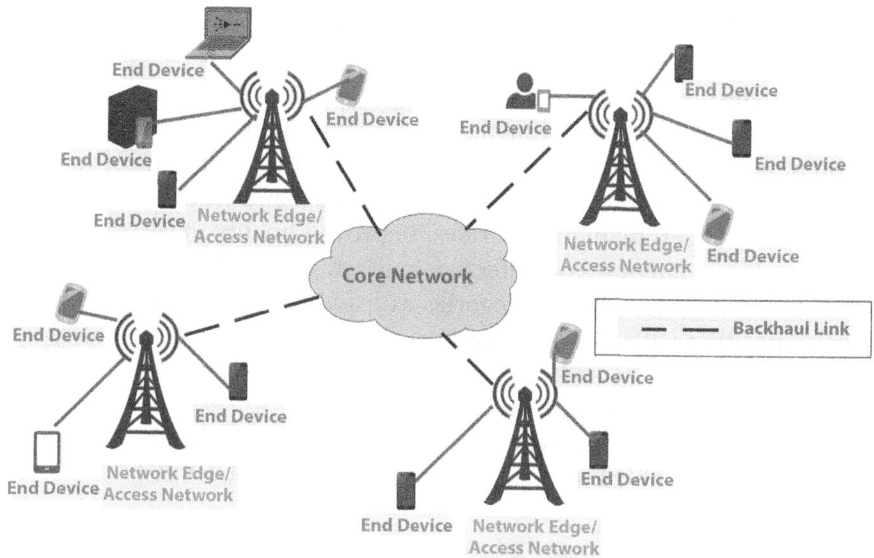

Figure 15.4 Communication system showing backhaul network.

5G also poses new interdependencies between previous straight forward network performance criterion. Till 4th generation, mobile networks were basically about people and data but 5G brings up another added dimension of things and services to the existing 4G. 5G is about people and things, data rate and latency [10].

15.2.5 Security Concerns

While 5G promises very high speed, low latency, and universal connectivity, it comes with a modified network architecture and adoption of a handful of new technologies. With the introduction of 5G, the number of devices connected to the network has increased exponentially along with numerous services which openly exposes unguarded data to anyone. The security solutions of 3G and 4G, therefore, cannot suffice for 5G [11].

5G supports virtualization and allows the usage of same network infrastructure for different applications. The preciously devised security architecture is not enough sophisticated for 5G. Some technologies like computing and virtualization, adopted to attain better efficiency, has their own security flaws.

15.3 Classification of 5G Services

5G is the latest mobile cellular technology implemented and is comprised of novel features like controlling devices objects and machines in addition to interconnecting people. As the basic cause for the evolutions of various cellular generations were to provide better speed, bandwidth, and services, 5G is characterized by high data rates, improved quality of service, larger coverage, lower latency, cost effective and reliability. 5G services can be broadly classified into three categories as listed below. Figure 15.5 shows the classification with its salient features.

- The extreme Mobile Broadband (eMBB)
- Massive Machine Type Communication (eMTC) and
- Ultra Reliable Low Latency Communication (URLLC)

To meet with the promises guaranteed by 5G, several new technologies were proposed and implemented.

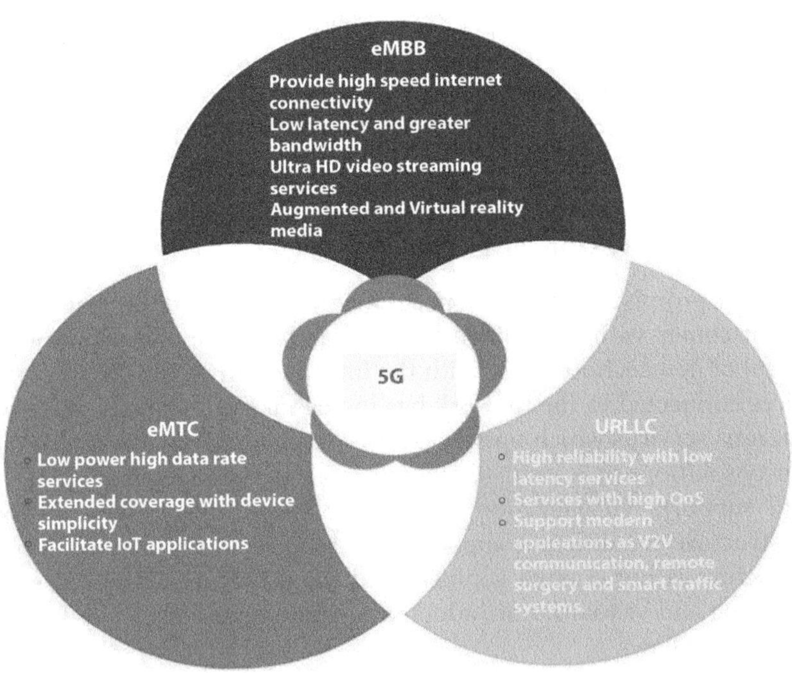

Figure 15.5 5G service categories.

15.4 Massive MIMO for 5G

Many MIMO systems were studied and proposed for modern wireless applications with 2-elements, 4-elements or 16-elements. MIMO was originally proposed for 4G implementation to achieve high bandwidth and higher throughput. As 5G bandwidth and throughput demands are much higher than that of 4G, simple MIMO system could hardly serve the purpose. Massive MIMO (Massive Multiple-Input Multiple-Output) is a cutting-edge extension of MIMO technology that exploits a significantly larger number of antennas at the base station or antenna panel in wireless communication systems. It has 16,32,64 or as large as 128 antenna elements. By doing so, it significantly inflates data throughput and augments the utilization of the existing spectrum. Figure 15.6 shows an architecture of massive MIMO [12–14]. The capability of multiplying antenna links capacity has positioned massive MIMO as a crucial component in modern wireless standards.

According to IEEE, moving from MIMO to massive MIMO involves making "a clean break with current practice using a large excess of service antennas over active terminals and time-division duplex operation. Extra antennas help by focusing energy into ever smaller regions of space to bring

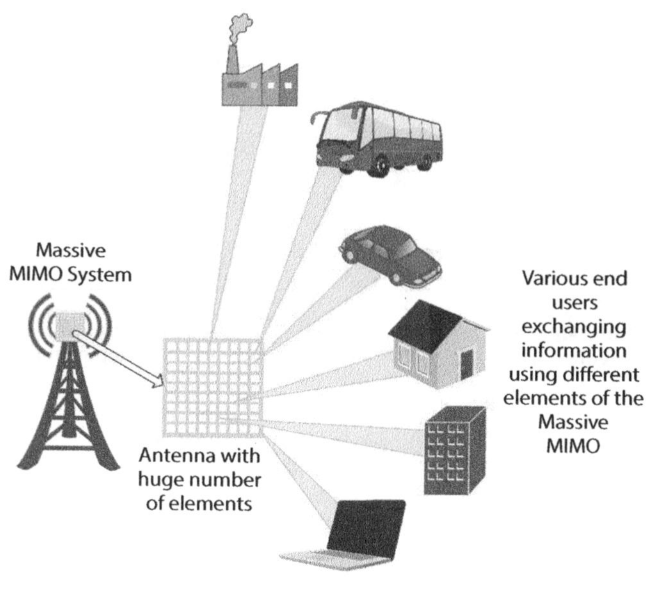

Figure 15.6 A MIMO system.

huge improvements in throughput and radiated energy efficiency." The group calls out other benefits including cheaper parts, lower latency, "simplification of the MAC layer, and robustness against intentional jamming."

Massive MIMO's deployment in 5G networks revolutionizes wireless technology by utilizing many antennas and advanced beamforming techniques. It significantly improves efficiency, capacity, and user experience, cementing its critical role in enabling the widespread adoption and success of 5G networks. Figure 15.7 lists some key points about Massive MIMO for 5G.

- **Increased Antenna Array:** Massive MIMO surpasses traditional MIMO systems, which typically have few antennas, by employing a significantly larger number of antennas—ranging from tens to thousands—at the base station. This multitude of antennas allows for simultaneous serving of multiple users, thereby improving system performance.
- **Spatial Multiplexing:** Leveraging spatial multiplexing techniques, Massive MIMO can transmit multiple independent data streams concurrently within the same frequency and

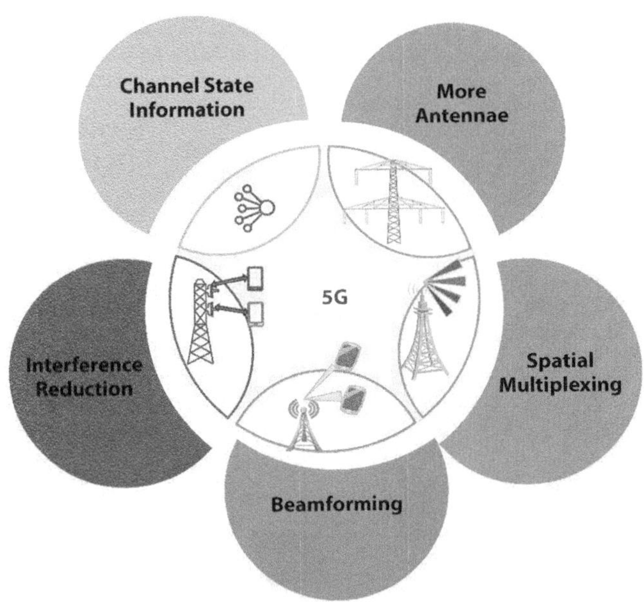

Figure 15.7 Some key aspects of massive MIMO.

time resources. By utilizing the spatial dimension, it achieves parallel serving of multiple users, leading to increased capacity and throughput.

- **Beamforming:** An essential aspect of Massive MIMO is beamforming, wherein the base station employs multiple antennas to create focused beams directed toward individual users [15]. This targeted approach enhances signal strength, reduces interference, and improves coverage, resulting in more efficient and reliable communication.
- **Interference Reduction:** With its extensive antenna array, Massive MIMO facilitates spatial separation of users, effectively reducing inter-user interference. By adaptively assigning distinct beams to each user and cancelling out interference from others, it enhances signal quality, overall network capacity, and performance.
- **Channel State Information (CSI):** Accurate channel state information is crucial for optimizing beamforming and interference cancellation in Massive MIMO. The base station constantly measures the channel conditions between antennas and users, enabling intelligent transmission beamforming and adaptive resource allocation.
- **Performance Benefits:** Massive MIMO brings numerous advantages to 5G networks, including increased capacity, improved spectral efficiency, wider coverage, and reduced latency. It excels in serving a larger number of simultaneous users and significantly enhances network performance in high-density areas, urban environments, stadiums, and other scenarios with high user demands [16].

Benefits of Massive MIMO include [17, 18]:

- **Enables Spatial Multiplexing:** Use of a large group of antenna helps to initiate spatial multiplexing to enhance the capacity/throughput, which will in turn assist to achieve high data rates and accommodate more users. Spatial Multiplexing is a technique used in wireless systems to physically separate the radio beams as shown in Figure 15.8. These radio beams can be focused to different users at the same time improving the spectral efficiency of the system.

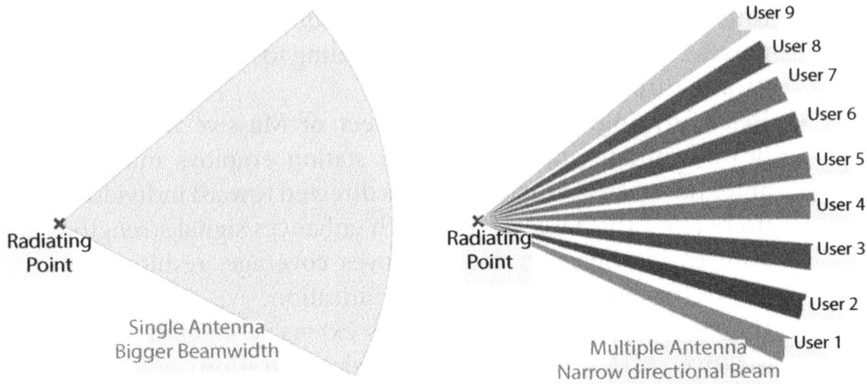

Figure 15.8 Comparison of beam with and without spatial multiplexing.

The beams so formed will be narrower and more directional in nature.

- **Low Power Consumption and Energy Efficiency:** Beamforming and beam steering achieved through massive MIMO help in focused beams which reduce the required power to be radiated. The system can thus achieve energy efficiency. The use of ultra-low power amplifiers in massive MIMO significantly reduces the required power and thus eliminated the need for other high power consuming electronic devices.
- **Higher Data Rate:** 5G promises 10 to 100 times more data rate and capacity than 4G. Massive MIMO technique helps in achieving array gain and spatial multiplexing which plays a major role in achieving better data rates.
- **Tracking Facilities and Improved Security:** Another important characteristic of Massive MIMO is beamforming. Massive MIMO employs multiple beams which are more focused. These narrow beams ensure more reliability, precise tracking, and improved security.
- **Minimized Fading:** The employment of massive number of antennae also acts against fading effects. Multiple receiver antennae receive the same signal through multiple paths and this ensures minimum loss of signal through fast or slow fading.

- **<u>Reduced Latency</u>:** Another welcoming benefit of 5G is the lower latency with respect to what it precedes. Latency can be explained as the reaction time or in other words, the time required to see the reaction for some action. 5G promises a latency as low as 1ms where 4G was offering a latency of 50ms which can be compared to the blink of an eye.
- **<u>Improved Robustness</u>:** 5G supported by massive MIMO offers high robustness against signal fading and jamming [19, 20]. The employment of multiple antennae also ensures quality of transmission even when an antenna fails.
- **<u>Simpler Linear Processing</u>:** The higher number of access station antennas in massive MIMO allows for simpler signal detectors and precoders, reducing overall system complexity.

Challenges of Massive MIMO include:

- **<u>Pilot contamination:</u>** This refers to the interference caused by reuse of pilot signals among different cells, which can degrade system performance and limit the achievable capacity due to correlated channel estimation errors.
- **<u>Channel estimation</u>:** Channel estimation in massive MIMO presents several challenges due to many antennas and users. These challenges include increased pilot contamination, pilot overhead, high-dimensional channel estimation, and limited coherence time, which require advanced algorithms and techniques to mitigate the impact of these factors and achieve accurate channel estimation.
- **<u>Precoding</u>:** The transmission of multiple streams in a multi-antenna system is achieved by the precoding technique. Precoding algorithms must efficiently optimize beamforming weights for multiple users while mitigating interference, requiring advanced techniques to handle the increased dimensionality and ensure efficient and effective precoding in massive MIMO systems.
- **<u>Muliti-user interference</u>:** Using massive MIMO technique, multiple users are supported by multiple antennas at the base station at the same time. This simultaneous communications between different users can cause multi-user interference, and hence degrade the total throughput performance of the system.

15.5 Conclusion

The article is a study of the major challenges faced in the implementation of 5th generation of mobile technology. A brief introduction is given on the different mobile technology generations continued by the identification of various challenges in the 5G technology. The chapter studies in detail, the 5G enabling , massive MIMO technology, its features, benefits and challenges.

References

1. Dangi, R., Lalwani, P., Choudhary, G., You, I., Pau, G., Study and investigation on 5G technology: A systematic review. *Sensors*, *22*, 1, 26, 2021.
2. Al-Falahy, N. and Alani, O.Y., Technologies for 5G networks: Challenges and opportunities. *IT Prof.*, *19*, 1, 12–20, 2017.
3. Guevara, L. and Auat Cheein, F., The role of 5G technologies: Challenges in smart cities and intelligent transportation systems. *Sustainability*, *12*, 16, 6469, 2020.
4. Taheribakhsh, M., Jafari, A., Peiro, M.M., Kazemifard, N., 5g implementation: Major issues and challenges, in: *2020 25th International Computer Conference, Computer Society of Iran (CSICC)*, 2020, January, IEEE, pp. 1–5.
5. Gustavsson, U., Frenger, P., Fager, C., Eriksson, T., Zirath, H., Dielacher, F., Carvalho, N.B., Implementation challenges and opportunities in beyond-5G and 6G communication. *IEEE J. Microwaves*, *1*, 1, 86–100, 2021.
6. Hausl, C., Emmert, J., Mielke, M., Mehlhorn, B., Rowell, C., Mobile network testing of 5G NR FR1 and FR2 networks: Challenges and solutions, in: *2022 16th European Conference on Antennas and Propagation (EuCAP)*, 2022, March, IEEE, pp. 1–5.
7. Yang, Y., Xu, J., Shi, G., Wang, C.X., *5G wireless systems*, Springer International Publishing, 2018.
8. Jaber, M., Imran, M.A., Tafazolli, R., Tukmanov, A., 5G backhaul challenges and emerging research directions: A survey. *IEEE Access*, *4*, 1743–1766, 2016.
9. Sawad, I., Nilavalan, R., Al-Raweshidy, H., Backhaul in 5G systems for developing countries: A literature review. *IET Commun.*, *17*, 6, 659–669, 2023.
10. Mowla, M.M., Ahmad, I., Habibi, D., Phung, Q.V., Energy efficient backhauling for 5G small cell networks. *IEEE Trans. Sustain. Comput.*, *4*, 3, 279–292, 2018.
11. Khan, R., Kumar, P., Jayakody, D.N.K., Liyanage, M., A survey on security and privacy of 5G technologies: Potential solutions, recent advancements, and future directions. *IEEE Commun. Surv. Tutor.*, *22*, 1, 196–248, 2019.

12. Zhang, J., Björnson, E., Matthaiou, M., Ng, D.W.K., Yang, H., Love, D.J., Prospective multiple antenna technologies for beyond 5G. *IEEE J. Sel. Areas Commun.*, *38*, 8, 1637–1660, 2020.
13. Samdanis, K. and Taleb, T., The road beyond 5G: A vision and insight of the key technologies. *IEEE Netw.*, *34*, 2, 135–141, 2020.
14. Siddiqi, M.A., Yu, H., Joung, J., 5G ultra-reliable low-latency communication implementation challenges and operational issues with IoT devices. *Electronics*, *8*, 9, 981, 2019.
15. Abdullah, Q., Salh, A., MohdShah, N.S., Abdullah, N., Audah, L., Hamzah, S.A., Nordin, S., A brief survey and investigation of hybrid beamforming for millimeter waves in 5G massive MIMO systems, arXiv preprint arXiv:2105.00180, 2021.
16. Belgiovine, M., Sankhe, K., Bocanegra, C., Roy, D., Chowdhury, K.R., Deep learning at the edge for channel estimation in beyond-5G massive MIMO. *IEEE Wireless Commun.*, *28*, 2, 19–25, 2021.
17. de Figueiredo, F.A.P., An Overview of Massive MIMO for 5G and 6G. *IEEE Lat. Am. Trans.*, *20*, 6, 931–940, 2022.
18. Shaikh, A. and Kaur, M.J., Comprehensive survey of massive MIMO for 5G communications, in: *2019 Advances in Science and Engineering Technology International Conferences (ASET)*, 2019, March, IEEE, pp. 1–5.
19. Avasthi, S., Tripathi, S.L., Sanwal, T., Mahmud, M., Deep Learning Models for Image Annotation Application in a 6G Network Environment, in: *Development of 6G Networks and Technology*, pp. 201–221, 2024.
20. Pragati, Tripathi, S.L., Patre, S.R., Singh, S., Singh, S.P., Triple-band microstrip patch antenna with improved gain. *2016 International Conference on Emerging Trends in Electrical Electronics & Sustainable Energy Systems (ICETEESES)*, Sultanpur, India, pp. 106–110, 2016, doi: 10.1109/ICETEESES.2016.7581362.

16

Smart Nanomaterials: Revolutionizing Drug Delivery Strategies

Jujhaar Singh Aidhen[1], Arjun Vitthal Chambarge[1], Chavan Aniket Navnath[1], Atharv Mohan Patil[1], Vedant Dnyandev Arjun[1], Jupinder Kaur[1]* and Rajan Vohra[2]

[1]*Department of Engineering Sciences, Vishwakarma University, Pune, Maharashtra, India*
[2]*School of Electronics and Electrical Engineering (SEEE), Lovely Professional University, Punjab, India*

Abstract

This review comprehensively explores advancements in nanomedicine, with a primary focus on developing and applying cutting-edge drug delivery systems for medical and pharmaceutical progress. Emphasizing key aspects such as structural stability, encapsulation efficiency, and targeted therapy approaches, the investigation spans various nanomaterials, including carbon-based systems, chitosan, graphene quantum dots, metal-organic frameworks, and superparamagnetic iron oxide nanoparticles. The paper delves into nanocomposites, smart drug delivery systems, and stimuli-responsive strategies, thoroughly examining these materials' synthesis, optimization, and biocompatibility for specific medical applications, particularly in cancer treatment. Highlighting breakthroughs in understanding tumor microenvironments, exploring various targeting ligands, and elucidating the role of nanomaterials in responsive drug delivery, the research extends to diverse synthesis methods and functionalization approaches, aiming to optimize drug loading capacity, release behavior, and therapeutic outcomes. Additionally, the review explores the integration of computational models, magnetic nanoparticles, and advanced imaging techniques in drug delivery systems. The future trajectory involves the development of biodegradable nanomaterials, biosensors, and green nanotechnology, all while emphasizing collaboration, standardization, and addressing safety concerns in medical nanotechnology.

**Corresponding author*: jupinder.kaur@vupune.ac.in

Suman Lata Tripathi, Vrinda Gupta, Sobhit Saxena, Shipra Upadhyay and Sudip Ghosh (eds.) Micro and Nano Semiconductor Devices for Digital, Analog and Sensor Design, (289–324) © 2026 Scrivener Publishing LLC

Keywords: Nanomedicine, drug delivery systems, nanomaterials, targeted therapy, stimuli-responsive, cancer treatment, biocompatibility

16.1 Introduction

Drugs play an important role in modern medicine, offering solutions to many health issues and giving pain relief from treating complex diseases. The safety and efficiency of these drugs are highly significant. In the drug delivery field side effects are minimized. Drug delivery involves the utilization of nanoparticles which are small in size and measured in nanometers. These nanoparticles serve as carriers for medications and are assimilated within the body. When nanoparticles are designed for performing drug delivery applications, it gives advantages over traditional drug delivery methods. These drugs improve the pharmacokinetics of drugs, enhance their bioavailability, target specific cells, and reduce systemic toxicity. In drug delivery, nanoparticles are taken in different forms like liposomes, polymeric nanoparticles, metallic nanoparticles, and dendrimers. These carriers can protect them from degradation and also allow for controlled release and sustained therapeutic effects. To navigate the complex biological barriers within the body nanoparticles are tailored. It improves the efficient delivery of drugs to the specific location where they are intended to take effect.

Drug delivery systems (DDS) play a pivotal role in the management of diverse medical conditions. Liposomes, dendrimers, and polymers—these nanoparticle-based delivery systems are prominent due to their ability to enhance the drug stability, solubility, and targeted drug delivery. This study of nanoparticles focuses on carbonaceous nanomaterials, specifically carbon nanohorns (CNHs)as potential drug carriers. Carbon nanohorns are inherently hydrophobic and can be used for aqueous applications through the oxygen plasma treatment which introduces the functional groups of oxygen and enhances hydrophilicity. Some nanoparticles are used in cancer treatment like mesoporous silica nanoparticles. The significant challenges faced in cancer treatment primarily due to limitations of conventional chemotherapy include poor targeting and severe side effects. To minimize these issues, a drug delivery system has been developed, utilizing various nanoparticles like mesoporous silica nanoparticles (MSNs) which are gatekeepers to control drug release. The gatekeeper likes β-cyclodextrin (βCD), which can cap the pores of MSNs effectively. This system remained effectively inactive in normal cells, but the drugs were released in the tumor microenvironment with real-time imaging capabilities. This pH-responsive and targeted FA-CβCD-MSNs system holds efficient tools for diagnostic and therapeutic applications in cancer treatment. Magnetically responsive nanoparticles like iron oxide are used in drug delivery. Superparamagnetic iron oxide nanoparticles (SPIONs) showed great work in augmenting therapy and drug delivery in biomedical applications. These nanoparticles have features like ease of synthesis and biocompatibility, along with the ability to be remotely controlled using magnetic fields. This study focuses on using SPIONs under oscillating magnetic fields (OMFs) to improve drug transport and achieve the triggered drug release at low OMF strength and minimize the risk of tissue damage associated with temperatures and high field strengths. When these nanoparticles are broken down in biopolymers including DNA, it leads to improved drug penetration. The release system was developed for utilizing doxorubicin (DOX) and gold nanoparticles (GNPs) for potential cancer therapy. DOX is a widely used chemotherapy drug which is known for its side effects and this system aims to control and target DOX release. Partial funding for the research was also provided through the ExCITE program, backed by the National Institutes of Health. A groundbreaking study aims to revolutionize drug delivery with stimulus-responsive nanoparticles. The focus is on the designing of photosensitive polymeric nanoparticles, specifically photoresponsive nano-micelles (PNms), for precise drug release. When UV light is exposed, these nanoparticles undergo photoisomerization and allow for the controlled structural changes. The application of light offers immense potential for precise drug delivery. Further research may explore biocompatibility to advance their medical applicability and offer a promising avenue to enhance drug delivery specificity and efficacy while reducing the side effects.

Natural polymers can be used for drug delivery. The utilization of natural polymers like chitosan have gained significant attention in drug delivery systems with a focus on enhancing drug resistance time in the gastrointestinal tract. Due to electrostatic interactions chitosan exhibits bioadhesive medical applications in the GI tract, particularly in the mucosal tissues. Chitosan nanoparticles can control drug release and targeted delivery. The chitosan concentration, molecular weight, and the chitosan/tripolyphosphate (TPP) ratio, have already been investigated to achieve controlled and sustained drug release as these are the factors affecting the drug release. Polymers, particularly polysaccharides like chitosan and carrageenan, attract attention within the realm of drug delivery and biomedical applications because of their unique characteristics. Chitosan, obtained from chitin, possesses minimal toxicity alongside biocompatibility and biodegradability, rendering it a well-suited polymer for drug delivery mechanisms. In the context of delivering drugs, nanoparticles sensitive to temperature variations find applications, presenting inventive approaches for controlled drug release. These nanoparticles possess distinctive properties and the capacity for reversible phase change at the lower critical solution temperature, enabling precise control over drug delivery. Through synthesis and characterization, these nanoparticles have been evaluated for their drug delivery potential. biocompatibility. These results provide a foundation for targeted drug delivery for future applications and tissue engineering based on temperature sensitivity.

16.2 Disease Specific Drug Delivery

Chitosan and its derivative, carboxymethyl chitosan (CMC), are polymers known for their biocompatibility and biodegradability, positioning them to be promising options for applications in delivering drugs. Synthesis of CMC nanoparticles, utilizing a cross-linking method, was accompanied by characterization through SEM, FTIR, and magnetic measurements. The resulting nanoparticles exhibited a spherical morphology with diameters ranging from 200 to 250 nm. Importantly, these nanoparticles displayed no toxicity towards mouse fibroblast cells and demonstrated antibacterial activity against staphylococcus aureus.

For drug delivery purposes, CMC nanoparticles using the emulsion crosslinking method were filled with the anticancer drug 5-fluorouracil (5-Fu). The encapsulation efficiency for 5-Fu was found to be 78.3%. *In vitro* drug discharge studies indicated that the nanoparticles discharged 74% of drug concentration at pH 6.8 after 12 hours. Cell viability assays further revealed that 5-Fu-loaded CMC nanoparticles reduced the feasibility of KB cancer cells in a manner contingent on the dosage. Collectively, these results indicate the potential of CMC nanoparticles for efficient cancer drug delivery.

Additionally, FT-IR and SEM analyses confirmed the dimensions of CMC nanoparticles within the 200-250 nm range, making them promising for drug delivery applications. The magnetic behavior observed at low temperatures hints at their potential for targeted drug delivery. Cytotoxicity studies affirmed their non-toxicity towards mouse L929 cells, and remarkable antibacterial efficacy was noted in the presence of staphylococcus strains. Moreover, *in vitro* drug release research investigations demonstrated gradual and regulated dispensation of 5-Fu at pH 6.8, emphasizing their suitability for the purpose of regulating drug delivery. Notably, 5-Fu-incorporated CMC nanoparticles exhibited increased toxicity towards KB cells, correlating with elevated initial loading concentrations. Overall, these findings imply that CMC nanoparticles, with their unique magnetic properties, hold promise as carriers for cancer drugs, warranting further exploration in therapeutic applications [1]. The following figure will help you with better understanding (Figure 16.1).

Bacterial infections pose a significant global health challenge, often exhibiting resistance to traditional treatments like antibiotics. Nanomaterials-based antimicrobial strategies emerge as promising alternatives, offering targeted drug delivery and multiple passive antimicrobial mechanisms. Nevertheless, traditional nanomaterial-based treatments may result in the demise of beneficial bacteria and exhibit limited bactericidal effectiveness. On-demand and precise nanomaterials-based strategies address these challenges by minimizing damage to healthy tissue and reducing drug toxicity. This review delves into the design and applications of these strategies, highlighting their mechanisms, strengths, limitations, and future perspectives. Stimuli-responsive nanomaterials, activated by the infectious microenvironment, and targeting nanomaterials, loaded with targeting moieties, are two promising approaches. Each approach has its own advantages and disadvantages, which are discussed in detail. The article provides concrete instances of hypoxia-responsive, pH-responsive, ROS-responsive, GSH-responsive, toxin-responsive, and enzyme-responsive nanomaterials. These examples demonstrate their capability for precise and on-demand antibacterial therapy. On-demand and precise antibacterial therapies using nanomaterials hold immense promise for combating bacterial infections. MMP-responsive, lipase-responsive,

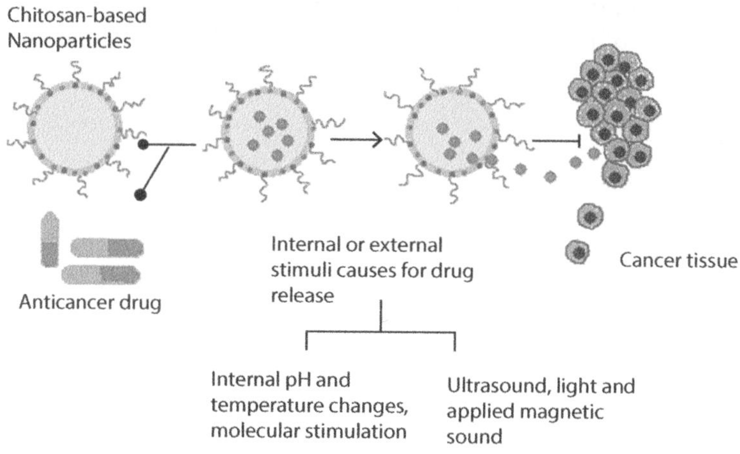

Figure 16.1 Anticancer drug with chitin and chitosan-based nanoparticles.

and gelatinase-responsive nanomaterials are under development, each targeting specific enzymes present at infection sites. Targeting bacteria through nonspecific binding, hydrophobic interactions, antigen-antibody reactions, and aptamer-responsive strategies offers precise antibacterial targeting. However, challenges like charge density optimization, anion-rich structure interference, and the development of broad-spectrum antibodies need to be addressed. Targeting infected cells using electrostatic interaction, apt-receptor interaction, and bacteriophage modification is promising but requires further research. Membrane-coated nanomaterials, including macrophage-, neutrophil-, dendritic cell-, and platelet-derived membranes, offer simple preparation, high biosafety, and rich functions. Hybrid membrane-coated nanomaterials combine the advantages of different cell membranes. Bacteria hunters and virus-based or biomimetic virus-structured nanomaterials provide natural targeting towards specific bacteria. The figure below will help you understand (Figure 16.2).

Despite the potential of on-demand and precise nanomaterials-mediated antibacterial therapy, challenges remain in developing more stable, reproducible, cost-effective nanomaterials with fewer side effects. Further research is crucial to translate these promising therapies into widespread clinical use [2].

This study delves into the capability of utilizing Ag-doped silicene co-doped with boron, phosphorus, sulfur and nitrogen as a targeted delivery system for drugs, treating advanced metastatic melanoma with Dacarbazine (DCB). Employing the first principles density functional theory, it further identifies dcb_S@AgSi as the most efficient adsorbent for DCB. The analysis uncovers enhanced electronic properties and significant alterations in bond length and energy gap, shedding light on the reactivity and stability of the molecular interactions. Further molecular analyses,

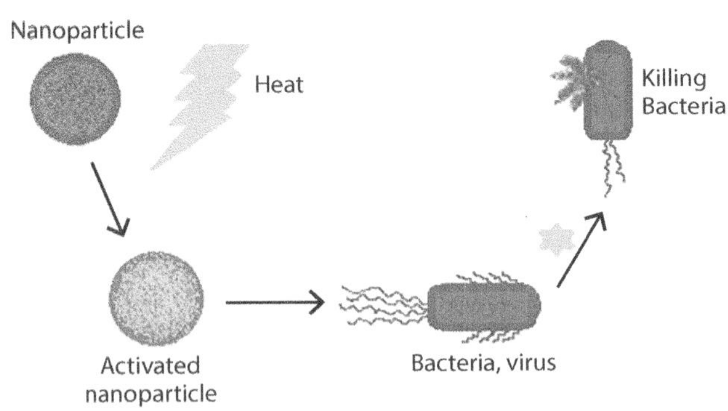

Figure 16.2 Nanomaterial working to kill the bacteria and viruses.

such as Density of States plots, Electron Localization Function analysis, non-covalent interaction analyses, and Quantum Theory of Atoms in Molecule contribute to a comprehensive understanding of the system's behavior. Adsorption studies confirm the chemical viability of the process, emphasizing the potential of dcb_S@AgSi as the preferred adsorbent for effective DCB delivery. The study's findings extend to the optimization of the systems, revealing post-optimization results through Gauss View 6.0.16 software. Electronic properties and quantum descriptors, determined through established equations, are complemented by Natural Bond Orbital (NBO) analysis and wave function analysis, enhancing the depth of insights into the molecular dynamics. Frontier Molecular Orbital (FMO) analysis provides valuable information on reactivity, stability, and conductivity. Notably, the study concludes with implications for the effective delivery of DCB for the management of advanced melanoma, highlighting the significance of dcb_S@AgSi in the above context. Consider the figure below (Figure 16.3). Dopants introduced into Ag-functionalized silicene nanomaterials influence their electronic properties and reactivity, affecting their efficacy as drug delivery vehicles for the anticancer drug dacarbazine (DCB). The introduction of DCB generally decreased the energy gap (Eg) of the doped surfaces, except for dcb_S@AgSi, which exhibited increased stability and lower reactivity. Global hardness (η) and softness (σ) are inversely related to reactivity, with higher η suggesting lower reactivity and vice versa. HOMO and LUMO analyses revealed electron transfer from the adsorbents to DCB during adsorption. Solvent polarity significantly affects Eg and global quantum descriptors. Density of states (DOS) analysis showed varying HOMO levels and highlighted the importance of Si atoms in the electronic structure of these systems.

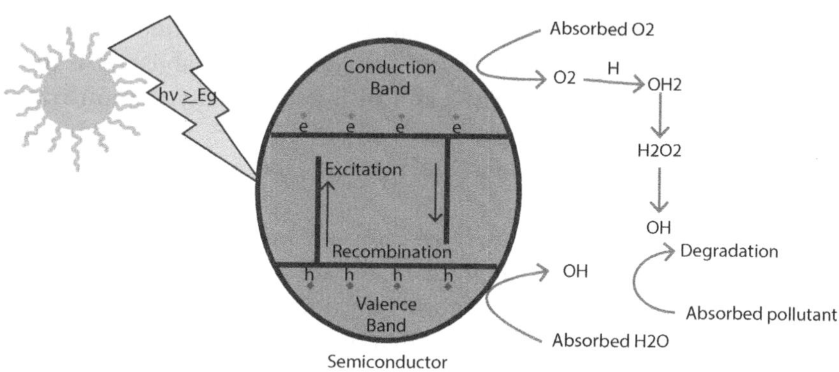

Figure 16.3 Multifunctional graphene based nanocomposites for drug delivery.

ELF analysis reveals increased electrophilicity in Ag-doped silicene and nucleophilicity in Si and H atoms. According to NBO analysis, the dcb_S@AgSi system is identified as the most stable. DFT calculations show robust adsorption in dcb_S@AgSi, indicating its potential for drug delivery. QTAIM analysis confirms stability and electrophilicity in dcb_S_Ag@Si. NCI analysis favors dcb_N_Ag@Si and dcb_B_Ag@Si for delivery of DCB, with the latter showing the weakest interaction. Van der Waals forces that were weak contribute to stability, and N-co-doped silicene stands out as a promising adsorbent for DCB delivery with strong adsorption and weak intermolecular interactions [3].

Researchers have developed Lipopolysaccharide-modified tumor-derived microparticles (TMPs-PEI-LPS) using polyethylenimine as a linking together agent to improve the immunogenicity of TMPs while mitigating the high toxicity associated with Lipopolysaccharide (LPS). Characterization using various techniques confirmed that TMPs-PEI-LPS exhibited heightened uptake by dendritic cells compared to unmodified TMPs. Moreover, they demonstrated an upregulation in the expression of CD86, CD80, and TNF-α in dendritic cells and facilitated M1 polarization of macrophages. *In vivo* studies further illustrated that TMPs-PEI-LPS induced immune stimulation, resulting in reduced tumor size and growth rate, along with an increase in CD8+T cells in mice with tumors. Crucially, TMPs-PEI-LPS did not induce toxicity or promote tumor growth. In summary, TMPs-PEI-LPS emerge as a promising antigen delivery platform for cancer immunotherapy due to their enhanced immunogenicity, reduced toxicity, and potential for personalization [4].

Magnetic Resonance Navigation utilizes Magnetic Nanoparticles incorporated into microrobots or microcarriers, allowing precise navigation through blood vessels by leveraging magnetic gradients and MRI tracking. These MNPs also serve the purpose of inducing localized hyperthermia, temporarily disrupting the Blood-Brain Barrier and facilitating targeted transportation of drugs to the brain. Research has established a direct relationship between hyperthermia and BBB permeability, demonstrating controlled opening and subsequent recovery of the BBB. This non-invasive methodology presents considerable promise for addressing disorders related to the brain [7].

Melanoma, recognized as the most intensely invasive form of skin cancer, presents formidable challenges in its advanced stages. Traditional treatments like radiotherapy, chemotherapy, immunotherapy and targeted therapy, often encounter limitations, marked by the development of drug resistance and poor response rates. Nanotechnology emerges as a promising avenue, enhancing drug delivery, refining treatment efficacy, and targeting specific signaling pathways integral to melanoma progression. A variety of nanoparticles like inorganic nanoparticles, carbon based nanoparticles, encompassing liposomes, protein-based nanoparticles, polymersomes and dendrimers, have been explored for treating melanoma. These nanoparticles play a pivotal role in safeguarding and encapsulating drugs, delivering them precisely to tumor sites, and facilitating sustained drug release. Nanotechnology has been integrated into conventional therapies such as targeted therapy, chemotherapy, photodynamic therapy and immunotherapy. Notably, the nanoparticle-mediated delivery of chemotherapeutic agents demonstrates heightened efficacy while minimizing side effects. Targeted therapy utilizing nanoparticles to deliver inhibitors against specific signaling pathways, such as the PI3K/Akt and MAPK pathways, has exhibited promising outcomes. Furthermore, nanotechnology holds potential in directing its focus towards the mitochondrial apoptotic pathway, a critical player in melanoma progression. By inducing apoptosis in melanoma cells, approaches rooted in nanotechnology could substantially enhance treatment outcomes. Overall, nanotechnology offers a promising avenue for improving melanoma treatment by enhancing drug delivery, targeting specific signaling pathways, and inducing apoptosis in melanoma cells. Check figures to learn treatments (Figure 16.4). Nanotechnology has

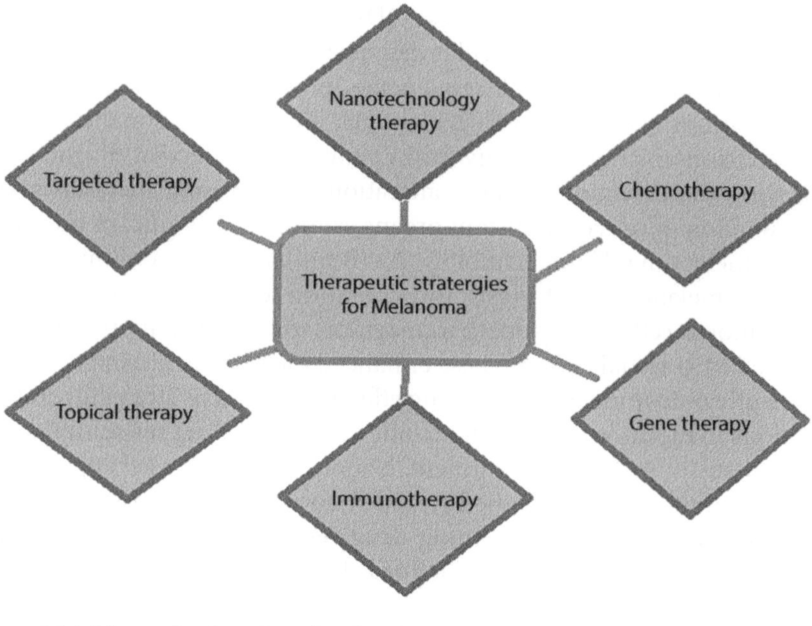

Figure 16.4 Nanotechnology-based and melanoma-based treatments.

emerged as a promising tool in the fight against melanoma, a type of skin cancer known for its aggressive nature and resistance to conventional therapies. Utilizing their distinctive properties and capacity to selectively target specific cells, nanoparticles present a spectrum of advantages in the context of melanoma treatment, diagnosis, and prospective research avenues. In the domain of treatment, nanoparticles have proven instrumental in augmenting the effectiveness while mitigating the side effects associated with diverse therapeutic approaches. For instance, chemotherapy drugs like Doxorubicin (Dox) have been encapsulated in gold nanoparticles or cholesterol-rich nano-emulsions, improving their delivery and reducing systemic toxicity. Similarly, nanoparticles carrying siRNAs against BRAF, Akt3, and c-Myc have shown promise in targeted therapy, effectively inhibiting signaling pathways involved in melanoma progression. Nanoparticle-based immunotherapy has also demonstrated potential, delivering agents like IL-2, IFN-alpha, and CpG-loaded virus-like particles to boost the immune response against melanoma cells. Additionally, gold nanoparticles conjugated with photosensitizers have enhanced the effectiveness of photodynamic therapy (PDT) in melanoma treatment.

Early diagnosis is crucial for improving melanoma survival rates, and nanotechnology plays a significant role in this area. Nanoparticle quantum dots (QDs), with their unique photophysical properties, can be conjugated with cancer-specific molecules to recognize and distinguish melanoma cells from normal melanocytes. These QDs enable early detection and facilitate timely intervention. Molecular diagnosis, using techniques like tissue arrays, proteomics, and DNA sequencing, provides insights into molecular alterations associated with melanoma, aiding in personalized treatment decisions. Nanotechnology further enhances molecular diagnosis by improving the detection of these alterations. Addressing metastasis, the dissemination of cancer to distant organs, stands as a significant hurdle in treating melanoma. Nanotechnology has been harnessed to refine animal models of melanoma metastasis, incorporating techniques such as tracking melanoma cells using superparamagnetic iron oxide nanoparticles for enhanced MRI visualization. These enhanced models can aid in predicting therapeutic outcomes and developing effective treatment strategies. Future directions in melanoma research include addressing drug resistance, a significant issue in melanoma treatment. Nanotechnology-based approaches, such as theranostic nanomedicine, which combines diagnosis and treatment, hold promise in overcoming drug resistance and providing real-time monitoring of therapeutic actions. In conclusion, nanotechnology offers a versatile and promising approach to improve melanoma treatment, diagnosis, and future research directions, paving the way for more effective and personalized therapies for this challenging disease [6].

Nanomaterials, with their unique properties and diverse applications, have revolutionized the field of biomedical science, particularly in targeted drug delivery and antibacterial therapy. These materials, ranging from MXene and polymers to graphene and CNTs, offer exceptional strength, durability, and diffusivity, making them ideal for controlled drug release and combating bacterial infections. Biodegradable nanomaterials are favored over non-biodegradable ones due to their biocompatibility and reduced environmental impact. Advanced characterization techniques like SEM, FTIR, and XRD provide insights into the morphology and structure of nanomaterials, enabling researchers to assess their suitability for specific applications. Combining nanomaterials with strong and conductive counterparts further enhances their performance, expanding their potential in biomedical applications. Nanomaterials like carbon nanotubes (CNTs) and graphene oxide (GO) have shown promise in treating brain tumors, enhancing drug delivery, and exhibiting antibacterial properties. MXene, a novel 2D nanomaterial, exhibits excellent conductivity and stability, making it suitable for cancer therapy and tissue regeneration. Layered double hydroxides function as carriers for drug delivery and enhance implant hydrophilicity. Natural and synthetic polymers including silk fibroin, gelatin, collagen, PCL, PTMC, PLA, PLGA, and PGA, find extensive applications in regenerative medicine and tissue engineering, owing to their biocompatibility and adjustable characteristics. Metal and metal-based nanomaterials find applications in joint replacements and dental implants, while conductive materials like CNTs and graphene are incorporated into polymers for nerve and cardiac tissue engineering. Nanomaterials, with their vast potential and continuous advancements, are poised to transform biomedical applications, offering targeted therapies, enhanced tissue regeneration, and improved diagnostic capabilities. Gold nanoparticles have emerged as versatile tools in nanomedicine, enabling imaging, theranostics, and drug delivery applications. Their unique properties, combined with polymers, have shown promise in enhancing electrical conductivity and mechanical properties for cardiac tissue regeneration. Surface morphology, structure, geometry, dimension, molecular weight, and thermal/electrical conductivity are among the distinct physical and chemical characteristics exhibited by nanomaterials. Characterization techniques play a crucial role in identifying these properties, which are essential for evaluating the performance and safety of nanomedicines. Reliable characterization methodologies are crucial for assessing the toxicity, quality, and safety of nanostructures. Biological compatibility research using these techniques

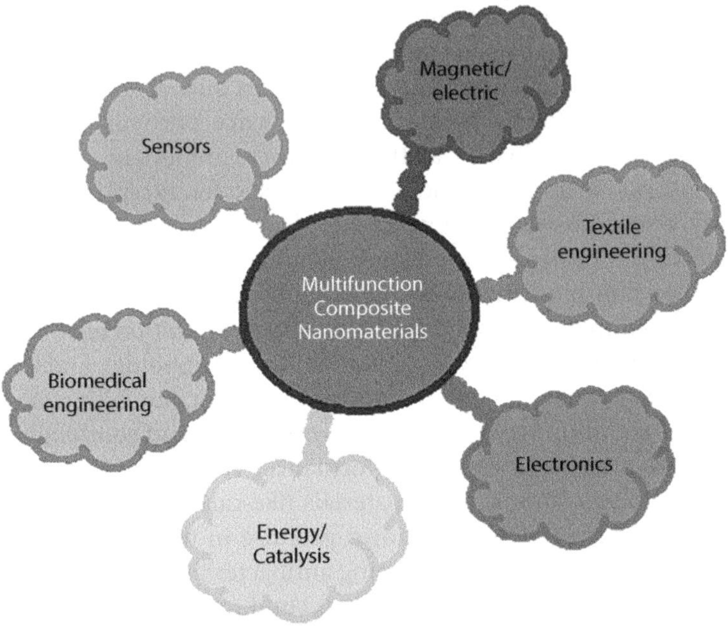

Figure 16.5 Applications of multifunctional composite nanomaterials.

is paramount for healthcare applications of nanoparticles. Specialized methods are needed to provide structural and chemical investigations with appropriate spatial resolutions for various forms of nanostructured substances. Characterization methods are employed to investigate the interior geometry, structure, morphology, and chemical properties of nanomaterials. These tools help determine the presence of different chemicals inside nanoparticles and chemical functionalities on their surfaces. Transmission electron microscopy (TEM), with its high spatial resolution (0.08 nm), is a primary method for characterizing nanomaterials. It provides information about particle size, morphology, and structural details through imaging, spectroscopy, and diffraction data. TEM's ability to capture high-resolution images of individual nanomaterials makes it suitable for analyzing crystallinity and structural details.

Refer to the figure above (Figure 16.5) for potential applications. Scanning electron microscopy (SEM) stands as a widely employed method for the examination and analysis of solid objects, particularly through micro- and nanoparticle imaging. With a resolution reaching 10 nm, it proves suitable for investigating particle sizes. SEM scrutinizes the surface of the specimen using a beam of electrons, generating signals that unveil topographical and atomic structure details.

Atomic force microscopy (AFM), known for its environmental friendliness, furnishes functional, mechanical and high-resolution structural data, on nanomaterials. Its operation involves a sharp needle and a cantilever for detecting surface features, providing genuine 3D surface characteristics of samples.

Magnetic resonance force microscopy (MRFM) endeavors to detect MRI signals emanating from nanoscale specimens with near-atomic resolution. Merging features of AFM and NMR, it delivers high-resolution 3D images of biological materials.

X-ray diffraction (XRD) finds application in identifying crystalline forms within nanomaterials and measuring their structural characteristics. Unlike assessing, XRD evaluates characteristics and chemical composition linked to crystal structure. It proves valuable in detecting mineral phases in environmental samples [8].

Stroke, a prominent cause of mortality and disability, faces limitations in treatment options due to the challenges posed by the blood-brain barrier and the necessity to address multiple therapeutic targets. In response to these challenges, researchers have introduced a cutting-edge class of antioxidant nanoparticles (NPs) tailored for stroke treatment, with a specific emphasis on those derived from betulinic acid (BA). The BA NPs underwent a chemical transformation, evolving into betulinic amine (BAM) NPs, notable for their accelerated drug release in acidic conditions. Subsequently, these BAM NPs were further refined by integrating AMD3100, recognized as a CXCR4 antagonist, onto their surface, resulting in the formation of AMD3100-conjugated BAM NPs (A BAM NP's). This sophisticated modification demonstrated heightened targeting precision and improved delivering of drugs to the ischemic brain in murine models. Impressively, A-BAM NPs exhibited therapeutic efficacy in stroke treatment, both as standalone agents and carriers for NA1, a fusion peptide renowned for its neuroprotective qualities. Consequently, A BAM NPs emerge as a promising pathway for precise drug delivery and advanced stroke therapy [5].

Cancer remains a formidable global health challenge, with existing treatment options often falling short in various aspects. Nano Drug Carriers (NDCs) present a promising avenue for advancing cancer therapy by addressing prevalent issues like suboptimal efficacy, elevated costs, and non-specific toxicity. Polymeric nanoparticles (PNPs) stand out among NDCs due to their advantageous features, encompassing a controlled molecular weight, cost-effective synthesis, biocompatibility, customizable chemical structure, facile functionalization and biodegradability. This study delved into a novel approach, employing metal-free ring-opening copolymerization (ROCOP), to craft a low molecular weight, fully alternating semi-aromatic polyester (SAP) using commercially available monomers and metal-free activators.

The synthesized SAP formed the basis for fabricating curcumin (CUR)-loaded nanoparticles (NPs) tailored for enhanced cancer therapy. Outcomes highlighted the successful production of the biocompatible, biodegradable, and hemocompatible poly(cyclohexene oxide-alt-phthalic anhydride) (poly(CHO-alt-PA))copolymer. These engineered poly(CHO-alt-PA)CUR nanoparticles showcased a regulated and prolonged release of CUR, coupled with heightened effectiveness in anti-cancer therapy across *ex vivo* and *in vitro* models. Crucially, rigorous assessments in BALB/c mice revealed no discernible acute or systemic toxicity associated with the poly(CHO-alt-PA) nanomaterial.

Essentially, this study highlights the capability of metal-free ROCOP-based SAP nanoparticles as a sturdy foundation for successful and secure cancer therapy. Polymeric nanomaterials, especially copolymers, have garnered prominence in biomedical applications. However, conventional synthesis methods, entailing metal catalysts, raise concerns regarding toxicity and cost. In response, the researchers innovatively introduced an approach without metal to synthesize a low molecular weight, fully alternating poly-(cyclohexene oxide-alt-phthalic anhydride)(poly(CHO-alt-PA))copolymer. This copolymer, characterized by biocompatibility, biodegradability, and hemocompatibility, emerges as a fitting candidate for diverse biomedical applications.

Leveraging this copolymer, the researchers formulated curcumin (CUR)-loaded nanoparticles (NPs) tailored for potential cancer therapy. The resultant poly(CHO-alt-PA)CUR NPs exhibited enhanced thermal stability, thorough drug encapsulation, and a controlled release pattern. Furthermore, these nanoparticles exhibited increased internalization of the drug within cells and increased cytotoxicity against various cancer cells in comparison to free CUR. In tumor spheroid models, the poly(CHO-alt-PA)CUR nanoparticles demonstrated efficacy in reducing tumor spheroid size, indicating their promise as an anticancer drug delivery system. Rigorous *in vivo* studies involving BALB/c mice affirmed the copolymer's safety profile, solidifying its potential for further development.

Overall, the synthesized poly(CHO-alt-PA)copolymer, devoid of metal, distinguishes itself as a promising candidate for biomedical applications, particularly as a carrier for anticancer drugs like CUR. Its favorable attributes, including biocompatibility, biodegradability, and hemocompatibility, coupled with its demonstrated efficacy in drug delivery and absence of toxicity, position it as a compelling avenue for future exploration and development [11].

Cancer, a leading cause of mortality, imposes substantial physical and emotional burdens on individuals. Traditional treatment modalities, such as surgery, radiation therapy, and chemotherapy, often fall short in achieving complete eradication of tumor cells and are accompanied by severe side effects. In response to these challenges, researchers are turning to nanotechnology to develop intelligent drug delivery systems that precisely target tumor sites, mitigating damage to healthy tissues. These systems leverage the distinctive features of the tumor microenvironment (TME), including its acidic pH, heightened redox potential, and overexpression of specific receptors, to trigger controlled drug release.

Among the array of nanomaterial-based carriers, black phosphorus, gold nanoparticles, graphene oxide and mesoporous silica nanoparticles, have gained prominence due to their versatility and customizable properties. MSNs, characterized by their porous structure and expansive surface area, can effectively encapsulate diverse therapeutic agents, spanning chemical drugs, proteins, and nucleic acids. GO, with its distinctive π-π stacking interactions, provides a robust platform for efficient drug loading. BP, renowned for its exceptional drug loading capacity, emerges as a promising carrier for chemotherapy drugs. GNPs, renowned for their facile synthesis and functionalization, find applications in various drug delivery strategies.

These intelligent drug delivery systems, engineered to respond to both endogenous and exogenous stimuli, present a compelling avenue for enhancing cancer treatment outcomes. By selectively targeting tumor cells and minimizing systemic toxicity, these systems hold the potential to usher in a paradigm shift in cancer therapy [9].

Chemotherapy, a foundational method in cancer treatment, utilizes cytotoxic agents to hinder the proliferation of cancer cells, aiming for palliative or curative outcomes. Supramolecular nanotechnology, extensively explored in biomedical applications, has become a focal point of interest. This field leverages non-covalent interactions that include hydrophobic interactions, hydrogen bonding, host guest interactions and electrostatic interactions, to engineer intricate nanoscale structures with tailored properties.

Hydrophobic interactions induce the aggregation of non-polar molecules in aqueous environments, while hydrogen bonding emerges as a result of the attractive forces between molecules, involving interactions between hydrogen bond donors and acceptors through dipole-dipole attractions. Host-guest interactions, a hallmark of supramolecular chemistry, involve the formation of a complex between two structurally complementary components. Electrostatic interactions, governed by electromagnetic forces, dictate the repulsion or attraction between charged particles, macromolecules, or nanoparticles. Electrostatic self-assembly, particularly advantageous for delivering charged therapeutic substances like metal ions, nucleotides and polyelectrolytes, depends on the attraction between entities carrying opposite charges.

Aromatic systems, characterized by delocalized π electrons, engage in π–π stacking interactions, a crucial driving force for supramolecular assembly. Macrocyclic moieties, encompassing cyclodextrins, pillar[n]arenes, cucurbit[n]urils and calix[n]arenes, play a pivotal role in biomedical applications. For instance, α-cyclodextrins (α-CDs) can effectively close pore openings to avoid premature drug spilling from mesoporous silica nanocarriers, while folic acid-conjugated α-CDs impart aiming for specificity towards tumor cells that exhibit overexpression of folate receptors.

The design of drug delivery systems responsive to stimuli hinges on the alteration of charges within specific components, whether they be structural elements or the drugs themselves. This method facilitates controlled drug release in reaction to particular stimuli, ultimately enhancing therapeutic effectiveness and reducing off-target effects.

In the realm of cancer treatment, nanotechnology has revolutionized drug delivery systems by utilizing nano-scaled carriers to enhance drug efficacy and minimize side effects. However, conventional nano-drug carriers often grapple with challenges; examples of challenges include inadequate accumulation in tumors and limited penetration. To address these limitations, researchers have pioneered smart nano-drug carriers endowed with adaptive morphology and switchable surface ligands. These innovative carriers dynamically adjust their size and shape in response to the tumor microenvironment, thereby facilitating deep tumor penetration, enhancing cellular uptake, and improving drug retention. This groundbreaking approach holds significant promise for advancing cancer therapy [10].

Nanomaterials with morphological adaptability, typically below 600 nm in size, can navigate through leaky tumor vasculature and dynamically adjust their size or structure within the tumor microenvironment. This responsive behavior enhances their penetration into tumor tissues and facilitates cellular uptake. Conversely, within the tumor parenchyma, smaller nanomaterials may undergo aggregation or assembly, leading to prolonged retention. Innovative strategies, such as the conversion of 100-nm gelatin nanoparticles into 10-nm quantum dots initiated by enzymes specific to tumors, exemplify this approach, enabling deeper tumor penetration.

The morphological characteristics of nanomaterials assume a crucial role in cellular internalization and the controlled release of drugs within cells, providing precise spatial control in intracellular drug delivery. Both organic and inorganic smart materials contribute to the development of morphology-adaptive nanomaterials. Organic materials, including small molecules and polymers, respond to tumor-specific reactions, while inorganic nanoparticles can undergo *in vivo* aggregation, enhancing their accumulation at tumor sites. These advancements hold potential in addressing physiological barriers linked to anti-cancer drug delivery, including clearance by the reticuloendothelial system and mononuclear phagocytic system restricted intratumoral accumulation, and insufficient cellular internalization.

While the potential of morphology-adaptive nanomaterials is promising, challenges remain in optimizing drug delivery efficiency, ensuring successful clinical translation, and addressing the heterogeneous nature of the enhanced permeability and retention phenomenon in human subjects. Future research endeavors should prioritize the simplification, intelligence, and biocompatibility of these nanomaterials to fully unlock their potential in precision medicine [12].

16.3 Synthesis of Nanomaterials for Drug Delivery

In the context of burgeoning interest in natural polymers for drug delivery and biomedical applications, particularly polysaccharides, this research focused on the remarkable versatility of chitosan and carrageenan. Derived from chitin, chitosan exhibits promising attributes such as low toxicity, biodegradability, biocompatibility, and mucoadhesive properties, rendering it as an ideal candidate for drug delivery. Simultaneously, carrageenan, sourced from red seaweed, exhibited the ability to form hydrogels and nanoparticles through interactions with oppositely charged polyelectrolytes. The study thoroughly investigated the formulation and characterization of protein-loaded nanoparticles, utilizing carrageenan and chitosan as the materials for carriers. It also incorporated ovalbumin through ionic complexation to be used as a model protein.

The investigation meticulously examined the *in vitro* drug release dynamics, physicochemical properties, and morphology of these nanoparticles, with a comprehensive assessment of their biocompatibility using rat fibroblast cells. Materials, including chitosan, ovalbumin, carrageenan, and various chemicals, were sourced from reputable suppliers. The purification process of chitosan included dissolving it in a 1% acetic acid solution, followed by filtration, neutralization with NaOH, washing, and freeze-drying. The subsequent determination of its molecular weight was carried out using the Mark-Houwink equation.

The chitosan/carrageenan nanoparticles were methodically synthesized through ionic complexation, showcasing the versatility of these natural polymers. Exploring different CS/CRG ratios and incorporating ovalbumin yielded protein-loaded nanoparticles with varying protein content. The nanoparticle production yield was determined through gravimetry, and morphological analysis, conducted using transmission electron microscopy (TEM), offered insights into particle size, zeta potential, and how various components interacted with each other were explored using Fourier transform infrared spectroscopy.

Significantly, the research assessed the nanoparticles' compatibility with living tissues by employing a mouse fibroblast-like cell line (L929), illustrating non-cytotoxic behavior within the examined concentration range. The investigation thoroughly examined loading capacity, association efficiency, and the release kinetics of ovalbumin from the nanoparticles in an *in vitro* setting were investigated, unveiling sustained release profiles over a three-week period. Exhibiting minimal toxicity when interacting with cells resembling fibroblasts, these carrageenan and chitosan nanoparticles demonstrate potential as a promising foundation for achieving controlled and sustained release of drugs. This feature holds substantial potential for use in both tissue engineering and drug delivery applications [13].

The research advanced the field of temperature-sensitive nanoparticles, focusing on enhancing their applicability in biotechnology and medicine, specifically for controlled drug delivery. The nanoparticles underwent a reversible phase change at the lower critical solution temperature (LCST), a crucial feature for facilitating controlled drug release. The study addressed challenges associated with incorporating additional molecules, such as antibodies, onto NIPA-AAm nanoparticles without significant alterations to their LCST. To overcome this limitation, allylamine was introduced to NIPA-AAm nanoparticles, maintaining their LCST while providing amine groups for potential conjugation. The nanoparticles were synthesized through free radical polymerization and extensively characterized for size, chemical composition, and conjugation capability.

Fluorescent antibodies were successfully incorporated into the nanoparticles, and cytotoxicity was rigorously assessed using MTS assays and 3T3 fibroblast cells. Drug release studies, focusing on doxorubicin, were conducted at temperatures below and the LCST of the nanoparticles. The comprehensive characterization involved the use of various chemicals, including monomers, initiators, surfactants, and cross-linking agents.

The study employed cutting-edge techniques, including dynamic light scattering, transmission Fourier transform infrared spectroscopy, electron microscopy, UV-Vis spectrophotometry, enhanced optical microscopy, and nuclear magnetic resonance spectroscopy, for a comprehensive and detailed analysis.

The results showcased the successful synthesis of temperature-sensitive nanoparticles with an LCST above body temperature, featuring amine functional groups for potential biomolecule conjugation. The nanoparticles exhibited favorable biocompatibility, as demonstrated by cytotoxicity studies on 3T3 fibroblast cells. Moreover, drug release studies highlighted the temperature-responsive behavior of these nanoparticles, with enhanced release at the LCST. The findings suggested promising applications for these nanoparticles in controlled drug delivery and bioconjugation, paving the way for future investigations into their targeting capabilities and pharmacological effects in drug delivery applications [14].

The growing interest in natural polymers for drug delivery applications, particularly within the gastrointestinal tract, underscores the significance of this research focused on chitosan nanoparticles loaded with metronidazole (MZ). Chitosan, known for its mucoadhesive properties, holds promise for enhancing drug residence time in the gastrointestinal tract, crucial for effective drug delivery. The study systematically explores key factors influencing drug release, including chitosan concentration, molecular weight, and the chitosan-to-TPP (tripolyphosphate) ratio, with the goal of achieving controlled and prolonged drug release.

Utilizing chitosan with varying viscosities and a high deacetylation degree, the research successfully formulates MZ-loaded chitosan nanoparticles through ionic gelation with TPP. The thorough characterization includes the evaluation of particle size distribution, drug dissolution studies, and the scrutiny of nanoparticles using scanning electron microscopy. The findings emphasize the crucial influence of the chitosan-to-TPP ratio in determining nanoparticle size and distribution, revealing that a selected ratio of 4:1 for drug-loaded nanoparticles yields sizes below 300 nm.

Furthermore, the study delves into the mucoadhesive properties of the drug-loaded chitosan nanoparticles, evaluating their interaction with pig intestinal mucosa. The findings reveal that increased nanoparticle size corresponds to decreased mucoadhesive strength, emphasizing the intricate relationship between size and adhesive properties. The release profiles observed *in vitro* depict an initial rapid release, succeeded through a gradual release and controlled, which is attributed to the hydrophilic nature of chitosan and the diffusion of MZ through the chitosan matrix.

In summary, this research not only successfully formulates MZ-loaded chitosan nanoparticles but also systematically analyzes crucial parameters affecting their performance. The drug being released consistently throughout a 12-hour period positions this formulation as a promising option for various drug delivery applications, holding the potential for enhanced patient compliance and therapeutic effectiveness. The study's insights into the interplay of formulation variables contribute valuable knowledge to the design of mucoadhesive drug delivery systems using natural polymers [15].

The field of wound healing has experienced a transformative evolution with the emergence of advanced technologies, particularly the utilization of 3D bioprinting in the creation of cutting-edge wound dressings. Wounds, often posing significant healthcare challenges, necessitate meticulous care throughout the intricate phases of healing. While traditional wound dressings exhibit limitations, modern solutions leverage biocompatible polymers and nanomaterials to address shortcomings and optimize the wound healing process. The incorporation of 3D bioprinting technology facilitates the development of personalized and intricate wound dressings, providing benefits such as accelerated healing, customized options, and the incorporation of bioactive molecules and cells.

Bioinks play a crucial role in 3D bioprinting by creating a supportive environment for cells in the formation of scaffolds, thereby enhancing biocompatibility, biodegradability, and tensile strength. The incorporation of various nanomaterials, whether natural, synthetic, or composite, has propelled advancements in wound healing. Metallic nanoparticles, such as silver and copper, bring antimicrobial properties, while nanocomposites like titanium dioxide and zeolitic imidazolate nanoparticles exhibit antibacterial effects and improved mechanical properties in 3D printed wound healing scaffolds.

The use of smart hydrogels, responsive to stimuli like pH and temperature, further extends the capabilities of wound dressings by enabling controlled drug release. Stem cells delivered through 3D scaffolds hold promise in regenerative medicine, promoting angiogenesis and facilitating wound healing. Despite these remarkable strides, challenges persist, including the limited availability of suitable materials for 3D printed medicines, quality control issues, and regulatory hurdles. Overcoming challenges related to biocompatibility, safety, and regulatory approval remains paramount for the widespread adoption of 3D bioprinting in wound care.

The future path involves filling regulatory framework gaps, transforming the pharmaceutical landscape for wound care, and progressing 4D printing systems that integrate the element of time into drug delivery. The integration of theranostic components within wound scaffolds, allowing for real-time monitoring of wound healing prognosis, represents a promising avenue for comprehensive and personalized wound care solutions. Success in these endeavors requires multidisciplinary research, collaborative efforts, and the establishment of agreed-upon scientific fabrication protocols to propel the field toward impactful clinical applications and commercial viability [18].

The growing complexities in early cancer detection and treatment, coupled with the high mortality rates associated with cancer, have underscored the critical need for personalized and effective therapeutic approaches. Existing therapies, frequently reliant on nonselective drug administration, pose the potential hazard of inducing toxic effects in healthy tissues. Consequently, there is a compelling need to devise tailored approaches for cancer treatment. Within this context, the integration of customizable multifunctional nanoparticles has emerged as a promising avenue. These nanoparticles, characterized by PLGA cores, phospholipid outer shells, and the incorporation of both hydrophilic and hydrophobic drugs, offer a versatile platform for personalized cancer therapy. The utilization of multicellular tumor spheroids (MCTS) to simulate tumor characteristics as an *in vitro* model has yielded valuable insights into the behavior of these nanoparticles.

Through the hanging droplet method, multifunctional hybrid-polymer nanoparticles were synthesized, with various PLGA/DSPE ratios controlling their characteristics. Surface modification using DOTA-NHS-ester allowed for further functionalization, including the incorporation of gadolinium for MRI imaging. The research rigorously analyzed the nanoparticles using a range of analytical methods, encompassing measurements of particle size and zeta potential. UV-Vis spectroscopy was employed to assess drug loading, while imaging was conducted using both scanning and transmission electron microscopy.

In vitro studies, involving fluorescence uptake and spheroid treatment, illuminated the nanoparticles' behavior over time. Flow cytometry analysis provided insights into cellular responses after treatment with nanoparticles, emphasizing the importance of nanoparticle size in their interaction with tumor spheroid tissues. *In vitro* release investigations scrutinized the regulated discharge of Paclitaxel from nanoparticles across various media.

Transitioning to *in vivo* experiments involving mice with MIA PaCa-2 tumors, the research showcased the prolonged and selective retention of 110 nm nanoparticles within the tumors. Fluorescence imaging and MRI further supported the efficacy of these nanoparticles in enhancing imaging contrast within tumor tissues. The findings consistently highlighted the potential of these multifunctional nanoparticles for both therapeutic and imaging applications.

The study concludes by emphasizing the versatility of hybrid-polymeric nanoparticles for diverse applications in cancer treatment, providing a foundation for further advancements, such as the integration of smaller nanoparticles for combined MRI and heat generation. This research represents a significant step toward the development of advanced, personalized cancer therapies that can address the evolving nature of cancer biology and enhance treatment efficacy [17].

The research centers on the creation and analysis of mesoporous silica nanoparticles (MSNs) obtained from byproducts generated during the production of LCD panels. The primary emphasis lies in assessing their safety, biocompatibility, and their potential utility as carriers for drug delivery. The waste powder obtained from LCD panel production serves as the precursor for MSN synthesis, showcasing a sustainable approach to repurpose industrial by-products. The MSNs exhibit promising attributes for drug delivery, characterized by a large surface area and porous structure. Comprehensive analyses include BET method for surface area determination, t-plot method for pore volume measurement, X-ray diffraction for crystal structure assessment, and high-performance liquid chromatography (HPLC) for quantifying drug entrapment efficiency. Biocompatibility evaluations encompass examinations with ovarian cancer (ES-2) and human lung fibroblast (WI-38) cells. Exploration of potential inflammation-related effects is to be noted, taking into account LDH release and the repercussions on both normal and cancer cells. The results underscore the biocompatible attributes of MSNs, emphasizing their suitability as carriers for drug delivery. Furthermore, the research explores the entrapment of chemotherapeutic agents, methotrexate (MTX), and camptothecin (CPT), revealing physicochemical dependencies governing drug adsorption onto MSNs. The study contributes valuable insights into the synthesis, safety, and application potential of MSNs, paving the way for their utilization as effective drug delivery carriers with biodegradable properties and defined dose limitations [16].

16.4 Location Specific Drug Delivery

Nanomaterials, with unique surface properties and porous structures, offer a promising approach for various applications, especially in drug delivery. Table 16.1 shows the list of smart materials, key features and applications. One interesting aspect of these materials is their available magnetism, which can be utilized for specific purposes. For example, iron oxide magnetic nanoparticles can be used to target and release mutant cells. The field

Table 16.1 Smart materials, key features and applications.

Smart material	Key features	Applications
Chitosan	Environmentally friendly, low toxicity, binds well to mucus, and easily absorbed by the body.	Supports gradual drug release, particularly in digestive tract applications; widely used for encapsulating proteins and other therapeutic agents.
Carrageenan	Forms gels and nanoparticles, derived from red algae, compatible with oppositely charged polymers.	Effective as a carrier for nanoparticles that provide steady release of therapeutics, aiding in tissue engineering and prolonged drug activity.
NIPA-AAm (Temperature-sensitive polymer)	Sensitive to specific temperature thresholds (LCST), allows for biomolecule attachment.	Facilitates drug delivery that activates at certain temperatures; advantageous in targeting specific body areas, especially for localized cancer therapy.
Metallic Nanoparticles (Silver, Copper)	Antimicrobial, used in infection prevention; can respond to thermal and magnetic fields.	Incorporated into wound dressings that release antimicrobial compounds upon exposure to stimuli, aiding in infection control and wound care.
Mesoporous Silica Nanoparticles (MSNs)	High surface area and porosity; sourced from recycled materials; suitable for carrying drug molecules.	Serve as vehicles for gradual and sustained drug release, especially for chemotherapies and targeted cancer treatments.

(*Continued*)

Table 16.1 Smart materials, key features and applications. (*Continued*)

Smart material	Key features	Applications
Hybrid Polymer Nanoparticles (PLGA-based)	Biocompatible, adaptable structure; can carry both water- and fat-soluble drugs.	Used for dual-function applications such as imaging and treatment; effective for retaining therapeutic agents in targeted tumor areas.
Stimuli-Responsive Hydrogels	Adjustable to changes in pH or temperature; versatile for encapsulating bioactive agents.	Delivers therapeutic agents such as antibiotics or growth factors on-demand, beneficial in wound healing and tissue repair.
Nanomaterial-Enhanced Bioinks for 3D Bioprinting	Customizable for creating biological scaffolds; includes nanocomposites like titanium and zinc.	Enhances wound healing with controlled bioactive release and provides structural support in regenerative medicine applications.

of drug delivery, in particular, gains advantages from diverse characteristics of nanomaterials, including biocompatibility, biodegradability, substantial drug-loading capacity, and added controllability.

However, the fabrication of nanostructures with all these desirable properties is a challenging task. Instead, we tend to design different nanostructures and use them for specific purposes. Examples of biodegradable biocompatibility alternatives include liposomes, phospholipids, and polymeric nanostructures, which are well suited for drug delivery. Inorganic nanoparticles such as magnetic nanoparticles, Au nanoparticles and silica nanoparticles are also valuable tools in this regard.

Carbon nanomaterials like carbon nanohorns, carbon nanotubes, and graphene display distinctive structures and characteristics. These materials have found applications in drug delivery systems and exhibit different drug release behaviors depending on their specific properties. As an illustration, graphene oxide, characterized by its planar structure and functional groups, features the highest drug loading capacity. In contrast, carbon nanotubes exhibit the lowest capacity, attributed to their surface morphology for several reasons.

Experimental methods involve loading Rhodamine B (RB) on carbon nanomaterials by mixing and sonication process, followed by centrifugation and rinsing The RB loading capacity was measured by UV-vis spectroscopyas shown in Figure 16.6. Drug dispensing tests include real-time cash flow [19].

Metal-organic frameworks (MOFs) have become versatile materials for delivering active molecules, allowing for the customization of MOFs to achieve stimuli-controlled drug release. This is facilitated by their self-assembly properties, high surface area, and cavity size. Furthermore, their adaptable nature, owing to modular synthesis, makes them suitable for various applications. A particular class of MOFs, zeolitic imidazolate frameworks (ZIFs) stand out for small pores and large cavities, allowing high loading and slow release properties Furthermore, there are ZIFs known for their stability, biocompatibility and pH-responsive behavior, making them promising candidates for drug delivery. Among ZIFs, ZIF-8 is popular because of its unique properties.

However, ZIF-8 has a low shear index and emits white light, exhibiting attractive properties such as near-infrared phosphorus. While volatiles release active molecules with changes in temperature, their high temperature (above 45°C) limit limits their physiological applications To overcome this, materials are a core shell composite seed in order to regulate the temperature and make ZIF-8 temperature responsive. In these composites,

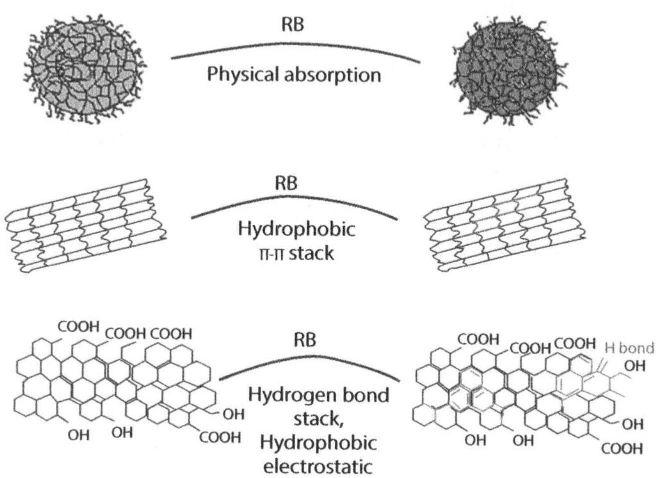

Figure 16.6 Illustrative diagram depicting the interplay among RB and CNT, GO, and CB.

metal nanoparticles in the core-shell structure act as controllable nano-heaters, and trivalent lanthanide ions act as nanoscale thermometers They are illuminated by lanthanide ions such as Eu3+ and Tb3 + in use to monitor temperature changes in core-shell composites.

(ZIF-8@AuNP) core-shell composites represent an alternative for thermos-controlled drug delivery. These composites utilize the thermal properties of gold nanoparticles (AuNPs) to induce chemical release from ZIF-8, while lanthanide ions regulate temperature Notably, this approach enables delivery [20].

A system incorporating doxorubicin (DOX) and gold nanoparticles (GNP) has been created as a solution to the requirement for targeted drug delivery, aiming to minimize side effects commonly associated with traditional chemotherapeutic agents. DOX, an extensively utilized chemotherapeutic agent recognized for its cardiotoxicity and susceptibility to drug resistance, possesses dual properties, enabling interaction with nucleic acids and showcasing potential in both chemotherapy and immunotherapy.

The design incorporates gold nanoparticles (GNPs) coated with a thiolated anticancer oligonucleotide named AS1411, featuring the control sequence CRO. The deposition of doxorubicin (DOX) onto the coated GNPs occurs under specific conditions. The release of DOX from GNPs can be induced either by reducing the oligonucleotide-thiol bond using dithiothreitol (DTT) or exposing it to gamma radiation.

To ascertain the stability and size of the GNP-oligonucleotide complexes, measurements of zeta potential and dynamic light scattering were conducted. The subsequent release of DOX from the GNP-oligonucleotide complexes following the incorporation were studied using fluorescence techniques.

The principal goal of the investigation was to investigate how different oligonucleotide sequences affect the loading and release of DOX from GNP-oligo colloidal dispersions. GNP exhibits distance-dependent fluorescence quenching effects on organic/inorganic dyes within a specific size range. Notably, fluorescence quenching of DOX was observed in GNP-oligo colloids, suggesting non-covalent intercalation with the proximity of DOX to AS1411 and/or CRO aptamer bands around the golden core.

Partial funding for this study was provided by the ExCITE program, backed by the National Heart, Lung, and Blood Institute of the National Institutes of Health, with the award number U01HL127518 associated with this backing, as suggested by the study results [21]. The research is centered on exploring the dynamic behavior of nano-scale magnets within biomaterials under the influence of external magnetic fields. While magnetic nanoparticles show significant potential in biomedicine, their interactions with surrounding solvent molecules remain incompletely understood. This review seeks to establish foundational insights into the movement of these nano-scale magnets within diverse materials.

To study this, researchers use an intense photon beam at the SPring-8 facility to track the flow of magnetic fluid and understand how nano-scale magnets respond to an external magnetic field as it is found under various materials such as agarose gel, eggs, plant leaves and oils.

The study employs theoretical models to accurately compute the velocity of 0.5 mm magnetic clusters, and the results align well with experimental data obtained in oil. These findings suggest that external magnetic fields can precisely control the movement of magnetic fluid within various materials, opening avenues for diverse medical applications.

Magnetic nanoparticles, as a part of nanotechnologies, show significant promise in therapeutic applications. These applications include enhanced magnetic resonance imaging, hyperthermal treatment of cancer cells, targeted drug delivery and the modification of cell membranes. Through experiments, the researchers seek to comprehend the behavior of magnetic nanoparticles suspended in the fluid environments within the human body [22].

This research aims to enhance drug delivery strategies through stimuli-reactive nanoparticles, with the goal of mitigating side effects and improving specificity and efficacy. The primary focus of the study revolves around the polymeric development of light-sensitive nanoparticles designed for localized drug delivery.

The initial step involves utilizing PPAPE(poly-2-[4-phenylazo phenoxy] ethyl acrylate co acrylic acid), a photo isomerized amphiphilic copolymer, as the key component in the synthesis of nanoparticles. Subsequently, micellarization of these PPAPE molecules into nanoparticles is achieved through the nanoprecipitation method.

Various characterization techniques such as X-ray photoelectron spectroscopy, Fourier transform infrared spectroscopy, and proton nuclear magnetic resonance are employed to investigate polymer properties. Additionally, light scattering is utilized to scrutinize the size distribution, and the resulting nanoparticle morphology is analyzed through electron microscopy (TEM).

The photostimulation response of these nanoparticles is confirmed by UV-VIS spectroscopy, where photoisomerization properties were demonstrated, which are important for controlled drug release Clinical studies have shown that polymeric nanoparticles, when used for drug delivery, can improve drug resistance, bioavailability, time of blood circulation, tumor site accumulation, and reduce side effects especially stimuli responsive micelles, their biocompatibility, And because of their hydrophobic properties, they are of great interest. These intelligent polymeric micelles can respond to external stimuli, including alterations in pH, temperature, exposure to light, or interaction with biological agents such as enzymes.

Light as an external stimulus is particularly attractive to induce dominance [23].

The subsequent content outlines a technique for identifying nonlinear discrete-time systems without measurement noise and with additive process noise. The ensuing points encapsulate the key features and contributions of this research.

1. Determination of nonlinear systems: The focus of this note is the challenging task of identifying nonlinear systems based on a finite set of time series data points. Nonlinear system identification involves retrieving the underlying nonlinear equations describing the system dynamics from the observed time series data.
2. Sparse Linear Regression and Bayesian Theory: The suggested approach frames the problem of identifying nonlinear systems as a sparse linear regression challenge, employing Bayesian principles. This methodology facilitates the recognition of nonlinear functional patterns and their corresponding parameters.
3. Convex Process: Although the identification problem contains non-convex optimizations, the notation introduces a convex process to better solve the problem. This leveling process helps to find a better solution.
4. Iterative Reweighted L1-Minimization Algorithm: The identification process employs an iterative reweighted L1-minimization algorithm to address problems related to sparse linear regression. This algorithm promotes sparsity in the designated parameters.
5. Sparsity-inducing priors: A sparsity-inducing prior and marginal likelihood maximization are applied to the parameters of the system to improve the efficiency of the detection method.
6. Applied to various industries: The effectiveness of the proposed detection method is demonstrated through examples from various industries including natural and physical systems [24].

Magnetic nanoparticles (MNPs) have revolutionized biomedical applications by offering a versatile platform with unique properties. Due to their small size and size, MNPs are widely applicable in various areas of diagnostics, molecular imaging, and therapeutics. Metal-oxide-based magnetic nanoparticles (MNPs), in particular, have become pivotal in the field of biomedicine, thanks to their attributes such as biocompatibility, adjustable surface functionality, and they being utilized as contrast agents for magnetic resonance imaging, among other roles, which not only prevent aggregation but also allow bioconjugation. High-magnetizing MNPs (HMNPs) with superparamagnetic behavior open up new possibilities for drug delivery when aqueous coatings with specific groups are provided to facilitate drug conjugation and targeting. They can be activated, leading to interesting controlled drug release strategies, especially in cancer therapy. Furthermore, MNPs have been demonstrated to help hyperthermia magnetic liquids for cancer therapy, providing controlled thermoregulation in tumors has been achieved. Their applications extend to magnetic separation techniques for the removal of target cells, enzyme immobilization for biocatalysis, and selective isolation of nucleic acids for clinical screening MNPs as a side tool many forms offer the potential to revolutionize disease diagnosis, imaging and treatment in biomedical applications [25].

Studies done recently have shown great promise in the use of statins to stimulate new bone formation, offering a new approach to osteoporosis treatment Although current therapies for osteoporosis focus primarily by preventing bone resorption though there are few options that promote bone density (osteoclast formation). Low-dose statins have been found to be able to promote bone growth. However, commonly used statins in clinical studies often present inconsistencies due to metabolites primarily in the liver, resulting in low levels in bone. Animal studies have shown promising results with high statin doses in to stimulate bone growth, but is also associated with systemic side effects. Other approaches such as drug delivery based on nanotechnology are being explored. In this context, tetracycline-loaded PLGA nanoparticles (TC–PLGA NPs) have been developed for efficient bone targeting and simvastatin (SIM) delivery, offering a potential solution for osteoporosis therapy. It utilizes a wide range of approaches, from nanoparticle synthesis and characterization to *in vitro* and *in vivo* assays, yielding promising materials for orthopedic targeted drug delivery [26].

Nanoparticles have emerged as a promising tool in cancer chemotherapy, mainly due to their ability to efficiently target tumor tissues around the reticuloendothelial system of the body. Various nanoparticles such as nanospheres, micelles, nanogels have been developed as drug carriers, using techniques such as oil/water emulsion, self-assembly and micro-emulsion polymerization but the main challenge in nanoparticle-based drug delivery is drug delivery take high-quality materials, encapsulation -To meet this challenge, which is essential for effective chemotherapeutics, researchers searched for new materials, including mesoporous silicon nanoparticles, graphene and carbon nanotubes, which provide specific sites for including high surface area and chemical absorption capacity.

Another approach is host-guest interaction, exemplified by polyrotaxanes. This unique structure is achieved by interactions between cyclodextrin-polymer chains, allowing nanoparticles to self-assemble α-cyclodextrin (α-CD), a cyclic oligosaccharide with exceptional biocompatibility, acts as a promising host molecule at polyrotaxanes into drug delivery due to its ability to form stable inclusion complexes with guest molecules ethylene glycol) (PEG) and other biodegradable materials. When combined with polyrotaxanes, it provides an attractive route for drug carriers. Furthermore, by immobilizing the cation fragments in polyrotaxanes [27].

Nanomedicine has surfaced as an extraordinary field with the capacity to transform medicine, particularly in addressing intricate challenges like the blood-brain barrier (BBB). Infusing sustainability and green practices into this challenge offers a promising solution. Nanoparticles, leveraging their distinctive features like high solubility, precise targeting capabilities, and the ability to enhance BBB penetration, have emerged as key contributors in endeavors to surmount this barrier.

In this context, nanoparticle drug delivery systems (NDDS) have taken center stage, offering a strategy to further enhance drug delivery by taking advantage of organic and inorganic drug carriers. Natural NDDS can improve drug delivery and cell delivery, with special focus on harvesting potential by BBB. This holds particular promise for the therapy of central nervous system (CNS) diseases.

The research you referred to conducted a comprehensive examination of data on blood-brain barrier (BBB)-crossing nanomedicine drug delivery systems (NDDS) retrieved from the Web of Science database. Several key insights into developments and trends in this area were published. Research in the NDDS for the BBB approach has been divided into two parts, and the number of published articles has increased significantly from 2008 to 2022. These articles highlight both regional and international partnerships, which are found in journals and books as it spans disciplines from chemistry to medicine to biomedicine [28].

Handling renal cell carcinoma presents a multifaceted challenge. RCC comprises various subtypes, such as clear chromophobe RCC, papillary

RCC and cell RCC, comprising around 3% of the total cancer cases, also demonstrating the greatest occurrence within Western nations. Traditional approaches to RCC involve surgical interventions like radical nephrectomy and partial nephrectomy, procedures that can notably affect renal function.

Phase III trials have shown that radical nephrectomy (RN) and nephron-sparing surgery (NSS) can reduce renal function, emphasizing the importance of renal function preservation in RCC treatment in addition to correlation between distinct links between chronic kidney disease (CKD) and cardiovascular risk.

Therapeutic interventions used in RCC, including contrast agents and anticoagulants, may inadvertently harm renal function. Furthermore, some patients undergoing stroke develop distant metastasis, emphasizing the need for effective systemic therapy.

Various treatments, including cytokines, radiotherapy, chemotherapy, vaccines, and recent innovations like immune checkpoint inhibitors and tyrosine kinase inhibitors, have been explored for renal cell carcinoma (RCC). However, combination therapies, especially pairing chemotherapy with immunotherapy, raise concerns regarding systemic side effects and impacts on immune function.

Nanoparticles emerge as a promising strategy for cancer therapy, offering secure drug encapsulation and minimizing potential toxicity in normal cells. Notably, sunitinib-loaded micellar nanocomplexes (SU-MNCs) present an alternative approach for targeted drug delivery with low toxicity [29].

Nanotechnology is a versatile field that combines chemical and physical properties, making it versatile. It has evolved into four distinct generations of materials: passive nanostructures, active nanostructures, nanosystems, and molecular nanosystems, each offering unique capabilities for different applications.

One of the major areas where nanotechnology has had a significant impact is in the development of nanomaterials. These materials are characterized by compactness, high performance, surface area, and performance. As a result, nanomaterials have found applications in various industries such as energy, pharmaceuticals, oil exploration and drug delivery.

In nanomedicine, drug delivery systems (DDS) have made remarkable progress. These systems have become increasingly effective in targeting and delivering drug conjugates to specific microsites in the body. Delivery methods, such as capsules, tablets and injections, have inherent limitations and challenges.

Another approach to drug delivery is the use of "nano gated materials" using nanocarriers equipped with stimuli-responsive gates that open in response to specific internal or external stimuli. Mesoporous silica nanoparticles (MSNs) emerge as key nanomaterials at these gates this packaging and acting as drug couriers. MSNs offer many advantages, including surface-modifying ability, biocompatibility, high drug-loading capacity, and drug inertness.

MSNs can be generated under various conditions and exhibit the ability to release drugs as gate structures change in response to specific stimuli [30–32].

16.5 Future Scope

The future of drug delivery systems research holds immense promise, propelled by innovative approaches and cutting-edge technologies. One area of investigation focuses on the advancement of intricate drug delivery systems, particularly tackling the hurdles related to ensuring structural stability and achieving efficient encapsulation of hydrophilic drugs. Simultaneously, research is intensifying on the refinement of carbon nanomaterial-based delivery systems, aiming to enhance fluorescent stability, prevent agglomeration, and enable pH-sensitive drug release. Natural polysaccharides, particularly the versatile chitosan, are being thoroughly investigated for their potential in creating intelligent, pH-regulated drug-controlled release systems. Evolving targeted therapy approaches leverage nanoparticle-drug conjugates with specific ligands, ensuring preferential accumulation in cancer cells while minimizing harm to normal cells. Research efforts are also focused on elevating drug delivery efficiency and selectivity, particularly in developing sustained drug release mechanisms for prolonged cytotoxic effects in cancer cells. This multifaceted exploration into novel materials, precision medicine strategies, and stimuli-responsive drug delivery systems sets the stage for transformative advancements in the field.

16.6 Conclusion

In conclusion, the panorama of nanomaterial-based drug delivery systems depicted in this review unveils a tapestry of innovation with profound implications for biomedical applications. From the refinement of carbon nanomaterials to the exploration of natural polysaccharides like chitosan, the field is marked by a diverse array of materials and strategies. Targeted therapy, exemplified by nanoparticle-drug conjugates and sustained release mechanisms, stands out as a pivotal approach, particularly in minimizing collateral damage in cancer treatment. Novel materials, including graphene quantum dots, metal-organic frameworks, and superparamagnetic iron oxide nanoparticles, signal a transformative era in precision medicine. As the exploration delves into stimuli-responsive systems, computational modeling, and the intricate dynamics of the tumor microenvironment, the collaborative interdisciplinary research emerges as imperative. The collective commitment to optimizing nanomaterial properties and navigating challenges in clinical translation underscores the potential for nanomaterial-based treatments to revolutionize drug delivery, diagnostic imaging, and therapeutic precision, shaping the future landscape of biomedical science.

References

1. Dev, A., Mohan, J.C., Sreeja, V., Tamura, H., Patzke, G.R., Hussain, F., Weyeneth, S., Nair, S.V., Jayakumar, R., Novel carboxymethyl chitin nanoparticles for cancer drug delivery applications. *Carbohydr. Polym.*, 2009, 10.1016/j.carbpol.2009.10.038.
2. Cheng, S., Wang, Q., Qi, M., Sun, W., Wang, K., Li, W., Lin, J., Dong, B., Wang, L., Nanomaterials-mediated on-demand and precise antibacterial therapies. *Mater. Des.*, 2023, 10.1016/j.matdes.2023.111982.
3. Bassey, V.M., Okon, E.E.D., Louis, H., Benjamin, I., Chukwuemeka, K., Gber, T.E., Ezekiel, M.C., Qader, S.W., Adeyinka, A.S., Nano-enhanced Drug Delivery of Dacarbazine using Heteroatoms (B, N, P, S) doped Ag-Functionalized Silicene Nanomaterials: Insight from Density Functional Theory. *Chem. Phys. Impact*, 2023, 10.1016/j.chphi.2023.100297.
4. Tan, Q., Yang, Z., Bu, S., Chen, J., Chen, W., Geng, W., Huang, Q., Duan, L., Guo, M., Wu, Y., Deng, J., Zhou, E., Li, M., Wu, F., Jin, Y., Tumor-derived microparticles-based nanomaterial as platform for delivery of tumor antigens to enhance immunogenicity. *Chem. Eng. J.*, 2023, 10.1016/j.cej.2023.142497.
5. Zhang, S., Peng, B., Chen, Z., Yu, J., Deng, G., Bao, Y., Ma, C., Du, F., Sheu, W.C., Kimberly, W.T., Simard, J.M., Coman, D., Chen, Q., Hyder, F., Zhou, J., Sheth, K.N., Brain-targeting, acid-responsive antioxidant nanoparticles for stroke treatment and drug delivery. *Bioact. Mater.*, 2022, 10.1016/j.bioactmat.2022.02.033.
6. Gupta, P.K., Tripathi, S.K., Pappuru, S., Chabattula, S.C., Govarthanana, K., Gupta, S., Biswal, B.K., Chakraborty, D., Verm, R.S., Metal-free semi-aromatic polyester as nanodrug carrier: A novel tumor targeting drug delivery vehicle for potential clinical application. *Mater. Sci. Eng. C*, 2019, 10.1016/j.msec.2019.110285.
7. Tabatabaei, S.N., Duchemin, S., Girouard, H., Martel, S., Towards MR-navigable Nanorobotic Carriers for Drug Delivery into the Brain. *Int Conf Robot Autom*, 2012, 10.1109/ICRA.2012.6225041.

8. Chen, J., Shao, C.R., Zhang, X.D., Chen, C., Applications of nanotechnology for melanoma treatment, diagnosis, and theranostics. *Int. J. Nanomed.*, 2012, 10.2147/IJN.S45429.
9. Shi, Z., Zhou, Y., Fan, T., Lin, Y., Zhang, H., Mei, L., Inorganic nano-carriers based smart drug delivery systems for tumor therapy. *Smart Mater. Med.*, 2020, 10.1016/j.smaim.2020.05.002.
10. Liu, Y., Wu, Y., Luo, Z., Li, M., Designing supramolecular self-assembly nanomaterials as stimuli-responsive drug delivery platforms for cancer therapy. *iScience*, 2023, 10.1016/j.isci.2023.106279.
11. Solangi, N.H., Karri, R.R., Mubarak, N.M., Mazari, S.A., Mechanism of polymer composite-based nanomaterial for biomedical applications. *Adv. Ind. Eng. Polym. Res.*, 2023, 10.1016/j.aiepr.2023.09.002.
12. Cheng, D.B., Qiao, Z.Y., Xuan, L., Wang, H., Recent advances of morphology adaptive nanomaterials for anti-cancer drug delivery. *Prog. Nat. Sci.: Mater. Int.*, 2020, 10.1016/j.pnsc.2020.09.002.
13. Grenha, A., Gomes, M.E., Rodrigues, M., Santo, V.E., Mano, J.F., Neves, N.M., Reis, R.L., Development of new chitosan/carrageenan nanoparticles for drug delivery applications. *J. Biomed. Mater. Res. Part A*, 2009, 10.1002/jbm.a.32466.
14. Rahimi, M., Kilaru, S., Sleiman, G.E.H., Saleh, A., Rudkevich, D., Nguyen, K., Synthesis and Characterization of Thermo-Sensitive Nanoparticles for Drug Delivery Applications. *J. Biomed. Nanotechnol.*, 2008, 10.1166/jbn.2008.014.
15. Elzatahry, A.A. and Eldin, M.S.M., Preparation and characterization of metronidazole-loaded chitosan nanoparticles for drug delivery application. *Polym. Adv. Technol.*, 2007, 10.1002/pat.1195.
16. Lin, Y.C., Lin, L.Y., Gao, M.Y., Fang, Y.P., Mesoporous silica nanoparticles synthesized from liquid crystal display manufacturing extracts: A potential candidate for a drug delivery carrier. *Int. J. Nanomed.*, 2013, 10.2147/IJN.S50991.
17. Tng, D.J.H., Song, P., Lin, G., Soehartono, A.M., Synthesis and characterization of multifunctional hybrid-polymeric nanoparticles for drug delivery and multimodal imaging of cancer. *Int. J. Nanomed.*, 2015, 10.2147/IJN.S86468.
18. Yayehrad, A.T., Siraj, E.A., Matsabisa, M., Birhanu, G., 3D printed drug loaded nanomaterials for wound healing applications. *Regener. Ther.*, 2023, 10.1016/j.reth.2023.08.007.
19. Zhang, R. and Olin, H., Carbon nanomaterials as drug carriers: Real time drug release investigation. *Mater. Sci. Eng. C*, 2012, 10.1016/j.msec.2012.03.016.
20. Silva, J.Y.R., Proenza, Y.G., da Luz, L.L., Araújo, S.S., Gomes Filho, M.A., Junior, S.A., Thermo-responsive adsorbent-heater-thermometer nanomaterial for controlled drug release: (ZIF-8,EuxTby)@AuNP core-shell. *Mater. Sci. Eng. C*, 2019, 10.1016/j.msec.2019.04.078.
21. Chauhan, R., Allen, N.C., Keynton, R.S., Bates, P.J., Malik, M.T., O'Toole, M.G., Engineering sequence and stimuli dependent doxorubicin release

from anti-nucleolin aptamer coated gold nanoparticles. *International Symposium on Signal Processing and Information Technology*, 2018, 10.1109/ISSPIT.2018.8642659.

22. Nakano, M., Matsuura, H., JU, D.Y., Uozumi, Y., Koide, K., Noda, N., BIAN, P., Makino, K., Dynamics of Magnetic Fluid Aimed for Local Drug Delivery Systems. *Second International Conference on Innovative Computing, Information and Control (ICICIC 2007)*, 2007, 10.1109/ICICIC.2007.276.
23. Pérez-Buitrago, S., Mena-Giraldo, P., Pinal, R., Hoyos-Palacio, L., Azopolymer based nanoparticles for phototriggered drug delivery. *Engineering in Medicine and Biology Society (EMBC)*, 2019, 10.1109/EMBC.2019.8856432.
24. Nakagawa, H. and Ohuchi, M., Sparse Bayesian Approach to the Identification of Nonlinear State-Space Systems. *Trans. Magn.*, 2015, 10.1109/TAC.2015.2426291.
25. Yang, H.W., Hua, M.Y., Liu, H.L., Huang, C.Y., Wei, K.C., Potential of magnetic nanoparticles for targeted drug delivery. *Nanotechnol. Sci. Appl.*, 2012, 10.2147/NSA.S35506.
26. Wang, H., Liu, J., Tao, S., Chai, G., Wang, J., Hu, F.Q., Yuan, H., Tetracycline-grafted PLGA nanoparticles: Bone-targeting drug delivery system. *Int. J. Nanomed.*, 2015, 10.2147/IJN.S88798.
27. Lai, Y., He, B., Li, Y., Wang, G., Chang, S., Gu, Z., Supramolecular nanoparticles generated by the self-assembly of polyrotaxanes for antitumor drug delivery. *Int. J. Nanomed.*, 2012, 10.2147/IJN.S33649.
28. Zhong, D., Gan, Z., Zheng, M., Li, L., Cheng, H., Luo, H., Liu, D., Liu, P., Li, H., Ou, X., Deng, Y., Knowledge mapping of nano drug delivery systems across blood-brain barrier from 1996 to 2022: A bibliometric analysis. *Heliyon*, 2023, 10.1016/j.heliyon.2023.e15828.
29. Li, J., Wu, K., Zhang, J., Gao, H., Xu, X., Progress in the treatment of drug-loaded nanomaterials in renal cell carcinoma. *Biomed. Pharmacother.*, 2023, 10.1016/j.biopha.2023.115444.
30. Isroni, M., Sagita, F., Culsum, N.T.U., Kadja, G.T.M., Molecule gated mesoporous silica for on-command drug delivery: A review. *Results Chem.*, 2023, 10.1016/j.rechem.2023.101053.
31. Tripathi, S.L., Prakash, K.B., Balas, V.E., Mohapatra, S.K., Nayak, J. (Eds.), *Electronic devices, circuits, and systems for biomedical applications: Challenges and intelligent approach*, Academic Press, Amsterdam, Netherlands, Elsevier, 2021, https://doi.org/10.1016/C2020-0-02289-9.
32. Tripathi, S.L., Kumar, A., Rao, K.S., Mudimela, P.R., *Nanodevices for Integrated Circuit Design*, Wiley & Scrivener Publishing LLC, USA, October-2023, DOI:10.1002/9781394186396.

About the Editors

Suman Lata Tripathi, PhD is a Professor at Lovely Professional University with more than 22 years of experience in academics and research. She has published more than 125 research papers in refereed science journals, conference proceedings, and e-books, edited and authored more than 27 books, 14 Indian patents, and four copyrights. Her areas of expertise include microelectronics device modeling and characterization, low-power VLSI circuit design, advanced FET design for IoT, and embedded system design.

Vrinda Gupta, PhD is an Associate Professor in the Department of Electronics and Communication Engineering at the National Institute of Technology Kurukshetra. She has more than 75 publications in international journals, national and international conferences, and book chapters. Her research interests are in the fields of computer communications, network and information security, wireless communications and networking, embedded systems design, and Internet of Things.

Sobhit Saxena, PhD is a as Professor in the School of Electronics and Electrical Engineering at Lovely Professional University with more than 14 years of teaching experience. He has published more than 35 research papers in international journals and conferences, two book chapters, and two books, as well as filed three patents. His areas of expertise include nanomaterial synthesis and characterization, and electrochemical analysis.

Shipra Upadhyay, PhD is an Assistant Professor at the Ramaiah Institute of Technology. She has published many papers in peer-reviewed international journals, conferences, and books. Her research interests include, custom analog IC design, field programmable gate array programming, nanoelectronics, and low-power circuit design.

Sudip Ghosh, PhD is an Assistant Professor in the School of VLSI Technology at the Indian Institute of Engineering Science and Technology. He has more than 60 publications in international journals and conferences. His areas of interest include digital image and video watermarking systems design, logic synthesis and verification of digital circuits, VLSI physical design, and VLSI testing.

Index

Also of Interest

By the same editors

Hydrogen Energy Systems: Advancing Sustainable Power Solutions, Edited by Krishan Arora, Himanshu Sharma, Suman Lata Tripathi, and Sandesh S. Chougule, ISBN: 9781394358373. Drive the global shift toward sustainability with this comprehensive guide that explores the latest advancements in renewable energy technologies, smart grids, and energy storage and providing the expert policy and economic insights needed to navigate the green energy transition.

Smart Chips for Smart Devices: VLSI Design for Next-Generation IoT Solutions, Edited by Abhishek Kumar and Suman Lata Tripathi, ISBN: 9781394335800. Master the future of semiconductor technology with this essential guide, which provides the advanced circuit design strategies necessary to develop the high-performance, low-power VLSI chips that drive the modern IoT revolution.

Waste Management and Utilization for a Sustainable Environment, Edited by Amit Srivastava and Suman Lata Tripathi, ISBN: 9781394292974. *Waste Management and Utilization for a Sustainable Environment* is an essential guide to understanding how cutting-edge technologies and innovative engineering solutions can transform waste management into a cornerstone of a sustainable future.

Energy-Efficient Communication Networks, Edited by Shakti Raj Chopra, Krishan Arora, Suman Lata Tripathi, and Vikram Kumar, ISBN: 9781394271658. *Energy-Efficient Communication Networks* is essential for anyone looking to understand and implement cutting-edge energy optimization strategies for communication systems, ensuring they meet growing energy demands while seamlessly integrating renewable energy sources and enhancing battery life in embedded applications.

Smart Textiles and Wearables for Health and Fitness, Edited by Jyotirmoy Pathak, Abhishek Kumar, Suman Lata Tripathi, and Balwinder Raj, ISBN: 9781394302949. *Smart Textiles and Wearables for Health: The Flexible Electronics Revolution* provides an in-depth exploration of how innovative technologies and materials are reshaping healthcare, making it an essential resource for anyone looking to understand the transformative power of smart textiles and wearables in patient monitoring, diagnosis, and rehabilitation.

Development of 6G Networks and Technology, Edited by Suman Lata Tripathi, Mufti Mahmood, C. Narmadha, and S. Albert Alexander, ISBN: 9781394230655. *Development of 6G Networks and Technology* provides an in-depth exploration of the potential impact of 6G networks on various industries, including healthcare, agriculture, transport, and national security, making it an essential resource for researchers, scholars, and students working in the field of wireless networks and high-speed data processing systems.

INTEGRATED DEVICES FOR ARTIFICIAL INTELLIGENCE AND VLSI: VLSI Design, Simulation and Applications, Edited by Balwinder Raj, Suman Lata Tripathi, Tarun Chaudhary, K. Srinivasa Rao, and Mandeep Singh, ISBN: 9781394204359. With its in-depth exploration of the close connection between microelectronics, AI, and VLSI technology, this book offers valuable insights into the cutting-edge techniques and tools used in VLSI design automation, making it an essential resource for anyone seeking to stay ahead in the rapidly evolving field of VLSI design.

DECENTRALIZED SYSTEMS AND DISTRIBUTED COMPUTING, Edited by Sandhya Avasthi, Suman Lata Tripathi, Namrata Dhanda, and Satya Bhushan Verma, ISBN: 978139420436. This book provides a comprehensive exploration of next-generation internet, distributed systems, and distributed computing, offering valuable insights into their impact on society and the future of technology.

ELECTRIC VEHICLE DESIGN: Design, Simulation and Applications, Edited by Krishan Arora, Suman Lata Tripathi, and Himashu Sharma, ISBN: 9781394204373. This book will serve as a definitive guide to conceptual and practical knowledge about the design of hybrid electrical vehicles (HEV), battery electrical vehicles (BEV), fuel cell electrical vehicles (FCEV), plug-in hybrid electrical vehicles (PHEV), and efficient EV charging techniques with advanced tools and methodologies for students, engineers, and academics alike.

INDUSTRIAL CONTROL SYSTEMS, Edited by Vipin Chandra Pal, Suman Lata Tripathi, and Souvik Ganguli, ISBN: 9781119829256. This volume serves as a comprehensive guide in the journey of industrial control systems with a multidisciplinary approach to the key engineering problems in the 21st century.

MODERN AUTOMOTIVE ELECTRICAL SYSTEMS: Theory and Applications, Edited by Pedram Asef, Sanjeevikumar Padmanaban, and Andrew Lapthorn, ISBN: 1119801047. Presenting the concepts and advances of modern automotive electrical systems, this volume, written and edited by a global team of experts, also goes into the practical applications for the engineer, student, and other industry professionals.

NANODEVICES FOR INTEGRATED CIRCUIT DESIGN, Edited by Suman Lata Tripathi, Abhishek Kumar, K. Srinivasa Rao, and Prasantha R. Mudimela, ISBN: 9781394185788. Written and edited by a team of experts in the field, this important new volume broadly covers the design of nano-devices and their integrated applications in digital and analog integrated circuits (IC) design.

EXPLAINABLE MACHINE LEARNING MODELS AND ARCHITECTURES: Real-Time System Implementation, Edited by Suman Lata Tripathi and Mufti Mahmud, ISBN: 9781394185849. This cutting-edge new volume covers the hardware architecture implementation, the software implementation approach, and the efficient hardware of machine learning applications.

MACHINE LEARNING TECHNIQUES FOR VLSI CHIP DESIGN, Edited by Abhishek Kumar, Suman Lata Tripathi, and K. Srinivasa Rao, ISBN: 9781119910398. This cutting-edge new volume covers the hardware architecture implementation, the software implementation approach, and the efficient hardware of machine learning applications with FPGA or CMOS circuits, and many other aspects and applications of machine learning techniques for VLSI chip design.

INTELLIGENT GREEN TECHNOLOGIES FOR SMART CITIES, Edited by Suman Lata Tripathi, Souvik Ganguli, Abhishek Kumar, and Tengiz Magradze, ISBN: 9781119816065. Presenting the concepts and fundamentals of smart cities and developing "green" technologies, this volume, written and edited by a global team of experts, also goes into the practical applications that can be utilized across multiple disciplines and industries, for both the engineer and the student.

DESIGN AND DEVELOPMENT OF EFFICIENT ENERGY SYSTEMS, Edited by Suman Lata Tripathi, Dushyant Kumar Singh, Sanjeevikumar Padmanaban, and P. Raja, ISBN: 9781119761631. Covering the concepts and fundamentals of efficient energy systems, this volume, written and edited by a global team of experts, also goes into the practical applications that can be utilized across multiple industries, for both the engineer and the student.

Electrical and Electronic Devices, Circuits, and Materials: Technical Challenges and Solutions, Edited by Suman Lata Tripathi, Parvej Ahmad Alvi, and Umashankar Subramaniam, ISBN: 9781119750369. Covering every aspect of the design and improvement needed for solid-state electronic devices and circuit and their reliability issues, this new volume also includes overall system design for all kinds of analog and digital applications and developments in power systems.

Green Energy: Solar Energy, Photovoltaics, and Smart Cities, Edited by Suman Lata Tripathi and Sanjeevikumar Padmanaban, ISBN: 9781119760764. Covering the concepts and fundamentals of green energy, this volume, written and edited by a global team of experts, also goes into the practical applications that can be utilized across multiple industries, for both the engineer and the student.

Check out these other related titles from Scrivener Publishing

HYDROGEN ELECTRICAL VEHICLES, Edited by Mehmet Sankir and Nurdan Demirci Sankir, ISBN: 9781394166381. Hydrogen electrical vehicles are an essential component of the "Green New Deal" and this book covers cutting-edge technologies designed for fuel-cell-powered cars.

PAVEMENT ASSET MANAGEMENT, By Ralph Haas, W.R. Hudson, and Lynne Cowe Falls, ISBN: 9781119038702. Whether you are new to managing pavement assets or a seasoned practitioner, this book provides the most up-to-date information in the field on the history, technology, practice and look ahead.

Printed and bound by CPI Group (UK) Ltd, Croydon, CR0 4YY
19/09/2026
14965093-0002